0198541128X

D1685932

B

Lectures on Biostatistics

D. COLQUHOUN

Lectures on Biostatistics

An Introduction to Statistics with Applications in Biology and Medicine

CLARENDON PRESS · OXFORD · 1971

Oxford University Press, Ely House, London W. 1

GLASGOW   NEW YORK   TORONTO   MELBOURNE   WELLINGTON
CAPE TOWN   IBADAN   NAIROBI   DAR ES SALAAM   LUSAKA   ADDIS ABABA
DELHI   BOMBAY   CALCUTTA   MADRAS   KARACHI   LAHORE   DACCA
KUALA LUMPUR   SINGAPORE   HONG KONG   TOKYO

PRINTED IN GREAT BRITAIN AT THE PITMAN PRESS, BATH

Preface

'In statistics, just like in the industry of consumer goods, there are producers and consumers. The goods are statistical methods. These come in various kinds and 'brands' and in great and often confusing variety. For the consumer, the applier of statistical methods, the choice between alternative methods is often difficult and too often depends on personal and irrational factors.

The advice of producers cannot always be trusted implicitly. They are apt—as is only natural—to praise their own wares. The advice of consumers—based on experience and personal impressions—cannot be trusted either. It is well known among applied statisticians that in many fields of applied science, e.g. in industry, experience, especially 'experience of a lifetime', compares unfavourably with objective scientific research: tradition and aversion from innovation are usually strong impediments for the introduction of new methods, even if these are better than the old ones. This also holds for statistics'

(J. HEMELRIJK 1961).

DURING the preparation of the courses for final year students, mostly of pharmacology, in Edinburgh and London on which this book is based, I have often been struck by the extent to which most textbooks, on the flimsiest of evidence, will dismiss the substitution of assumptions for real knowledge as unimportant if it happens to be mathematically convenient to do so. Very few books seem to be frank about, or perhaps even aware of, how little the experimenter actually *knows* about the distribution of errors of his observations, and about facts that are assumed to be known for the purposes of making statistical calculations. Considering that the purpose of statistics is supposed to be to help in the making of inferences about nature, many texts seem, to the experimenter, to take a surprisingly deductive approach (*if* assumptions *a*, *b*, and *c* were true *then* we could infer such and such). It is also noticeable that in the statistical literature, as opposed to elementary textbooks, a vast number of methods have been *proposed*, but remarkably few have been *assessed* to see how they behave under the conditions (small samples, unknown distribution of errors, etc.) in which they are likely to be used.

These considerations, which are discussed at greater length in the text, have helped to determine the content and emphasis of the methods in this book. Where possible, methods have been advocated that involve a minimum of untested assumptions. These methods, which occur mostly in Chapters 7–11, have the secondary advantage that they are much easier to understand at the present level than the methods, such as Student's *t* test and the chi-squared test (also described and exemplified in this book) which have, until recently, been the main sources of misery for students doing first courses in statistics.

In Chapter 12 and also in § 2.7 an attempt has been made to deal with non-linear problems, as well as with the conventional linear ones. Statistics is heavily dominated by linear models of one sort and another, mainly for reasons of mathematical convenience. But the majority of physical relationships to be tested in practical science are not straight lines, and not linear in the wider sense described in § 12.7, and attempts to make them straight may lead to unforeseen hazards (see § 12.8), so it is unrealistic not to discuss them, even in an elementary book.

In Chapters 13 and 14, calibration curves and assays are discussed. The step by step description of parallel line assays is intended to bridge the gap between elementary books and standard works such as that of Finney (1964).

In Chapter 5 and Appendix 2 some aspects of random ('stochastic') processes are discussed. These are of rapidly increasing importance to the practical scientist, but the textbooks on the subject have always seemed to me to be among the most incomprehensible of all statistics books, partly, perhaps, because there are not really any elementary ones. Again I have tried to bridge the gap.

The basic ideas are described in Chapters 1–4. They may be boring, but the ideas in them are referred to constantly in the later chapters when these ideas are applied to real problems, so the reader is earnestly advised to study them.

There is still much disagreement about the fundamental principles of inference, but most statisticians, presented with the problems described, would arrive at answers similar to those presented here, even if they justified them differently, so I have felt free to choose the justifications that make the most sense to the experimenter.

I have been greatly influenced by the writing of Professor Donald Mainland. His *Elementary medical statistics* (1963), which is much more concerned with statistical thinking than statistical arithmetic, should be read not only by every medical practitioner, but by everyone who has to interpret observations of any sort. If the influence of Professor Mainland's wisdom were visible in this book, despite my greater concern with methods, I should be very happy.

I am very grateful to many statisticians who have patiently put up with my pestering over the last few years. If I may, invidiously, distinguish two in particular they would be Professor Mervyn Stone who read most of the typescript and Dr. A. G. Hawkes who helped particularly with stochastic processes. I have also been greatly helped by Professor D. R. Cox, Mr. I. D. Hill, Professor D. V. Lindley, and Mr. N. W. Please, as well as many others. Needless to say, none of these people has any responsibilities for errors of judgment or fact that I have doubtless persisted with, in spite of their best efforts. I am also very grateful to Professor C. R. Oakley for permission to quote extensively from his paper on the purity-in-heart index in § 7.8.

University College London   D.C.
April 1970

STATISTICAL TABLES FOR USE WITH THIS BOOK

The appendix contains those tables referred to in the text that are not easily available elsewhere. Standard tables, such as normal distribution, Student's *t*, variance ratio, and random sampling numbers, are so widely available that they have not been included. Any tables should do. Those most referred to in the text are Fisher and Yates *Statistical tables for biological, agricultural and medical research* (6th edn 1963, Oliver and Boyd), and Pearson and Hartley *Biometrika tables for statisticians* (Vol. 1, 3rd edn 1966, Cambridge University Press). The former has more about experimental designs; the latter has tables of the Fisher exact text for a 2×2 contingency table (see § 8.2), but anyone doing many of these should get the full tables: Finney, Latscha, Bennett, and Hsu *Tables for testing significance in a* 2×2 *table* (1963, Cambridge University Press). The *Cambridge elementary statistical tables* (Lindley and Miller, 1968, Cambridge University Press) give the normal, *t*, chi-squared, and variance ratio distributions, and some random sampling numbers.

Contents

TABLES

Index of symbols

Reference is to the main section in which the symbol is used, explained, or defined.

Mathematical symbols and operations

$=$	is equal to
$\equiv$	is equivalent to, is defined as
$\simeq$	is approximately equal to
$>$	is greater than
$\gg$	is much greater than
$\geqslant$	is equal to or greater than
$a < P < b$	P is between a and b (greater than a and less than b)
$x^{1/n}$	$\sqrt[n]{x}$, nth root of x
x^{-a}	$1/x^a$
$x\,!$	factorial x (§ 2.1)
$\log x$	logarithm of x. The power to which the base must be raised to equal x. If the base is important it is inserted (e.g. $\log_e x$, $\log_{10} x$), otherwise the expression holds for any base
antilog$_a x$	$= a^x$ where a is the base of the logarithms
Σ	add up all terms like the following (§ 2.1)
Π	multiply together all terms like the following (§ 2.1)
and	logical and (§ 2.4)
or	logical or (§ 2.4)
$\left.\begin{array}{l}\mathrm{d}y/\mathrm{d}a \\ \partial y/\partial x \\ \int \end{array}\right\}$	see Thompson (1965) or Massey and Kestelman (1964)

Roman symbols

a	a constant (§ 2.7)
a	entry in a 2×2 table (§ 8.2)
a	estimate of α (value of y when $x = \bar{x}$) (§ 12.2)
a'	estimate of value of y when $x = 0$ (§ 12.2)
a or $\hat{a}$	least squares value of a (§§ 12.2, 12.7)
A	total in 2×2 table (§ 8.2)
and	logical 'and' (§ 2.4)
b	estimate of β, slope of straight line (§ 12.2)
b or $\hat{b}$	least squares estimate of β (§§ 12.2, 12.7)
b_s	slope for standard line (§ 13.4)
b_u	slope for unknown line (§ 13.4)
b	denominator of a ratio (§ 13.5)
B	total in 2×2 table (§ 8.2)
c	a constant (§ 2.7)
c	concentration (§ A2.4)
$\mathscr{C}$	population value of coefficient of variation (§ 2.6)
C	sample estimate of $\mathscr{C}$ (§ 2.6)

$var(.)$	population variance of variable in brackets. Same as $\sigma^2(.)$. (§§ 2.6, 2.7)
var(.)	Sample estimate of $var(.)$. Same as $s^2(.)$ (§§ 2.6, 2.7)
$\mathscr{V}$	population (true) maximum value of y (§ 12.8)
V	a sample estimate of $\mathscr{V}$ (§ 12.8)
$\hat{V}$	least squares estimate of $\mathscr{V}$ (§ 12.8)
w	weight (§ 2.5)
x	any variable (§§ 2.1, 2.7, 4.1)
$\tilde{x}$	x considered as a random variable, x denoting a particular value of $\tilde{x}$ (§§ 4.1, A1.1)
$\tilde{x}$	geometric mean of x values (§ 2.5)
x	independent variable in curve fitting problems (§§ 12.1, 12.2)
x	log z (Chapter 13)
x_o, x_e	observed frequency (integer) and expected frequency (§ 8.5)
$\bar{y}_i.$, etc.	means of observations (§ 2.1)
y	observed value of dependent variable in curve-fitting problems (§§ 12.1, 12.2)
Y	value of dependent variable read off fitted curve (§ 12.2)
z_S, z_U	doses of standard and unknown (§ 13.1)
z'	doses giving equal responses (§ 13.3)

Greek symbols

α	(alpha) probability of an error of the first kind (§ 6.1)
α	population value of y when $x = \bar{x}$, estimated by a (§§ 12.2, 12.7)
α	population value of numerator of ratio (§ 13.5)
α	orthogonal coefficient (§ 13.8)
β	(beta) probability of an error of the second kind (§ 6.1)
β	population value of slope, estimated by b (§§ 12.2, 12.7)
β_1	block effect for ith block in model observation (§ 11.2)
β	population value for denominator of ratio (§ 13.5)
Δ	(delta) change in, interval in, value of following variable (§§ 5.1, A2.2)
λ	(lambda) population (true) mean number of events in unit time (§§ 5.1, A2.2)
λ	measure of probability of catabolism, disintegration, adsorption in a short time-interval (§§ A2.3–A2.6)
μ	(mu) population mean (§§ 2.5, 4.2, 12.2, A1.1)
μ	population (true) value of ratio (§ 13.5)
μ	measure of probability of desorption (§ A2.4)
π	$3 \cdot 141593 \ldots$
Π	(capital pi) multiply the following § 2.1
$\sigma^2(.)$	(sigma) population (true) variance of variable in brackets. Same as $var(.)$ (§§ 2.1, 2.6, 2.7, A1.2)
Σ	(sigma) add up the following (§ 2.1)
τ_j	(tau) treatment effect for jth treatment in model observation (§ 11.2)
$\chi^2_{(f)}$	(chi) chi-squared statistic with f degrees of freedom (§ 8.5)
χ^2_{rank}	rank statistic distributed approximately as χ^2 (§ 11.7)
ω^2	(omega) interblock standard deviation (§ 11.2)

1. Is the statistical way of thinking worth bothering about?

'I wish to propose for the reader's favourable consideration a doctrine which may, I fear, appear wildly paradoxical and subversive. The doctrine in question is this: that it is undesirable to believe a proposition when there is no ground whatever for supposing it true. I must of course, admit that if such an opinion became common it would completely transform our social life and our political system: since both are at present faultless, this must weigh against it. I am also aware (what is more serious) that it would tend to diminish the incomes of clairvoyants, bookmakers, bishops and others who live on the irrational hopes of those who have done nothing to deserve good fortune here or hereafter. In spite of these grave arguments, I maintain that a case can be made out for my paradox, and I shall try to set it forth.'

BERTRAND RUSSELL, 1935
(*On the Value of Scepticism*)

1.1. How to avoid making a fool of yourself. The role of statistics

IT is widely held by non-statisticians, like the author, that if you do good experiments statistics are not necessary. They are quite right. At least they are right as long as one makes an exception of the important branch of statistics that deals with processes that are *inherently* statistical in nature, so-called 'stochastic' processes (see Chapters 3 and 5 and Appendix 2). The snag, of course, is that doing good experiments is difficult. Most people need all the help they can get to prevent them making fools of themselves by claiming that their favourite theory is substantiated by observations that do nothing of the sort. And the main function of that section of statistics that deals with tests of significance is to prevent people making fools of themselves. From this point of view, the function of significance tests is to prevent people publishing experiments, not to encourage them. Ideally, indeed, significance tests should never appear in print, having been used, if at all, in the preliminary stages to detect inadequate experiments, so that the final experiments are so clear that no justification is needed.

The main aim of this book is to produce a critical way of thinking about experimentation. This is particularly necessary when attempting to measure abstract quantities such as pain, intelligence, or purity in heart (§ 7.8). As Mainland (1964) points out, most of us find arithmetic easier than thinking. A particular effort has therefore been made to explain the rational basis of as many methods as possible. This has been made much easier by starting with the randomization approach to significance testing (Chapters 6–11), because this approach is easy to understand, before going on to tests like Student's t test. The numerical examples have been made as self-contained as possible for the benefit of those who are not interested in the rational basis.

Although it is difficult to achieve these aims without a certain amount of arithmetic, all the mathematical ideas needed will have been learned by the age of 15. The only difficulty may be the occasional use of longer formulae than the reader may have encountered previously, but for the vast majority of what follows you do not need to be able to do anything but add up and multiply. Adding up is so frequent that a special notation for it is described in detail in § 2.1. You may find this very dull and boring until familiarity has revealed its beauty and power, but do not on any account miss out this section. In a few sections some elementary calculus is used, though anything at all daunting has been confined to the appendices. These parts can be omitted without affecting understanding of most of the book. If you know no calculus at all, and there are far more important reasons for no biologist being in this position than the ability to understand the method of least squares, try Silvanus P. Thompson's *Calculus made Easy* (1965).

A list of the uses and scope of statistical methods in laboratory and clinical experimentation is necessarily arbitrary and personal. Here is mine.

(1) Statistical prudence (Lancelot Hogben's phrase) encourages the design of experiments in a way that allows conclusions to be drawn from them. Some of the ideas, such as the central importance of randomization (see §§ 2.3, 6.3, and Chapters 8–11) are far from intuitively obvious to most people at first.

(2) Some processes are inherently probabilistic in nature. There is no alternative to a statistical approach in these cases (see Chapter 5 and Appendix 2).

(3) Statistical methods allow an estimate (usually optimistic, see § 7.2) of the uncertainty of the conclusions drawn from inexact observations. When results are assessed by hopeful intuition it is not uncommon for more to be inferred from them than they really imply. For example,

Schor and Karten (1966) found that, in no less than 72 per cent of a sample of 149 articles selected from 10 highly regarded medical journals. conclusions were drawn that were not justified by the results presented. The most common single error was to make a general inference from results that could quite easily have arisen by chance.

(4) Statistical methods can only cope with random errors and in real experiments systematic errors (bias) may be quite as important as random ones. No amount of statistics will reveal whether the pipette used throughout an experiment was wrongly calibrated. Tippett (1944) put it thus: 'I prefer to regard a set of experimental results as a biased sample from a population, the extent of the bias varying from one kind of experiment and method of observation to another, from one experimenter to another, and, for any one experimenter, from time to time.' It is for this reason, and because the assumptions made in statistical analysis are not likely to be exactly true, that Mainland (1964) emphasizes that the great value of statistical analysis, and in particular of the confidence limits discussed in Chapter 7, is that 'they provide a kind of minimum estimate of error, because they show how little a particular sample would tell us about its population, even if it were a strictly random sample.'

(5) Even if the observations were unbiased, the method of calculating the results from them may introduce bias, as discussed in §§ 2.6 and 12.8 and Appendix 1. For example, some of the methods used by biochemists to calculate the Michaelis constant from observations of the initial velocity of enzymic reactions give a biased result even from unbiased observations (see § 12.8). This is essentially a statistical phenomenon. It would not happen if the observations were exact.

(6) The important point to realize is that by their nature statistical methods can never *prove* anything. The answer always comes out as a probability. And exactly the same applies to the assessment of results by intuition, except that the probability is not calculated but guessed.

1.2. What is an experiment? Some basic ideas

Statistics originally meant state records (births, deaths, etc.) and its popular meaning is still much the same. However, as is often the case, the scientific meaning of the word is much narrower. It may be illustrated by an example.

Imagine a solution containing an unknown concentration of a drug. If the solution is assayed many times, the resulting estimate of concentration will, in general, be different at every attempt. An unknown true value such as the unknown true concentration of the drug is called a *parameter*. The mean value from all the assays gives an *estimate* of this parameter. An approximate experimental estimate (the mean in this example) of a parameter is called a *statistic* It is calculated from a *sample* of observations from the *population* of all possible observations.

In the example just discussed the individual assay results differed from the parameter value only because of experimental error. However, there is another slightly different situation, one that is particularly common in the biological sciences. For example, if identical doses of a drug are given to a series of people and in each case the fall in blood sugar level is measured then, as before, each observation will be different. But in this case it is likely that most of the difference is real. Different individuals really do have different falls in blood sugar level, and the scatter of the results will result largely from this fact and only to a minor extent from experimental errors in the determination of the blood sugar level. The average fall of blood sugar level may still be of interest if, for example, it is wished to compare the effects of two different hypoglycaemic drugs. But in this case, unlike, the first, the parameter of which this average is an estimate, the *true* fall in blood sugar level, is no longer a physical reality, whereas the true concentration was. Nevertheless, it is still perfectly all right to use this average as an estimate of a parameter (the value that the mean fall in blood sugar level would approach if the sample size were increased indefinitely) that is used simply to define the distribution (see §§ 3.1 and 4.1) of the observations. Whereas in the first case the average of all the assays was the only thing of interest, the individual values being unimportant, in the second case it is the individual values that are of importance, and the average of these values is only of interest in so far as it can be used, in conjunction with their scatter, to make predictions about individuals.

In short, there are two problems, the older one of estimating a true value by imperfect methods, and the now common problem of measuring effects that are really variable (e.g. in different people) by relatively very accurate methods. Both these problems can be treated by the same statistical methods, but the interpretation of the results may be different for each.

With few exceptions, scientific methods were applied in medicine and biology only in the nineteenth century and in education and the social sciences only very recently. It is necessary to distinguish two sorts of scientific method often called the *observational* method and the *experimental* method. Claude Bernard wrote: 'we give the name observer to the man who applies methods of investigation, whether simple or complex, to the study of phenomena which he does not vary and which he therefore gathers as nature offers them. We give the name experimenter to the man who applies methods of investigation, whether simple or complex, so as to make natural phenomena vary.' In more modern terms Mainland (1964) writes: 'the distinctive feature of an experiment, in the strict sense, is that the investigator, wishing to compare the effects of two or more factors (independent variables) assigns them *himself* to the individuals (e.g. human beings, animals or batches of a chemical substance) that comprise his test material.' For example, the type and dose of a drug, or the temperature of an enzyme system, are *independent variables*.

The observational method, or survey method as Mainland calls it, usually leads to a *correlation*; for example, a correlation between smoking habits and death from lung cancer, or between educational attainment and type of school. But the correlation, however perfect it may be, does not give any information at all about *causation*,† such as whether smoking *causes* lung cancer. The method lends itself only too easily to the confusion of sequence with consequence. 'It is the *post hoc, ergo propter hoc* of the doctors, into which we may very easily let ourselves be led' (Claude Bernard).

This very important distinction is discussed further in §§ 12.7 and 12.9. Probably the most useful precaution against the wrong interpretation of correlations is to imagine the experiment that might in principle be carried out to decide the issue. It can then be seen that bias in the results is controlled by the *randomization* process inherent in experiments. If all that is known is that pupils from type A schools do better than those from type B schools it could well have nothing to do with the type of school but merely, for example, be that children of educated parents go to type A school and those of uneducated parents to type B schools. If proper experimental methods were applied in the situations mentioned above the first step would be to divide the population (or a random sample from it) by a *random* process, into two groups. One group would be *instructed* to smoke (or to go to a particular sort of school), the other group would be *instructed* not to smoke (or go to

† It is not even necessarily true that zero correlation rules out causation, because lack of correlation does not necessarily imply independence (see § 12.9).

a different sort of school). The difficulty in the medical and social sciences is usually that an experiment may be considered unethical. Since it can hardly be assumed *a priori* that there is an equal chance of smoking having good or bad effects on health, it is not possible to *instruct* a group of people to smoke, though it should be perfectly acceptable to leave one randomly selected group to its normal habits (including smoking by some of the group) and to instruct the other group to stop smoking.

Often the situation is not as bad as this, however. There is genuine doubt about the relative merits of different sorts of school, and, very often, about different sorts of therapy, so in these cases it is not merely ethical to do a proper experiment, but it would be unethical, though not unusual, *not* to do the experiment.

1.3. The nature of scientific inference

We are concerned with the establishment of new knowledge about the real world. Therefore it will do no harm to mention something of the logical foundations of inference before starting on methods.

The earliest natural philosophers based their work largely on *deductive* arguments from *axioms*. The only criterion for a valid set of axioms is that it should not be possible to deduce contradictory conclusions from them, i.e. that the axioms should be *consistent*. Even if this is so it has no bearing on whether or not the axioms are *true*.

Later it came to be supposed that knowledge of the natural world could only be obtained by *induction* of general theories from particular observations, and not, as had previously been assumed, by *deduction* of the particular case from a general axiom.

The process of induction must clearly be subject to uncertainties, but it was not until much later that these uncertainties were investigated and attempts made to measure them. During the seventeenth century the study of *probability theory* was started by Fermat and Pascal. This was, and still is, a branch of mathematics, wholly deductive in nature.

Probability theory and experimental method grew up alongside each other, but largely separately. One of the first attempts at a synthesis came when astronomers wanted to find out whether the stars were distributed randomly, or in some sort of order. What was needed was a method for determining the *probability of a hypothesis being true* given some experimental observations relevant to it. For example: (1) the hypothesis that the stars are randomly distributed; (2) the hypothesis that morphine is a better analgesic than aspirin; or (3) the hypothesis that state schools provide a better education than private schools.

The use to which natural scientists wanted to put probability theory was, it seems, of a quite different kind from that for which the theory was designed. All that probability theory would answer were questions such as: Given certain premises about the thorough shuffling of the pack and the honesty of the players, what is the probability of drawing four consecutive aces? This is a statement of the probability of making some observations, given an hypothesis (that the cards are well shuffled, and the players honest), a deductive statement of *direct probability*. What was needed was a statement of the probability of the hypothesis, *given* some observations—an inductive statement of *inverse probability*.

An answer to the problem was provided by the Rev. Thomas Bayes in his *Essay towards solving a Problem in the Doctrine of Chances* published in 1763, two years after his death. Bayes' theorem states:

posterior probability of a hypothesis = constant × likelihood
of hypothesis × prior probability of the hypothesis   (1.3.1.)

In this equation prior (or *a priori*) probability means the probability of the hypothesis being true *before* making the observations under consideration, the posterior (or *a posteriori*) probability is the probability *after* making the observations, and the *likelihood of the hypothesis* is defined as the probability of making the given observations *if* the hypothesis under consideration were in fact true. This technical definition of likelihood will be encountered again later.

The wrangle about the interpretation of Bayes' theorem continues to the present day. Is 'the probability of an hypothesis being true' a meaningful idea? The great mathematician Laplace assumed that if nothing were known of the merits of rival hypotheses then their prior probabilities should be considered equal ('the equipartition of ignorance'). Later it was suggested that Bayes' theorem was not really applicable except in a small proportion of cases in which valid prior probabilities were known. This view is still probably the most common, but there is now a strong school of thought that believes the only sound method of inference is Bayesian. An uncontroversial use of Bayes' theorem, in medical diagnosis, is mentioned in § 2.4.

Fortunately, in most, though not all, cases the practical results are the same whatever viewpoint is adopted. If the prior probabilities of several mutually exclusive hypotheses are known or assumed to be equal then the hypothesis with the maximum posterior probability will also be that with the maximum likelihood. In fact a popular procedure is to ignore the prior probability altogether and to select the hypothesis with the maximum likelihood. This procedure avoids altogether the making of statements of inverse probability that many people think to be invalid, but loses something in interpretability. The probability considered is the *probability of the observations* calculated assuming the hypothesis in question to be true—a statement of direct probability.

It has been argued strongly by Karl Popper that scientific inference is a wholly deductive process. A hypothesis is framed by inspired guesswork. It consequences are *deduced* and then tested experimentally. This is certainly just how things should be done. But, as A. J. Ayer points out, the experiment is only useful if it is supposed that it will give the same result when it is repeated, and the argument leading to this supposition is the sort of inductive inference with which much of statistics is concerned.

2. Fundamental operations and definitions

'Considering how many fools can calculate, it is surprising that it should be thought either a difficult or a tedious task for any other fool to learn how to master the same tricks. SILVANUS P. THOMPSON

2.1. Functions and operators. A beautiful notation for adding up

Functional notation

IF the value of one variable, say y, depends on the value of another, say x, then y is said to be a function of x. For example, the response to a drug is a function of the dose. The usual algebraic way of saying this is $y = f(x)$ where f denotes the function. This equation is read 'y equals a function *of* x'. If it is required to distinguish different functions of the same variable then different symbols are chosen to represent each function. For example, $y_1 = g(x)$, $y_2 = \phi(x)$. If the function f were the square root, g were the logarithm, and ϕ the tangent then the above equations could be written in a less abstract form as $y = \sqrt{x}$, $y_1 = \log x$, and $y_2 = \tan x$. This notation can be extended to several variables. If the value of y depends on the value of two different variables, x_1 and x_2 say, this could be denoted $y = f(x_1, x_2)$. An example of such a function is $y = x_1{}^2 + x_2$.

Needless to say, the symbols, f, g, and ϕ do not stand for numbers and, for example, it is very important to distinguish $y = f(x)$ from 'y equals f times x'. In the present case f, g, and ϕ stand for *operations* carried out on the argument x in just the same way as the symbol '$+$' stands for the *operation* of addition of the quantities on each side of the plus sign, or the symbol d/dx stands for 'find the differential coefficient with respect to x'.

In the following pages this operational notation is used frequently. For example, $s(x)$ will stand for 'the estimated standard deviation *of* x' (not 's *times* x').† The square of the standard deviation is called the

---

† See § 2.6 for the definitions. Although it is commonly used, this is not really a consistent use of the notation. The sample standard deviation, $s(x)$, is not a function of a single variable x, but of the whole set of x values making up the sample. And in the case of the population standard deviation, $\sigma(x)$, σ is really an operator on the probability distribution of x (see Appendix 1).

variance. The variance of x is thus $[s(x)]^2$, which is usually written $s^2(x)$. The situation may look even more confusing if it is wished to denote the estimated standard deviation of a quantity like $x_1 - x_2$, i.e. a measure of the scatter of the values of the quantity $x_1 - x_2$. Using the notation given above this number would be written $s(x_1 - x_2)$, but this is *not* the same as $s(x_1) - s(x_2)$; s is an operator not a number. To add to the difficulties it is quite common for $s(x)$ or $f(x)$ to be abbreviated to s and f, the argument, x, being understood. So in this case s and f *do* stand for numbers; the numbers $s(x)$ and $f(x)$. Brackets rather than parentheses are sometimes used to make the notation clearer so the standard deviation of $x_1 - x_2$ is written $s[x_1 - x_2]$.

Two important operators are those used to denote the formation of sums and products, viz. Σ and Π (Greek capitals, sigma and pi). For example, Σx means find the sum of all the values of x, and Πx means find the product of all the values of x. These operations occur often and are discussed in more detail below.

Factorial notation

Another operation that will occur in the following pages is written $n!$, which is read as 'factorial n'. When n is an integer this has the value $n(n-1)(n-2)...1$. For example, $4! = 4 \times 3 \times 2 \times 1 = 24$. A more general definition (the gamma function) is valid also for non-integers and occurs often in more advanced work than is dealt with here. In the light of this more general definition, as well as for reasons of convenience that will be apparent later, $0!$ (factorial zero) is defined as having the value 1.

The use of the summation operator

The operation of adding up occurs very often. The arithmetic is familiar, but the notation used may not be. In the following pages the summation operator is used often. Frequently it is written with a full panoply of superscript and subscripts. This makes the operation unambiguous, at the expense of looking a bit complicated. It is very well worth while (for far wider reasons than merely understanding this book) making sure that you can add up, so the temptation to skip this section should be resisted. The use of the product operator Π is analogous, $+$ being replaced by $\times$.

Given a set of observations, for example n replicate observations on the same animal of the fall in blood pressure in response to a drug, an observation can be denoted y_i. This symbol stands for the ith fall in blood pressure. There are n observations so in general an observation is y_i where $i = 1, 2,..., n$. If $n = 5$, for example, then the five observations are symbolized y_1, y_2, y_3, y_4, y_5. Note that the subscript i has not necessarily got any particular experimental significance. It is merely a method for counting, or labelling, the individual observations.

The observations can be laid out in a table thus:

$$y_1 \quad y_2 \quad y_3 \cdots y_n.$$

Mathematicians would refer to such a table as a *one-dimensional array* or a *vector*, but 'table' is a good enough name for now.

The instruction to add up all values of y from the first value to the nth value is written

$$\text{sum} = \sum_{i=1}^{i=n} y_i \quad \text{or, more briefly,} \quad \sum_{i=1}^{n} y_i.$$

This expression symbolizes the number

$$\text{sum} = y_1 + y_2 + y_3 + ... + y_n.$$

Similarly,

$$\sum_{i=3}^{i=5} y_i \text{ stands for the number } y_3 + y_4 + y_5.$$

Thus the arithmetic mean of n values of y is

$$\bar{y} = \frac{\sum_{i=1}^{i=n} y_i}{n}$$

Notice that after the summation operation, the counting, or subscripted, variable, i, does not appear in the result.

A slightly more complicated situation arises when the observations can be classified in two ways. For example, if n readings of blood pressure (y) were taken on each of k animals the results would probably be arranged in a table like this:

		\multicolumn{5}{c}{Animal (value of j)}
		1	2	3	....	k
	1	y_{11}	y_{12}	y_{13}	....	y_{1k}
	2	y_{21}	y_{22}	y_{23}	....	y_{2k}
observation	3	y_{31}	y_{32}	y_{33}	....	y_{3k}
(value of i)	⋮	⋮	⋮	⋮		⋮
	n	y_{n1}	y_{n2}	y_{n3}	....	y_{nk}

Two subscripts, say i and j, are now needed, one for keeping count of the observations and one for the animals. i takes the values 1, 2, 3,..., n; and j takes the values 1, 2, 3,..., k. The ith observation on the jth animal is thus represented by the symbol y_{ij}. In more general terms, y_{ij} stands for the value of y in the ith row and jth column of a table (or *two-dimensional array*, or *matrix*) such as that shown above.

For example, a table with 3 columns and 4 rows could be written

$$
\begin{array}{ccc}
y_{11} & y_{12} & y_{13} \\
y_{21} & y_{22} & y_{23} \\
y_{31} & y_{32} & y_{33} \\
y_{41} & y_{42} & y_{43}
\end{array}
\quad
\begin{array}{l}
\text{and a particular table} \\
\text{of this size could be}
\end{array}
\quad
\begin{array}{ccc}
2 & 4 & 3 \\
1 & 6 & 4 \\
5 & 7 & 6 \\
8 & 4 & 5
\end{array}
$$

In this case $n = 4$ and $k = 3$ and the $n \times k$ table contains $nk = 12$ observations.

The row and column totals and means can be represented by an extension of the notation used above. For example the total of the observations in the jth column, which may be called $T_{.j}$ for short, would be written

$$T_{.j} = \sum_{i=1}^{i=n} y_{ij} \equiv y_{1j} + y_{2j} + y_{3j} + \ldots + y_{nj}. \tag{2.1.1}$$

Thus the total of the first column is $T_{.1} = y_{11} + y_{21} + y_{31} + \ldots + y_{n1}$ ($=16$ in the example). The mean of the readings in the jth column (the mean fall in blood pressure in the jth animal in the example given above), which is usually called $\bar{y}_{.j}$, is thus

$$\bar{y}_{.j} = \frac{\sum\limits_{i=1}^{i=n} y_{ij}}{n} = \frac{T_{.j}}{n}. \tag{2.1.2}$$

Again notice that after summing over the values of i (i.e. adding up the numbers in a specified column) the answer does not involve i, but does still involve the specified column number j. The symbol i, the subscript operated on, is replaced by a dot in the symbols $T_{.j}$ and $\bar{y}_{.j}$.

In an exactly similar way the total for the ith row, $T_{i.}$, is written

$$T_{i.} = \sum_{j=1}^{j=k} y_{ij} \equiv y_{i1} + y_{i2} + y_{i3} + \ldots + y_{ik} \tag{2.1.3}$$

For example, for the second row $T_{2.} = y_{21}+y_{22}+\ldots+y_{2k}$ ($=11$ in the example). The mean value for the ith row is

$$\bar{y}_{i.} = \frac{T_{i.}}{k} = \frac{\sum\limits_{j=1}^{j=k} y_{ij}}{k}. \tag{2.1.4}$$

Using the numbers in the 4×3 table above, the totals and means are found to be

		Column number (value of j)			Row totals	Row means
		1	2	3	$T_{i.} = \sum\limits_{j=1}^{j=k} y_{ij}$	$\bar{y}_{i.} = T_{i.}/k$
Row	1	$y_{11} = 2$	$y_{12} = 4$	$y_{13} = 3$	$T_{1.} = \sum\limits_{j=1}^{j=k} y_{1j} = 9$	$\bar{y}_{1.} = 9/3$
number	2	$y_{21} = 1$	$y_{22} = 6$	$y_{23} = 4$	$T_{2.} = \sum\limits_{j=1}^{j=k} y_{2j} = 11$	$\bar{y}_{2.} = 11/3$
(value of i)	3	$y_{31} = 5$	$y_{32} = 7$	$y_{33} = 6$	$T_{3.} = \sum\limits_{j=1}^{j=k} y_{3j} = 18$	$\bar{y}_{3.} = 18/3$
	4	$y_{41} = 8$	$y_{42} = 4$	$y_{43} = 5$	$T_{4.} = \sum\limits_{j=1}^{j=k} y_{4j} = 17$	$\bar{y}_{4.} = 17/3$
Column totals $T_{.j} = \sum\limits_{i=1}^{i=n} y_{ij}$		$T_{.1} = \sum\limits_{i=1}^{i=n} y_{i1}$ $= 16$	$T_{.2} = \sum\limits_{i=1}^{i=n} y_{i2}$ $= 21$	$T_{.3} = \sum\limits_{i=1}^{i=n} y_{i3}$ $= 18$	Grand total $\sum\limits_{i=1}^{i=n} \sum\limits_{j=1}^{j=k} y_{ij} = 55$	
Column means $\bar{y}_{.j} = T_{.j}/n$		$\bar{y}_{.1} = 16/4$	$\bar{y}_{.2} = 21/4$	$\bar{y}_{.3} = 18/4$		

The grand total of all the observations in the table illustrates the meaning of a double summation sign. The grand total (G, say†) could be written as the sum of the row totals

$$G = \sum_{i=1}^{i=n} (T_{i.}).$$

Inserting the definition of $T_{i.}$ from (2.1.3) gives

$$G = \sum_{i=1}^{i=n} \left(\sum_{j=1}^{j=k} y_{ij} \right).$$

Equally, the grand total could be written as the sum of the column totals

$$G = \sum_{j=1}^{j=k} (T_{.j})$$

† It would be more consistent with earlier notation to replace both suffixes by dots and call the grand total $T..$, but the symbol G is often used instead.

which, inserting the definition of $T_{.j}$ from (2.1.1), becomes

$$G = \sum_{j=1}^{j=k} \left(\sum_{i=1}^{i=n} y_{ij} \right).$$

Since the grand total is the same whichever order the additions are carried out in, the parentheses are superfluous and the operation is usually symbolized

$$G = \sum_{i=1}^{i=n} \sum_{j=1}^{j=k} y_{ij} \text{ or } \sum_{j=1}^{j=k} \sum_{i=1}^{i=n} y_{ij} \text{ or simply } \Sigma\Sigma y.$$

What to do if you get stuck

If it is ever unclear how to manipulate the summation operator simply write out the sum term by term and apply the ordinary rules of algebra. For example, if k denotes a constant then

$$\sum_{i=1}^{n} kx_i = k\sum_{i=1}^{n} x_i \tag{2.1.5}$$

because the left-hand side, written out in full, is $kx_1 + kx_2 + ... + kx_n = k(x_1 + x_2 + ... + x_n)$, which is the right-hand side. Thus if k is the same for every x it can be 'taken outside the summation sign'. However $\Sigma k_i x_i$, in which each x is multiplied by a different constant, is $k_1 x_1 + k_2 x_2 + ... + k_n x_n$, which cannot be further simplified.

It follows from what has been said that if the quantities to be added do not contain the subscript then the summation becomes a simple multiplication. If all the $x_i = 1$ in (2.1.5) then

$$\sum_{i=1}^{i=n} k = k + k + ... + k = nk \tag{2.1.6}$$

and furthermore, if $k = 1$,

$$\sum_{i=1}^{i=n} 1 = n. \tag{2.1.7}$$

Another useful result is

$$\sum_{i=1}^{n} (x_i - y_i) \equiv (x_1 - y_1) + (x_2 - y_2) + ... + (x_n - y_n)$$

$$= (x_1 + x_2 + ... + x_n) - (y_1 + y_2 + ... + y_n)$$

$$= \sum_{i=1}^{n} x_i - \sum_{i=1}^{n} y_i. \tag{2.1.8}$$

These results will be used often in later sections.

2.2. Probability

The only rigorous definition of probability is a set of axioms defining its properties, but the following discussion will be limited to the less rigorous level that is usual among experimenters. For practical purposes the probability of an event is a number between zero (implying impossibility) and one (implying certainty). Although statisticians differ in the way they define and interpret probability, there is complete agreement about the rules of probability described in § 2.4. In most of this book probability will be interpreted as a *proportion* or *relative frequency*. An excellent discussion of the subject can be found in Lindley (1965, Chapter 1).

The simplest way of defining probability is as a *proportion*, viz. 'the ratio of the number of favourable cases to the total number of equiprobable cases'. This may be thought unsatisfactory because the concept to be defined is introduced as part of its own definition by the word 'equiprobable', though a non-numerical ordering of likeliness more primitive than probability would be sufficient to define 'equally likely', and hence 'random'. Nevertheless when the reference set of 'the total number of equiprobable cases' is finite this description is used and accepted in practice. For example if 55 per cent of the population of college students were male it would be asserted that the probability of a single individual chosen from this finite population being male is 0·55, provided that the probability of being chosen was the same for all individuals, i.e. provided that the choice was made at *random*.

When the reference population is infinite the ratio just discussed cannot be used. In this case the *frequency* definition of probability is more useful. This identifies the probability P of an event as the limiting value of the relative frequency of the event in a random sequence of trials when the number of trials becomes very large (tends towards infinity). For example, if an unbiased coin is tossed ten times it would not be expected that there would be exactly five heads. If it were tossed 100 times the proportion of heads would be expected to be rather closer to 0·5 and as the number of tosses was extended indefinitely the proportion of heads would be expected to converge on exactly 0·5. This type of definition seems reasonable, and is often invoked in practice, but again it is by no means satisfactory as a complete, objective definition. A random sequence cannot be proved to converge in the mathematical sense (and in fact *any* outcome of tossing a true coin a million times is possible), but it can be shown to converge in a statistical sense.

Degrees of belief

It can be argued persuasively (e.g. Lindley (1965, p. 29)) that it is valid and sometimes necessary to use a subjective definition of probability as a numerical measure of one's degree of belief or strength of conviction in a hypothesis ('personal probability'). This is required in many applications of Bayes' theorem, which is mentioned in §§ 1.3 and 2.4 (see also § 6.1, para. (7)). However the application of Bayes' theorem to medical diagnosis (§ 2.4) does not involve subjective probabilities, but only frequencies.

2.3. Randomization and random sampling

The selection of *random* samples from the population under study is the basis of the design of experiments, yet is an extraordinarily difficult job. Any sort of statistical analysis (*and any sort of intuitive analysis*) of observations depends on random selection and allocation having been properly done. The very fundamental place of randomization is particularly obvious in the randomization significance tests described in Chapters 8–11.

It should never be out of mind that all calculations (and all intuitive assessments) belong to an *entirely imaginary world* of perfect random selection, unbiased measurement, and often many other ideal properties (see § 11.2). The assumption that the real world resembles this imaginary one is an extrapolation outside the scope of statistics or mathematics. As mentioned in Chapter 1 it is safer to assume that samples have some unknown bias.

For example, an anti-diabetic drug should ideally be tested on a random sample of all diabetics in the world—or perhaps of all diabetics in the world with a specified form and severity of the disease, or all diabetics in countries where the drug is available. In fact, what is likely to be available are the diabetic patients of one, or a few, hospitals in one country. Selection should be done *strictly* at random (see below) from this restricted population, but extension of inferences from this population to a larger one is bound to be biased to an unknown extent.

It is, however, quite easy, having obtained a sample, to divide it *strictly randomly* (see below) into several groups (e.g. groups to receive new drug, old drug, and control dummy drug). This is, nevertheless, very often not done properly. The hospital numbers of the patients will not do, and neither will their order of appearance at a clinic. It is very important to realize that 'random' is not the same thing as 'haphazard'. If two treatments are to be compared on a group of patients it is not good enough for the experimenter, or even a neutral person, to allocate a patient haphazardly to a treatment group. It has been shown repeatedly that any method involving human decisions is non-random. For all practical purposes the following interpretation of randomness, given by R. A. Fisher (1951, p. 11), should be taken as a fundamental principle of experimentation: ' . . . not determined arbitrarily by human choice, but by the actual manipulation of the physical apparatus used in games of chance, cards, dice, roulettes, etc., or, more expeditiously, from a published collection of random sampling numbers purporting to give the actual results of such manipulation.'

Published random sampling numbers are, in practice, the only reliable method. Samples selected in this way (see below) will be referred to as selected *strictly at random*. Superb discussions of the crucial importance of, and the pitfalls involved in random sampling have been given by Fisher (1951, especially Chapters 2 and 3) and by Mainland (1963, especially Chapters 1–7). Every experimenter should have read these. They cannot be improved upon here.

How to select samples strictly at random using random number tables

This is, perhaps, the most important part of the book. There are various ways of using random number tables (see, for example, the introduction to the tables of Fisher and Yates (1963)). Two sorts of tables are commonly encountered, and those of Fisher and Yates (1963) will be used as examples. The first is a table of random digits in which the digits from 0 to 9 occur in random order. The digits are usually printed in groups of two to make them easier to read, but they can be taken as single digits or as two, three, etc. digit numbers. If taken in groups of three the integers from 000 to 999 will occur in random order in the tables. The second form of table is the table of random permutations. Fisher and Yates (1963) give random permutations of 10 and 20 integers. In the former the integers from 0 to 9, and in the latter the integers from 0 to 19, occur in random order, but each number appears once only in each permutation.

To divide a group of subjects into several sub-groups strictly at random the easiest method is to use the tables of random permutations, as long as the total number of subjects is not more than 20 (or whatever is the largest size of random permutation available). Suppose that 15 subjects are to be divided into group of size n_1, n_2, and n_3. First number the subjects 0 to 14 in any convenient way. Then obtain a random permutation of 15 by taking the first random permutation of 20 from the tables and deleting the numbers 15 to 19. (This permutation in the table should then be crossed out so that it is not used again— use each once only.) Then allocate the first n_1 of the subjects to the first group, the next n_2 to the second group, and the remainder to the third group. For example, if the random permutation of 15 turned out to be 1, 6, 8, 5, 10, 12, 11, 9, 2, 0, 3, 14, 7, 4, 13 (the first permutation from Fisher and Yates (1963), p. 142)) and the 15 subjects were to be divided randomly into groups of 5, 4, and 6 subjects then subjects 1, 6, 8, 5, and 10 would go in the first group, 12, 11, 9, and 2 in the second group, and the rest in the third group.

For larger numbers of subjects the tables of random digits must be used. For example, to divide 24 subjects into 4 groups of 6 the procedure is as follows. First number the subjects in any convenient way with the numbers 00 to 23. Take the digits in the table in groups of two. The table then gives the integers from 00 to 99 in random order. One procedure would be to delete all numbers from 24 to 99, but it is more economical to delete only 96, 97, 98, and 99 (i.e. those equal to or larger than 96, which is the biggest multiple of 24 that is not larger than 100). Now the remaining numbers are a random sequence of the integers from 00 to 95. From each number between 24 and 47 subtract 24; from each number between 48 and 71 subtract 48; and from each number between 72 and 95 subtract 72 (or, in other words, divide every number in the sequence by 24 and write down the remainder). For example, if the number in the table is 94 then write down 22; or in place of 55 write down 07. (The numbers from 96 to 99 must, of course, be omitted because their presence would give the numbers 00 to 03 a larger chance than the others of occurring.) Some numbers may appear several times but *repetitions are ignored*. If the final sequence were 21, 04, 07, 13, 02, 02, 04, 09, 00, 23, 14, 13, 11, etc., then subjects 21, 04, 07, 13, 02, 09 are allocated to the first group, subjects 00, 23, 14, 11, etc. are allocated to the second group, and so on.

The method is simpler for the random block experiments described in §§ 11.6 and 11.7. Blocks are never likely to contain more than 20 treatments so the order in which the treatments occur in each block is taken from a random permutation found from the tables of random permutations as above. For example, if there are four treatments in each block number them 0 to 3, and for each block obtain a random permutation of the numbers 0 to 3, by deleting 4 to 9 from the tabulated random permutations of 10, crossing out each permutation from the table as it is used.

The selection of a Latin square at random is more complicated (see § 11.8).

2.4. Three rules of probability

The words **and** and **or** are printed in bold type when they are being used in a restricted logical sense. For our purposes E_1 **and** E_2 means that both the event E_1 and the event E_2 occur, and E_1 **or** E_2 means that either E_1 or E_2 *or both* occur (in general, that *at least one* of several events occurs). More explanation and details of the following rules can be found, for example, in Mood and Graybill (1963), Brownlee (1965), or Lindley (1965).

(1) *The addition rule of probability*

This states that the probability of either or both of two events, E_1 and E_2, occurring is

$$P[E_1 \text{ or } E_2] = P[E_1] + P[E_2] - P[E_1 \text{ and } E_2]. \qquad (2.4.1)$$

If the events are *mutually exclusive*, i.e. if the probability of E_1 and E_2 *both* occurring is zero, $P[E_1 \text{ and } E_2] = 0$, then the rule reduces to

$$P[E_1 \text{ or } E_2] = P[E_1] + P[E_2] \qquad (2.4.2)$$

the sum of probabilities. Thus, if the probability that a drug will decrease the blood pressure is 0·9 and the probability that it will have no effect or increase it is 0·1 then the probability that it will *either* (a) increase the blood pressure *or* (b) have no effect or decrease it is, since the events are mutually exclusive, simply $0·9 + 0·1 = 1·0$. Because the events considered are *exhaustive*, the probabilities add up to 1·0. It is certain that one of them will occur. That is

$$P[E \text{ occurs}] = 1 - P[E \text{ does not occur}]. \qquad (2.4.3)$$

This example suggests that the rule can be extended to more than two events. For example the probability that the blood pressure will not change might be 0·04 and the probability that it will decrease might be 0·06. Thus

$$P[\text{no change or decrease}] = 0·04 + 0·06 = 0·1 \text{ as before,}$$
$$P[\text{no change or increase}] = 0·04 + 0·9 = 0·94,$$
$$P[\text{no change or decrease or increase}] = 0·04 + 0·06 + 0·9 = 1·0.$$

The simple addition rule holds because the events considered are mutually exclusive. In the last case only they are also exhaustive. An example of the use of the full equation (2.4.1) is given below.

(2) *The multiplication rule of probability*

It is possible that the probability of event E_1 happening depends on whether E_2 has happened or not. The *conditional probability* of E_1 happening given that E_2 has happened is written $P[E_1|E_2]$, which is usually read 'probability of E_1 given E_2'.

The probability that *both* E_1 *and* E_2 will happen is

$$P[E_1 \text{ and } E_2] = P[E_1].P[E_2|E_1]$$
$$= P[E_2].P[E_1|E_2]. \qquad (2.4.4)$$

If the events are independent in the probability sense (different from the functional independence) then, *by definition* of independence,

$$P[E_1|E_2] = P[E_1]$$

and similarly $P[E_2|E_1] = P[E_2]$ $\qquad (2.4.5)$

so the multiplication rule reduces to

$$P[E_1 \text{ and } E_2] = P[E_1].P[E_2], \qquad (2.4.6)$$

the product of the separate probabilities. Events obeying (2.4.5) are said to be independent. Independent events are necessarily uncorrelated but the converse is not necessarily true (see § 12.9).

A numerical illustration of the probability rules. If the probability that a British college student smokes is 0·3, and the probability that the student attends London University is 0·01, then the probability that a student, selected at random from the population of British college students, is both a smoker and attends London University can be found from (2.4.6) as†

$$P[\text{smoker and London student}] = P[\text{smoker}] \times P[\text{London student}]$$
$$= 0{\cdot}3 \times 0{\cdot}01 = 0{\cdot}003$$

as long as smoking and attendence at London University are independent so that, from (2.4.5),

$$P[\text{smoker}] = P[\text{smoker}|\text{London student}]$$

† Notice that $P[\text{smoker}]$, could be written as $P[\text{smoker} \mid \text{British college student}]$. All probabilities are really conditional (on membership of a specified population). See, for example, Lindley (1965).

or, equivalently,

$$P[\text{London student}] = P[\text{London student}|\text{smoker}].$$

The first of these conditions of independence can be interpreted in words as 'the probability of a student being a smoker equals the probability that a student is a smoker given that he attends London University', that is to say 'the proportion of smoking students in the whole population of British college students is the same as the proportion of smoking students at London University', which in turn implies that the proportion of smoking students is the same at London as at any other British University (which is, no doubt, not true).

Because smoking and attendance at London University are not mutually exclusive the full form of the addition rule, (2.4.1), must be used. This gives

$$P[\text{smoker or London student}] = 0{\cdot}3+0{\cdot}01-(0{\cdot}3\times0{\cdot}01) = 0{\cdot}307.$$

The meaning of this can be made clear by considering random samples, each of 1000 British college students. On the average there would be 300 smokers and 10 London students in each sample of 1000. There would be 3 students $(1000\times0{\cdot}003)$ who were both smokers and London students if the implausible condition of independence were met (see above). Therefore there would be 297 students $(300-3)$ who smoked but were not from London, and 7 students $(10-3)$ who were from London but did not smoke. Therefore the number of students who *either* smoked (but did not come from London), *or* came from London (but did not smoke), *or* both came from London *and* smoked, would be $297+7+3 = 307$, as calculated $(1000\times0{\cdot}307)$ from (2.4.1).

(3) *Bayes' theorem, illustrated by the problem of medical diagnosis*

Bayes' theorem has already been given in words as (1.3.1) (see § 1.3). The theorem applies to any series of events H_j, and is a simple consequence of the rules of probability already stated (see, for example, Lindley (1965, p. 19 et seq.)). The interesting applications arise when the events considered are *hypotheses*. If the jth hypothesis is denoted H_j and the observations are denoted Y then (1.3.1) can be written symbolically as

$$P[H_j|Y] \quad = \quad k\times P[Y|H_j] \quad \times \quad P[H_j] \quad , \qquad (2.4.7)$$

posterior probability of hypothesis j likelihood of hypothesis j prior probability of hypothesis j

where k is a proportionality constant. If the set of hypotheses considered is exhaustive (one of them must be true), and the hypotheses are mutually exclusive (not more than one can be true), the addition rule states that the probability of (hypothesis 1) **or** (hypothesis 2) **or** . . . (which must be equal to one, because one or another of the hypotheses is true) is given by the total of the individual probabilities. This allows the proportionality constant in (2.4.7) to be found. Thus

$$\sum_{all\ j} P[H_j|\,Y] = k\Sigma(P[\,Y|H_j].P[H_j]) = 1 \text{ and therefore}$$

$$k = \frac{1}{\sum\limits_{all\ j}(P[\,Y|H_j].P[H_j])}. \qquad (2.4.8)$$

Bayes' theorem has been used in medical diagnosis. This is an uncontroversial application of the theorem because all the probabilities can be interpreted as proportions. Subjective, or personal, probabilities are not needed in this case (see § 2.2)

If a patient has a set of symptoms **S** (the observations) then the probability that he is suffering from disease D (the hypothesis) is, from (2.4.7),

$$P[D|\textbf{S}] = k \times P[\textbf{S}|D] \times P[D]. \qquad (2.4.9)$$

In this equation the prior probability of a patient having disease D, $P[D]$, is found from the proportion of patients with this disease in the hospital records. In principle the likelihood of D, i.e. the probability of observing the set of symptoms **S** if a patient in fact has disease D, $P[\textbf{S}|D]$, could also be found from records of the proportion of patients with D showing the particular set of symptoms observed. However, if a realistic number of possible symptoms is considered the number of different possible sets of symptoms will be vast and the records are not likely to be extensive enough for $P[\textbf{S}|D]$ to be found in this way. This difficulty has been avoided by assuming that symptoms are independent of each other so that the simple multiplication rule (2.4.6) can be applied to find $P[\textbf{S}|D]$ as the product of the separate probabilities of patients with D having each individual symptom, i.e.

$$P[\textbf{S}|D] = P[S_1|D] \times P[S_2|D] \times ... \times P[S_n|D], \qquad (2.4.10)$$

where **S** stands for the set of n symptoms (S_1 **and** S_2 **and** . . . **and** S_n) and $P[S_1|D]$, for example, is found from the records as the proportion of patients with disease D who have symptom 1. Although the assumption of independence is very implausible this method seems to have given some good results (see, for example, Bailey (1967, Chapter 11).

A numerical example

The simplest (to the point of naivety) example of the above argument is the case when only one disease and one symptom is considered. The example is modified from Wallis and Roberts (1956).

Suppose that a diagnostic test for cancer has a probability of 0·96 of being positive when the patient does have cancer. If S stands for the event that the test is positive and $\tilde{S}$ for the event that it is negative (the data), and if D stands for the event that the patient has cancer, and $\bar{D}$ for the event that he has not (the two hypotheses) then in symbols $P[S|D] = 0·96$ (the *likelihood* of D if S observed). Because the test is either positive or not a slight extension of (2.4.3) gives $P[\tilde{S}|D] = 1 - P[S|D] = 0·04$ (the *likelihood* of D if $\tilde{S}$ is observed). The proportion of patients with cancer giving a negative test (false negatives) is 4 per cent. Suppose also that 95 per cent of patients without cancer give a negative test, $P[\tilde{S}|\bar{D}] = 0·95$. Similarly $P[S|\bar{D}] = 1 - 0·95 = 0·05$, i.e. 5 per cent of patients without cancer give a positive test (false positives). As diagnostic tests go, these proportions of false results are not outrageous. But now consider what happens if the test is applied to a population of patients of whom 1 in 200 (0·5 per cent) suffer from cancer, i.e. $P[D] = 0·005$ (the prior probability of D) and $P[\bar{D}] = 1 - 0·005 = 0·995$ (from (2.4.3) again). What is the probability that a patient reacting positively to the test actually has cancer? In symbols this is $P[D|S]$, the posterior probability of D after observing S, and from (2.4.7) or (2.4.9), and (2.4.8) it is, using the probabilities assumed above,

$$P[D|S] = \frac{P[S|D].P[D]}{P[S|D].P[D] + P[S|\bar{D}].P[\bar{D}]}$$

$$= \frac{0·96 \times 0·005}{(0·96 \times 0·005) + (0·05 \times 0·995)}$$

$$= \frac{0·0048}{0·0048 + 0·04975}$$

$$= \frac{0·0048}{0·05455}$$

$$= 0·0880. \qquad (2.4.11)$$

In other words, only 8·80 per cent of positive reactors actually have cancer, and $100 - 8·80 = 91·2$ per cent do not have cancer. Not such a good performance. It remains true that 96 per cent of those with cancer are detected by the test, but a great many more without cancer also give positive tests.

It is easy to see how this arises without the formality of Bayes' theorem. Suppose that 100000 patients are tested. On average 500 $(= 100000 \times 0·005)$ will have cancer and 99500 will not have cancer. Of the 500 with cancer, $500 \times 0·96 = 480$ will give positive reactions on average. Of the 99500 without cancer, $99500 \times 0·05 = 4975$ will give positive reactions on average (a much smaller proportion, but a much larger number than for the patients with cancer). Of the total number of positive reactors, $480 + 4975 = 5455$, the *number* with cancer is 480 and the *proportion* with cancer is $480/5455 = 0·0880$ as above. If these numbers are divided by the total number of patients, 100000, they are seen to coincide with the probabilities calculated by Bayes' theorem in (2.4.11).

2.5. Averages

If a number of replicate observations is made of a variable quantity it is commonly found that the observations tend to cluster round some central value. Some sort of average of the observations is taken as an estimate of the *true* or *population* value (see § 1.2) of the quantity that is being measured. Some of the possible sorts of average will be defined now. It can be seen that there is no logical reason for the automatic use of the ordinary unweighted arithmetic mean, (2.5.2). If the distribution of the observations is not symmetrical it may be quite inappropriate, and nonparametric methods usually use the median (see §§ 4.5, 6.2, and 7.3 and Chapters 9, 10, and 14).

The arithmetic mean

The general form is the *weighted arithmetic sample mean* (using the notation described in § 2.1),

$$\bar{x} = \frac{\Sigma w_i x_i}{\Sigma w_i}. \tag{2.5.1}$$

This provides an estimate, from a sample of observations, of the unknown *population mean* value of x (as long as the sample was taken strictly at random, see § 2.3). The population mean is the mean of all the values of x in the population from which the sample was taken and will be denoted μ (see § A1.1 for a more rigorous definition).

The weight of an observation. The weight, w_i, associated with the ith observation, x_i, is a measure of the relative importance of the observation in the final result. Usually the weight is taken as the reciprocal of the variance (see § 2.6 and (2.7.12)), so the observations with the smallest scatter are given the greatest weight. If the observations are uncorrelated, this procedure gives the best estimate of the population mean, i.e. an unbiased estimate with minimum variance (maximum precision). (See §§ 12.1 and 12.7, Appendix 1, and Brownlee (1965, p. 95).) From (2.7.8) it is seen that halving the variance is equivalent to doubling the number of observations. Both double the weight. See § 13.4 for an example.

Weights may also be arbitrarily decided degrees of relative importance. For example, if it is decided in examination marking that the mark for an essay paper should have twice the importance of the marks for the practical and oral examinations, a weighted average mark could be found by assigning the essay mark (say 70 per cent) of a weight of 2 and the practical and oral marks (say 30 and 40 per cent) a weight of one each. Thus

$$\bar{x} = \frac{(2 \times 70) + (1 \times 30) + (1 \times 40)}{2 + 1 + 1} = 52 \cdot 5.$$

If the weights had been chosen as 1, 0·5, and 0·5 the result would have been exactly the same.

The definition of the weighted mean has the following properties.

(a) If all the weights are the same, say $w_i = w$, then the ordinary unweighted arithmetic mean is found; (using (2.1.5) and (2.1.6)),

$$\bar{x} = \frac{\Sigma w x_i}{\Sigma w} = \frac{w \Sigma x_i}{N w} = \frac{\Sigma x_i}{N}. \tag{2.5.2}$$

In the above example the unweighted mean is $\Sigma x_i / N = (70 + 30 + 40)/3 = 46 \cdot 7$.

(b) If all the observations (values of x_i) are the same, then $\bar{x}$ has this value whatever the weights.

(c) If one value of x has a very large weight compared with the others then $\bar{x}$ approaches that value, and conversely if its weight is zero an observation is ignored.

The geometric mean

The unweighted geometric mean of N observations is defined as the Nth root of their product (cf. arithmetic mean which is the Nth part of their sum).

$$\tilde{x} = \left(\prod_{i=1}^{i=N} x_i \right)^{1/N}. \tag{2.5.3}$$

It will now be shown that this is the sort of mean found when the arithmetic mean of the logarithms of the observations is calculated (as, for example, in § 12.6), and the antilog of the result found.

Call the original observations z_i, and their logarithms x_i, so

$$x_i = \log z_i$$

$$\text{Arithmetic mean of log } z = \bar{x} = \overline{(\log z)} = \frac{\Sigma x_i}{N} = \frac{\Sigma(\log z_i)}{N}$$

$$= \frac{\log(\Pi z_i)}{N} \quad \begin{array}{l}\text{because the sum of the logs is the log of the} \\ \text{product}\end{array}$$

$$= \log \{\sqrt[N]{(\Pi z_i)}\}$$

$$= \log \text{ (geometric mean of } z)$$

or, taking antilogs,

$$\text{antilog (arithmetic mean of log } z) = \text{geometric mean of } z. \quad (2.5.4)$$

This relationship is the usual way of calculating the geometric mean. For the figures used above the unweighted geometric mean is the cube root of their product, $(70 \times 30 \times 40)^{1/3} = 43 \cdot 8$. The geometric mean of a set of figures is *always* less than their arithmetic mean, as in this case. If even a single observation is zero, the geometric mean will be zero.

The median

The population (or true) median is the value of the variable such that half the values in the population fall above it and half above it (i.e. it is the value bisecting the area under the distribution curve, see Chapters 3 and 4). It is not necessarily the same as the population mean (see § 4.5). The population median is estimated by the

$$\text{sample median} = \text{central observation}. \quad (2.5.5)$$

This is uniquely defined if the number of observations is odd. The median of the 5 observations 1, 4, 9, 7, 6, is seen, when they are ranked in order of increasing size giving, 1, 4, 6, 7, 9, to be 6. If there is an even number of observations the sample median is taken half-way between two central observations; for example the sample median of 1, 4, 6, 7, 9, 12 is $6\frac{1}{2}$.

The mode

The sample mode is the most frequently observed value of a variable. The population mode is the value corresponding to the peak of the population distribution curve (Chapters 3 and 4). It may be different from the mean and median (see § 4.5).

The arithmetic mean as a least squares estimate

This section anticipates the discussion of estimation in § 2.6, Chapter 12, and § A1.3. The arithmetic mean of a sample is said to be a least squares estimate (see Chapter 12) because it is the value that best represents the sample in the sense that the sum of the squares of the deviations of the observations from the arithmetic mean, $\Sigma(x_i - \bar{x})^2$, is smaller than the sum of the squares of the deviations from any other value. This can be shown without using calculus as follows.

Suppose, as above, that the sample consists of N observations, $x_1, x_2, ..., x_N$. It is required to find a value of m that makes $\Sigma(x_i - m)^2$ as small as possible. This follows immediately from the algebraic identity

$$\Sigma(x_i - m)^2 = \Sigma(x_i - \bar{x})^2 + N(\bar{x} - m)^2. \qquad (2.5.6)$$

The values of m that minimizes this is clearly $m = \bar{x}$, the arithmetic mean, because this makes the last term zero; as small at it can be. For the example following (2.5.2), the sum of squares

$$\Sigma(x_i - \bar{x})^2 = (70 - 46 \cdot 7)^2 + (30 - 46 \cdot 7)^2 + (40 - 46 \cdot 7)^2 = 866 \cdot 7,$$

and a few trials will show that inserting any value other than $46 \cdot 7$ makes the sum of squares larger than $866 \cdot 7$.

The intermediate steps in establishing (2.5.6) are easy. By definition of the arithmetic mean $N\bar{x} = \Sigma x_i$, so the right-hand side of (2.5.6) can be written, completing the squares and using (2.1.6), as

$$\Sigma(x_i^2 - 2x_i\bar{x} + \bar{x}^2) + N\bar{x}^2 - 2N\bar{x}m + Nm^2$$
$$= \Sigma x_i^2 - 2\bar{x}\Sigma x_i + N\bar{x}^2 + N\bar{x}^2 - 2N\bar{x}m + Nm^2$$
$$= \Sigma x_i^2 - 2N\bar{x}^2 + N\bar{x}^2 + N\bar{x}^2 - 2N\bar{x}m + Nm^2$$
$$= \Sigma x_i^2 - 2N\bar{x}m + Nm^2$$
$$= \Sigma(x_i^2 - 2mx_i + m^2) = \Sigma(x_i - m)^2$$

as stated in (2.5.6).

Using calculus the same result can be reached more elegantly. The usual way of finding a minimum in calculus is to differentiate and equate the result to zero (see Thompson (1965, p. 78)). This process is described in detail, and illustrated, in Chapter 12. In this case $\Sigma(x_i - m)^2$ is to be minimized with respect to m. Differentiating $\Sigma(x_i - m)^2 = \Sigma x_i^2 - 2m\Sigma x_i + Nm^2$, remembering that the x_i are constants for a given sample, and equating to zero, gives

$$\frac{d}{dm}[\Sigma(x_i - m)^2] = -2\Sigma x_i + 2Nm = 0. \qquad (2.5.7)$$

Therefore $2Nm = 2\Sigma x_i$

and $m = \Sigma x_i / N = \bar{x}$

as found above.

2.6. Measures of the variability of observations

When replicate observations are made of a variable quantity the scatter of the observations, or the extent to which they differ from each other, may be large or may be small. It is useful to have some quantitative measure of this scatter. See Chapter 7 for a discussion of the way this should be done. As there are many sorts of average (or 'measures of location'), so there are many measures of scatter. Again separate symbols will be used to distinguish the estimates of quantities calculated from (more or less small) samples of observations from the true values of these quantities, which could only be found if the whole population of possible observations were available.

The range

The difference between the largest and smallest observations is the simplest measure of scatter but it will not be used in this book.

The mean deviation

If the deviation of each observation from the mean of all observations is measured, then the sum of these deviations is easily proved (and this result will be needed later on) to be always zero. For example, consider the figures 5, 1, 2, and 4 with mean = 3. The deviations from the mean are respectively $+2$, -2, -1, and $+1$ so the total deviation is zero. In general (using (2.1.6), (2.1.8), and (2.5.2)),

$$\sum_{i=1}^{N} (x_i - \bar{x}) = \Sigma x_i - N\bar{x} = N\bar{x} - N\bar{x} = 0. \qquad (2.6.1)$$

If, however, the deviations are all taken as positive their sum (or mean) *is* a measure of scatter.

The standard deviation and variance

The standard deviation is also known, more descriptively, as the root mean square deviation. The population (or true) value will be denoted $\sigma(x)$. It is defined more exactly in § A1.2. The estimate of the population value calculated from a more or less small sample of, say, N observations, the *sample standard deviation*, will be denoted $s(x)$. The square of this quantity is the *estimated* (or *sample*) *variance* (or *mean square deviation*) of x, var(x) or $s^2(x)$. The population (or true) variance will be denoted $var(x)$ or $\sigma^2(x)$. The estimates are calculated as

$$s(x) = \sqrt{\left[\frac{\Sigma(x_i-\bar{x})^2}{N-1}\right]}$$

$$(2.6.2)$$

$$\text{var}(x) \text{ or } s^2(x) = \frac{\Sigma(x_i-\bar{x})^2}{N-1}.$$

The standard deviation and variance are said to have $N-1$ *degrees of freedom*. In calculating the mean value of $(x_i-\bar{x})^2$, $N-1$, rather than N, is used because the sample mean $\bar{x}$ has been used in place of the population mean μ. This would tend to make the estimate too small if N were used (the deviations of the observations from μ will tend to be larger than the deviations from $\bar{x}$; this can be seen by putting $m = \mu$ in (2.5.6)). It is not difficult to show that the use of $N-1$ corrects this tendency (see § A1.3).† It also shows that no information about scatter can be obtained from a single observation if μ is unknown (in this case the number of degrees of freedom, on which the accuracy of the estimate of the scatter depends, will be $N-1 = 0$). If μ were known even a single observation could give information about scatter, and, as expected from the foregoing remarks, the estimated variance would be a straightforward mean square deviation using N not $N-1$.

$$s^2(x) = \frac{\Sigma(x_i-\mu)^2}{N}.$$

$$(2.6.3)$$

A numerical example of the calculation of the sample standard deviation is provided by the following sample of $N = 4$ observations with arithmetic mean $\bar{x} = 12/4 = 3$.

---

† If the 'obvious' quantity, N, were used as the denominator in (2.6.2) the estimate of σ^2 would be *biased* even if the observations themselves were perfectly free of bias (systematic errors). This sort of bias results only from the way the observations are treated (another example occurs in § 12.8). Notice also that this implies that the mean of a very large number of values of $\Sigma(x-\bar{x})^2/N$ would tend towards too small a value, viz. $var(x) \times (N-1)/N$, as the number of values, each calculated from a small sample, increases; whereas the same formula applied to a single very large sample would tend towards $var(x)$ itself as the size of the sample (N) increases. These results are proved in § A1.3. It should be mentioned that unbiasedness is not the only criterion of a good statistic and other criteria give different divisors, for example N or $N+1$.

x_i	$x_i - \bar{x}$	$(x_i - \bar{x})^2$
5	+2	+4
1	−2	+4
2	−1	+1
4	+1	+1
Totals 12	0	10

Thus $\sum_{i=1}^{i=4} (x_i - \bar{x})^2 = 10$ so, from (2.6.2), $s(x) = \sqrt{(10/3)} = 1\cdot 83$.

The coefficient of variation

This is simply the standard deviation expressed as a proportion (or percentage) of the mean (as long as the mean is not zero of course).

$$C(x) = \frac{s(x)}{\bar{x}} \quad \text{(sample value)},$$

$$\mathscr{C}(x) = \frac{\sigma(x)}{\mu} \equiv \sqrt{\left[\frac{var(x)}{\mu^2}\right]} \quad \text{(population value)}, \qquad (2.6.4)$$

where μ is the population mean value of x (and of $\bar{x}$). $C(x)$ is an estimate, from a sample, of $\mathscr{C}(x)$.

Whereas the standard deviation has the same dimensions (seconds, metres, etc.) as the mean, the coefficient of variation is a dimensionless ratio and gives the *relative* size of the standard deviation. If the scatter of means (see § 2.7) rather than the scatter of individual observations, were of interest $C(\bar{x})$ would be calculated with $s(\bar{x})$ in the numerator.

In the numerical example above $C(x) = s(x)/\bar{x} = 1\cdot 83/3 = 0\cdot 61$, or $100\,C(x) = 100 \times 0\cdot 61 = 61$ per cent.

The working formula for the sum of squared deviations

When using a desk calculating machine it is inconvenient to form the individual deviations from the mean and the sum of squared deviations is usually found by using the following identity. Using (2.1.6) and (2.1.8),

$$\Sigma(x_i - \bar{x})^2 = \Sigma(x_i^2 - 2x_i\bar{x} + \bar{x}^2) = \Sigma x_i^2 - 2\bar{x}\Sigma x_i + N\bar{x}^2.$$

Now, since $\Sigma x_i = N\bar{x}$, this becomes

$$\Sigma x_i^2 - 2N\bar{x}^2 + N\bar{x}^2 = \Sigma x_i^2 - N\bar{x}^2$$

and thus

$$\sum_{i=1}^{i=N} (x_i - \bar{x})^2 = \sum_{i=1}^{i=N} x_i^2 - \frac{(\sum^N x_i)^2}{N} \qquad (2.6.5)$$

In the above example $\Sigma x_i^2 = 5^2 + 1^2 + 2^2 + 4^2 = 46$ and therefore $\Sigma(x_i - \bar{x})^2 = 46 - 12^2/4 = 46 - 36 = 10$, as found above.

The covariance

This quantity is a measure of the extent to which two variables are correlated. Uncorrelated events are defined as those that have zero covariance, and statistically independent events are necessarily uncorrelated (though uncorrelated events *may* not be independent—see § 12.9).

The true, or population, covariance of x with y will be denoted $cov(x,y)$, and the estimate of this quantity from a sample of observations is

$$cov(x,y) = \frac{\Sigma(x - \bar{x})(y - \bar{y})}{N - 1}. \qquad (2.6.6)$$

The numerator is called the *sum of products*. That the value of this expression will depend on the extent to which y increases with x is clear from Fig. 2.6.1 in which, for example, y might represent body weight and x calorie intake. Each point represents one pair of observations.

If the graph is divided into quadrants drawn through the point $\bar{x}, \bar{y}$ it can be seen that any point in the top right or in the bottom left quadrant will contribute a positive term $(x - \bar{x})(y - \bar{y})$ to the sum of products, whereas any point in the other two quadrants will contribute a negative value of $(x - \bar{x})(y - \bar{y})$. Therefore the points shown in Fig. 2.6.2(a) would have a large positive covariance, the points in Fig. 2.6.2.(b) would have a large negative covariance, and those in Fig. 2.6.2(c) would have near zero covariance.

The working formula for the sum of products

A more convenient expression for the sum of products can be found in a way exactly analogous to that used for the sum of squares (2.6.5). It is

$$\Sigma(x-\bar{x})(y-\bar{y}) = \Sigma xy - \frac{(\Sigma x)(\Sigma y)}{N}. \tag{2.6.7}$$

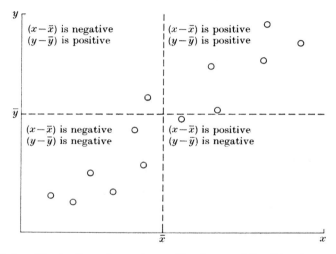

FIG. 2.6.1.   Illustration of covariance. For eleven of the thirteen observations, the product $(x_i-\bar{x})(y_i-\bar{y})$ is positive; for the other two it is negative.

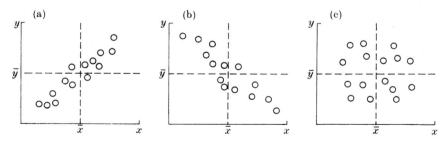

FIG. 2.6.2.   Illustration of covariance: (a) positive covariance as in Fig. 2.6.1; (b) negative covariance; (c) near zero covariance.

2.7. What is a standard error? Variances of functions of the observations. A reference list

Prediction of variances without direct measurement

A problem that recurs continually is the prediction of the scatter of some function calculated from the observations, using the internal evidence of one experiment.

Suppose, for example, that it is wished to know what degree of confidence can be placed in an observed sample mean, the mean, $\bar{x}$, of a single sample of, say, N values of x selected from some specified population (see § 1.2). A numerical example is given below. The sample mean, $\bar{x}$, is intended as an estimate of the population mean, μ. How good an estimate is it? The direct approach would be to repeat the experiment many times, each experiment consisting of (a) making N observations (i.e. selecting N values of x from the population) and (b) calculating their mean, $\bar{x}$. In this way a large set of means would be obtained. It would be expected that these means would agree with each other more closely (i.e. would be more closely grouped about the population mean μ) than a large set of single observations. And it would be expected that the larger the number (N) of observations averaged to find each mean, the more closely the means would agree with each other. If the set of means was large enough their distribution (see Chapter 4) could be plotted. Its mean would be μ, as for the x values (= 'means of samples of size $N = 1$'), but the standard deviation of the population of $\bar{x}$ values, $\sigma(\bar{x})$ say, would be less than the standard deviation, $\sigma(x)$, of the population of x values, as shown in Fig. 2.7.1. The closeness with which the means agree with each other is a measure of the confidence that could be placed in a single mean as an estimate of μ. And this closeness can be measured by calculating the variance of the set of sample means, using the means ($\bar{x}$ values) as the set of figures to which (2.6.2) is applied, giving var($\bar{x}$), or $s^2(\bar{x})$, as an estimate of $\sigma^2(\bar{x})$, as illustrated below.

If (2.6.2) was applied to a set of observations (x values), rather than a set of sample means, the result would be var(x), an estimate of the scatter of repeated observations. As it has been mentioned that a set of means would be expected to agree with each other more closely than a set of single observations, it would be expected that $var(\bar{x})$ would be smaller than $var(x)$, and this is shown to be so below ((2.7.8)). The standard deviation of the mean, $s(\bar{x}) = \sqrt{\text{var}(\bar{x})}$, is often called the *sample standard error* of the mean to distinguish it from $s(x)$, the sample

standard deviation of x, or 'sample standard deviation of the observations'. This term is unnecessary and *sample standard deviation of the
mean* is a preferable name for $s(\bar{x})$.

The sample standard deviation of the observations, $s(x)$, is the
quantity of interest if one wants to estimate the scatter of x values
(single observations from the population). In other words, it measures
the inherent variability of the population. Taking a larger sample

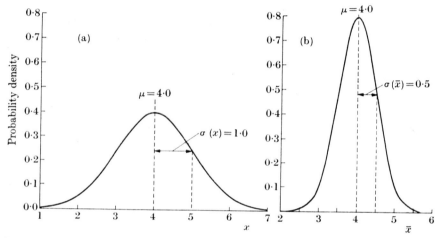

FIG. 2.7.1. (a) Distribution of observations (x values) in the population.
The area under the curve between any two x values is the probability of an
observation falling between these two values, so the total area under the curve is
1·0 (see Chapter 4 for details). This particular distribution is Gaussian but the
results in this chapter are valid for any distribution (though the standard deviation has a simple interpretation only when the distribution is Gaussian). The
mean value of x is 4·0 and the standard deviation, $\sigma(x)$, is 1·0. (b) The distribution
of $\bar{x}$ values. $\bar{x}$ is the mean of a sample of four x values from the population represented in (a). The area under this curve must be 1·0, like the distribution in (a).
To keep the area the same, the distribution must be taller, because it is narrower
(i.e. the $\bar{x}$ values have less scatter than the x values). The ordinate and abscissa
are drawn on the same scale in (a) and (b). The mean value of $\bar{x}$ is 4·0 and its
standard deviation, $\sigma(\bar{x})$ (the 'standard error of the mean'), is 0·5.

makes $s(x)$ a more *accurate* estimate of the population value, $\sigma(x)$.
On the other hand, the sample standard deviation of the mean, $s(\bar{x})$,
is the quantity of interest if one wants to estimate the accuracy of a
sample mean, $\bar{x}$. It is used if the object of making the observations is
to estimate the population mean, rather than to estimate the inherent
variability of the population. Taking a larger sample makes $s(\bar{x})$ smaller,

on the average, because it is an estimate of $\sigma(\bar{x})$ (the population 'standard error'), which is smaller than $\sigma(x)$.

The standard deviation of the mean may sometimes be measured, for special purposes, by making measurements of the sample mean (see Chapter 11, for example). It is the purpose of this section to show that the result of making repeated observations of the sample mean (or of any other function calculated from the sample of observations), for the purpose of measuring their scatter, can be predicted indirectly. If the scatter of the means of four observations were required it could be found by observing many such means and calculating the variance of the resulting figures from (2.6.2), but alternatively it could be predicted using 2.7.8) (below) even if there were only four observations altogether, giving only a single mean. An illustration follows.

A numerical example to illustrate the idea of the standard deviation of the mean

Suppose that one were interested in the precision of the mean found by averaging 4 observations. It could be found by determining the mean several times and seeing how closely the means agreed. Table 2.7.1 shows three sets of four observations. (It is not the purpose of this section to show how results of this sort would be analysed in practice.

TABLE 2.7.1.

Three random samples, each with $N = 4$ observations, from a population with mean $\mu = 4 \cdot 00$ and standard deviation $\sigma(x) = 1 \cdot 00$

	Sample 1	Sample 2	Sample 3	
x values	3·39	5·88	3·79	
	3·38	2·45	3·25	
	3·89	2·21	4·07	
	6·36	5·96	3·21	
Sample mean, $\bar{x}$	4·40	4·12	3·58	Grand mean = 4·04
Sample standard deviation, $s(x)$	1·33	2·08	0·420	

That is dealt with in Chapter 11.) The observations were all selected randomly from a population *known* (because it was synthetic, not experimentally observed) to have mean $\mu = 4 \cdot 00$ and standard deviation $\sigma(x) = 1 \cdot 00$, as shown in Fig. 2.7.1(a)

The sample means, $\bar{x}$, of the three samples, 4·40, 4·12, and 3·58, are all estimates of $\mu = 4\cdot00$. The grand sample mean, 4·04, is also an estimate of $\mu = 4\cdot00$ (see Appendix 1). The standard deviations, $s(x)$, of each of the three samples, 1·33, 2·08, and 0·420, are all estimates of the population standard deviation, $\sigma(x) = 1\cdot00$ (a better estimate could be found by averaging, or pooling, these three estimates as described in § 11.4). The population standard deviation of the mean can be estimated directly by calculating the standard deviation of the sample of three means (4·40, 4·12, 3·58) using (2.6.2). This gives

$$s(\bar{x}) = \sqrt{\left[\frac{(4\cdot40-4\cdot04)^2+(4\cdot12-4\cdot04)^2+(3\cdot58-4\cdot04)^2}{3-1}\right]}$$

$$= 0\cdot420.$$

Now according to (2.7.9) (see below), if we had an infinite number of $\bar{x}$ values instead of only 3, their standard deviation would be $\sigma(\bar{x}) = \sigma(x)/\sqrt{N} = 1\cdot00/\sqrt{4} = 0\cdot5$ (see Fig. 2.7.1(b)), and $s(\bar{x}) = 0\cdot420$ is a sample estimate of this quantity. And, furthermore, if we had only one sample of observations it would still be possible to estimate indirectly $\sigma(\bar{x}) = 0\cdot5$, by using (2.7.9). For example, with only the first group, $s(\bar{x}) = s(x)/\sqrt{N} = 1\cdot33/\sqrt{4} = 0\cdot665$ could be found as an estimate of $\sigma(\bar{x}) = 0\cdot500$, i.e. as a prediction of what the scatter of means *would* be *if* they were repeatedly determined. (This prediction refers to repeated samples from the *same* population. If the repeated samples were from different populations the prediction would be an under-estimate, as described in Chapter 11.)

A reference list

   The problem dealt with throughout this section has been that of predicting what the scatter of the values of various functions of the observations (such as the mean) would be *if* repeated samples were taken and the value of the function calculated from each. The aim is to predict this, given a single sample containing any number of observations that happen to be available (not fewer than two of course).

   The relationships listed below will be referred to frequently later on. The derivations should really be carried out using the definition of the population variance (§ A1.2 and Brownlee (1965, p. 57)), but it will, for now, be sufficient to use the sample variance (2.6.2). The results, however, are given properly in terms of population variances. The notation was defined in §§ 2.1 and 2.6.

Variance of the sum or difference of two variables. Given the variance of the values of a variable x, and of another variable y, what is the predicted variance of the figures found by adding an x value to a y value? From, (2.6.2),

$$\text{var}(x+y) = \frac{\sum\limits_{i=1}^{i=N}[(x_i+y_i)-\overline{(x+y)}]^2}{N-1}.$$

Now, since $\overline{(x+y)} = \Sigma(x_i+y_i)/N = \Sigma x_i/N + \Sigma y_i/N = \bar{x}+\bar{y}$ (from 2.1.8), this can be rearranged giving

$$\text{var}(x+y) = \frac{\Sigma[(x_i-\bar{x})+(y_i-\bar{y})]^2}{N-1}$$

$$= \frac{\Sigma[(x_i-\bar{x})^2+(y_i-\bar{y})^2+2(x_i-\bar{x})(y_i-\bar{y})]}{N-1}$$

$$= \frac{\Sigma(x_i-\bar{x})^2}{N-1}+\frac{\Sigma(y_i-\bar{y})^2}{N-1}+\frac{2\Sigma(x_i-\bar{x})(y_i-\bar{y})}{N-1}$$

suggesting, from (2.6.2) and (2.6.6), the general rule

$$var(x+y) = var(x)+var(y)+2\ cov(x,y). \tag{2.7.1}$$

By a similar argument the variance of the *difference* between two variables is found to be

$$var(x-y) = var(x)+var(y)-2\ cov(x,y). \tag{2.7.2}$$

Thus if the variables are uncorrelated, i.e. if $cov(x,y) = 0$, then the variance of either the sum or the difference is simply the *sum* of the separate variances

$$var(x+y) = var(x-y) = var(x)+var(y). \tag{2.7.3}$$

If variables are independent they are necessarily uncorrelated (see §§ 2.4 and 12.9), so (2.7.3) is valid for independent variables.

Variance of the sum of N variables. By a simple extension of the above argument for two variables, if $x_1, x_2, ..., x_N$ are N *uncorrelated* variables then (2.7.3) can be generalized giving

$$\begin{aligned} var(\Sigma x_i) &\equiv var(x_1+x_2+...+x_N) \\ &= var(x_1)+var(x_2)+...+var(x_N) \\ &= \Sigma var(x_i), \text{ or } N var(x), \end{aligned} \tag{2.7.4}$$

the second form being appropriate (cf. (2.1.6)) if all the x_i have the same variance, $var(x)$.

The effect of multiplying each x by a constant factor. If a is a constant then, by (2.6.2), the variance of the set of figures found by multiplying each of a set of x values by the same figure, a, will be

$$\text{var}(ax) = \frac{\Sigma(ax_i - \overline{ax})^2}{N-1} = \frac{\Sigma[a(x_i - \bar{x})]^2}{N-1} = \frac{a^2 \cdot \Sigma(x_i - \bar{x})^2}{N-1}$$

suggesting the general rule

$$var(ax) = a^2 var(x); \tag{2.7.5}$$

and similarly from (2.6.6),

$$cov(ax, by) = ab \; cov(x,y) \tag{2.7.6}$$

where a and b are constants.

The effect of adding a constant to each x. By similar arguments to those above it can be seen, from (2.6.2), that adding a constant has no effect on the scatter.

$$var(a+x) = var(x). \tag{2.7.7}$$

The variance of the mean of N observations and the standard error. This relationship, the answer to the problem of how to estimate indirectly the scatter of repeated observations of the mean discussed, with a numerical example, at the beginning of this section, follows from those already given.

$$var(\bar{x}) \equiv var\left(\frac{\Sigma x_i}{N}\right) = \frac{1}{N^2} var(\Sigma x_i) \qquad \text{(from (2.7.5))}$$

$$= \frac{N}{N^2} var(x) \qquad \text{(from (2.7.4))}$$

and therefore the variance of the mean is

$$var(\bar{x}) = \frac{var(x)}{N} \tag{2.7.8}$$

and the *standard deviation of the mean* (the standard error, see discussion above) is

$$\sigma(\bar{x}) = \sqrt{[var(\bar{x})]} = \frac{\sigma(x)}{\sqrt{N}}. \tag{2.7.9}$$

Notice that var(x), being an average (like $\bar{x}$), will be more or less the same whatever size sample is used to estimate it (though a larger sample will give a more *precise* value), whereas var($\bar{x}$) becomes smaller as the number of observations averaged increases, as expected from the discussion at the beginning of this section and from (2.7.8).

The variance of a linear function of the observations. A linear function of the observations x_1, x_2, ..., x_n is defined as

$$L = a_0 + a_1 x_1 + a_2 x_2 + ... + a_n x_n \equiv a_0 + \sum_{i=1}^{n} a_i x_i$$

where the a_i are constants. From (2.7.7) it can be seen that a_0 has no effect on the variance and can be ignored. Using (2.7.4) it can be seen that, *if* the observations are uncorrelated,

$$var(L) = var(\Sigma a_i x_i) = \Sigma var(a_i x_i)$$

and using (2.7.5) this becomes

$$var(L) = \Sigma(a_i^2 var(x_i)), \qquad (2.7.10)$$

$$\text{or } var(L) = var(x).\Sigma a_i^2$$

if the variances of all the x_i are the same, $var(x)$ say.

If the x_i do *not* have zero convariances (are not uncorrelated) a more general form is necessary. Using (2.7.5), (2.7.5), and an extension of (2.7.1) this is found to be

$$var(L) = \Sigma a_i^2 var(x_i) + 2 \sum_{\substack{i \\ i \neq j}} \sum_{j} a_i a_j \, cov(x_i, x_j), \qquad (2.7.11)$$

where the second term is the sum of all possible pairs of covariances. For example if $L = a_1 x_1 + a_2 x_2 + a_3 x_3$, then $var(L) = a_1^2 \, var(x_1) + a_2^2 \, var(x_2) + a_3^2 \, var(x_3) + 2a_1 a_2 \, cov(x_1,x_2) + 2a_1 a_3 \, cov(x_1,x_3) + 2a_2 a_3 \, cov(x_2,x_3)$.

The variance of the weighted arithmetic mean. The variance of the weighted mean, defined in (2.5.1), follows from (2.7.5) and (2.7.10).

$$var\left(\frac{\Sigma w_i x_i}{\Sigma w_i}\right) = \frac{1}{(\Sigma w_i)^2} var(\Sigma w_i x_i) = \frac{\Sigma[w^2 var(x_i)]}{(\Sigma w_i)^2}.$$

Now if $w_i = 1/var(x_i)$, as discussed in § 2.5, then $\Sigma[w^2 \, var(x_i)] = \Sigma w_i$ so

$$var\left(\frac{\Sigma w_i x_i}{\Sigma w_i}\right) = \frac{1}{\Sigma w_i}, \qquad (2.7.12)$$

and if all the weight (variances) are the same this reduces to (2.7.8).

The approximate variance of any function. The variance of any function $f(x_1, x_2, ..., x_n)$ of the *uncorrelated* variables x_1, x_2, ..., x_n is given *approximately* (taking only the linear terms in a Taylor series expansion of f) by

$$var(f) \simeq \left(\frac{\partial f}{\partial x_1}\right)^2 var(x_1) + \left(\frac{\partial f}{\partial x_2}\right)^2 var(x_2) + ... + \left(\frac{\partial f}{\partial x_n}\right)^2 var(x_n). \qquad (2.7.13)$$

if the variances are reasonably small relative to the means, so that the function can be represented approximately by a straight line in the range over which each x varies. The derivatives should be evaluated at the true mean value of the x variables. If the x variables are correlated then terms involving their covariances must be added as shown below. For discussion and the derivation see, for example, Lindley (1965, p. 134), Brownlee (1965, p. 144), and Kendall and Stuart (1963, p. 231).

If f is a linear function then (2.7.13) reduces to (2.7.10), which is exact. If f is not linear then the result is only approximate and furthermore f will not have the same distribution of errors as the x variables so if, for example, the x values were normally distributed (see § 4.2) f would not be normally distributed, so its variance, even if it were exact, could not be interpreted in any simple way.

The variance of $\log_e x$. If the true mean value of x is μ and then using the version of (2.7.13) for a single variable gives

$$var(\log_e x) \simeq \left(\frac{\mathrm{d}\,\log_e x}{\mathrm{d}x}\right)^2_{x=\mu} var(x) = \frac{var(x)}{\mu^2} = \mathscr{C}^2(x). \tag{2.7.14}$$

Therefore the standard deviation of $\log_e x$ is approximately equal to the coefficient of variation of x, $\mathscr{C}(x)$, defined by (2.6.4). If the standard deviation of x increases in proportion to the true mean value of x, so that the coefficient of variation of x is constant, the standard deviation of $\log_e x$ will be approximately constant (cf. §§ 11.2 and 12.2).

The variance of the product of two variables, $x_1 x_2$. In this case an exact result can be derived for the variance of values of $x_1 x_2$, given the variances of x_1 and of x_2. Suppose that x_1 and x_2 are independent of each other, and have population means μ_1 and μ_2 respectively. Then†

$$var(x_1 x_2) = var(x_1).var(x_2) + \mu_2^2\, var(x_1) + \mu_1^2\, var(x_2).$$

If this result is divided through by $(\mu_1\mu_2)^2$, it can be expressed in terms of coefficients of variation, defined in (2.6.4), as

$$\mathscr{C}^2(x_1 x_2) = \mathscr{C}^2(x_1).\mathscr{C}^2(x_2) + \mathscr{C}^2(x_1) + \mathscr{C}^2(x_2). \tag{2.7.15}$$

It is interesting to compare this with the result of applying the approximate formula, (2.7.13), viz.

$$var(x_1 x_2) \simeq \left(\frac{\partial x_1 x_2}{\partial x_1}\right)^2 var(x_1) + \left(\frac{\partial x_1 x_2}{\partial x_2}\right)^2 var(x_2)$$

$$= \mu_2^2\, var(x_1) + \mu_1^2\, var(x_2);$$

or, again dividing through by $(\mu_1\mu_2)^2$ to get the result in terms of coefficients of variation,

$$\mathscr{C}^2(x_1 x_2) \simeq \mathscr{C}^2(x_1) + \mathscr{C}^2(x_2).$$

† *Proof.* From appendix equation (A1.2.2), $var(x_1 x_2) = \mathrm{E}(x_1^2 x_2^2) - [\mathrm{E}(x_1 x_2)]^2$. Now $\mathrm{E}(x_1 x_2) = \mu_1 \mu_2$ and $\mathrm{E}(x_1^2 x_2^2) = \mathrm{E}(x_1^2).\mathrm{E}(x_2^2)$ if, as supposed, x_1 and x_2 are independent. Also, from (A1.2.2), $\mathrm{E}(x_1^2) = var(x_1) + \mu_1$, and similarly for x_2. Thus

$$var(x_1 x_2) = \mathrm{E}(x_1^2).\mathrm{E}(x_2^2) - \mu_1^2 \mu_2^2$$
$$= (var(x_1) + \mu_1^2)(var(x_2) + \mu_2^2) - \mu_1^2 \mu_2^2$$
$$= var(x_1).var(x_2) + \mu_2^2 var(x_1) + \mu_1^2 var(x_2)$$

as stated above.

By comparison with (2.7.15) it appears that use of the approximate formula involves neglecting the term $\mathscr{C}^2(x_1).\mathscr{C}^2(x_2)$. The approximation involved can be illustrated by two numerical examples.

First, suppose that both x_1 and x_2 have coefficients of variation of 50 per cent, i.e. $\mathscr{C}(x_1) = 0\cdot5$, $\mathscr{C}(x_2) = 0\cdot5$. In this case (2.7.15) gives

$$\mathscr{C}(x_1 x_2) = \sqrt{[(0\cdot5^2 \times 0\cdot5^2) + 0\cdot5^2 + 0\cdot5^2]} = \sqrt{(0\cdot0625 + 0\cdot25 + 0\cdot25)} = 0\cdot750,$$

i.e. 75·0 per cent. The approximate form gives

$$\mathscr{C}(x_1 x_2) \simeq \sqrt{(0\cdot5^2 + 0\cdot5^2)} = 0\cdot707,$$

i.e. 70·7 per cent. Secondly, consider more accurate observations, say 100 $\mathscr{C}(x_1)$ = 5 per cent and 100 $\mathscr{C}(x_2)$ = 5 per cent. Similar calculations show that (2.7.15) gives 100 $\mathscr{C}(x_1 x_2)$ = 7·075 per cent, whereas the approximate form gives 100 $\mathscr{C}(x_1 x_2) \simeq 7\cdot071$ per cent. The more accurate the observations, the better the approximate version will be.

The variance of the ratio of two variables, x_1/x_2. Using (2.7.13) gives, in terms of coefficients of variation, the approximate result

$$\mathscr{C}^2(x_1/x_2) \simeq \mathscr{C}^2(x_1) + \mathscr{C}^2(x_2). \qquad (2.7.16)$$

An exact treatment for the ratio of two normally distributed variables is given in § 13.5 and exemplified in §§ 13.11–13.15.

The variance of the reciprocal of a variable, $1/x$. According to (2.7.13),

$$var(1/x) \simeq \left(\frac{\mathrm{d}(1/x)}{\mathrm{d}x}\right)^2_{x=\mu} var(x) = \frac{var(x)}{\mu^4}. \qquad (2.7.17)$$

The weight (see § 2.5) to be attached to a value of $1/x$ is therefore approximately proportional to the fourth power of x if $var(x)$ is constant! This explains why plots involving reciprocal transformations may give bad results (see § 12.8 for details) if not correctly weighted.

Correlated variables. In the simplest case of two correlated variables, x_1 and x_2, the appropriate extension of (2.7.13) is

$$var[f(x_1,x_2)] \simeq \left(\frac{\partial f}{\partial x_1}\right)^2 var(x_1) + \left(\frac{\partial f}{\partial x_2}\right)^2 var(x_2) + 2\left(\frac{\partial f}{\partial x_1}\frac{\partial f}{\partial x_2}\right) cov(x_1,x_2). \qquad (2.7.18)$$

This relationship is referred to in § 13.5. For a linear function this reduces to the two variable case of (2.7.11). The n variable extension of (2.7.18) involves all possible pairs of x variables in the same way as (2.7.11).

Sum of a variable number of random variables. Let S denote the sum of a randomly variable number of random variables

$$S = \sum_{i=1}^{m} z_i,$$

where z_i are independent variables with coefficient of variation $\mathscr{C}(z)$ and m is a random variable with coefficient of variation $\mathscr{C}(m)$. If each S is made up of a random sample (of variable size) from the population of z values, then it is shown in § A1.4 that the coefficient of variation of S is

$$\mathscr{C}(S) = \sqrt{\left(\frac{\mathscr{C}^2(z)}{\mu_m} + \mathscr{C}^2(m)\right)}, \tag{2.7.19}$$

where μ_m is the population mean value of m (size of the sample). This result is illustrated on p. 58.

3. Theoretical distributions: binomial and Poisson

THE variability of experimental results is often assumed to be of a known mathematical form (or distribution) to make statistical analysis easy, though some methods of analysis need far more assumptions than others. These theoretical distributions are mathematical idealizations. The only reason for supposing that they will represent any phenomenon in the real world is comparison of real observations with theoretical predictions. This is only rarely done.

3.1. The idea of a distribution

If it were desired to discover the proportion of European adults who drink Scotch whisky then the population *involved* is the set of all European adults. From this population a *strictly random sample* (see § 2.3) of, say, twenty-five Europeans might be examined and the proportion of whisky drinkers in this sample taken as an estimate of the proportion in the population.

A similar statistical problem is encountered when, for example, the true concentration of a drug solution is estimated from the mean of a few experimental estimates of its concentration. Although it is convenient still to regard the experimental observations as samples from a population, it is apparent that in this case, unlike that discussed in the previous paragraph, the population has no physical reality but consists of the infinite set of all valid observations that might have been made.

The first example illustrates the idea of a *discontinuous probability distribution* (it is *not* meant to illustrate the way in which a single sample would be analysed). If very many samples, each of 25 Europeans, were examined it would not be expected that all the samples would contain exactly the same number of whisky drinkers. If the proportion of whisky drinkers in the whole population of European adults were 0·3 (i.e. 30 per cent) then it might reasonably be expected that samples containing about 7 or 8 cases would appear more frequently than samples containing any other number because $0·3 \times 25 = 7·5$. However samples containing about 5 or 10 cases would be frequent, and 3 (or fewer) or 13 (or more) drinkers would appear in roughly 1 in 20 samples. If a sufficient number of samples were taken it should be possible to discover something approaching the true proportion of samples containing r drinkers, r being any specified number between 0 and 25. These figures are called the probability distribution of the proportion of whisky drinkers in a sample of 25 and since this proportion is a discontinuous variable (the number of drinkers per sample must be a whole number) the distribution is described as discontinuous. The distribution is usually plotted as a block histogram as shown in Fig. 3.4.1 (p. 52), the block representing, say, 6 drinkers extending from 5·5 to 6·5 along the abscissa.

The second example, concerning the estimation of the true concentration of a drug solution, leads to the idea of a *continuous probability distribution*. If many estimates were made of the same concentration it would be expected that the estimates would not be identical. By analogy with the discontinuous case just discussed it should be possible, if a large enough number of estimates were made, to find the proportion of estimates having any given value. However, since the concentration is a continuous variable the problem is more difficult because the proportion of estimates having *exactly* any given value (e.g. *exactly* 12 μg/ml, that is 12·00000000...μg/ml) will obviously in principle be indefinitely small (in fact experimental difficulties will mean that the answer can only be given to, say, three significant figures so that in practice the concentration estimate will be a discontinuous variable). The way in which this difficulty is overcome is discussed in § 4.1.

3.2. Simple sampling and the derivation of the binomial distribution through examples

The binomial distribution predicts the probability, $P(r)$, of observing any specified number (r) of 'successes' in a series of n *independent trials* of an event, when the outcome of a trial can be of only two sorts ('success' or 'failure'), and when the probability of obtaining a 'success' is constant from trial to trial. If the conditions of independence and constant probability are fulfilled the process of taking a sample (of n trials) is described as *simple sampling*. When there are more than two possible outcomes a generalization of the binomial distribution known as the multinomial distribution is appropriate. Often it will not be possible *a priori* to assume that sampling is *simple* and when this is so it must be found out by experiment whether the observations are binomially distributed or not.

The example in this section is intended to illustrate the nature of the binomial distribution. It would not be a well-designed experiment to test a new drug because it does not include a control or placebo group. Suitable experimental designs are discussed in Chapters 8–11.

Suppose that n trials are made of a new drug. In this case 'one trial of an event' is one administration of the drug to a patient. After each trial it is recorded whether the patient's condition is apparently better (outcome B) or apparently worse (outcome W). It is assumed for the moment that the method of measurement is sensitive enough to rule out the possibility of no change being observed.

The derivation of the binomial distribution specifies that the probability of obtaining a success shall be the same at every trial. What exactly does this mean? If the n trials were all conducted on the same patient this would imply that the patient's reaction to the drug must not change with time, and the condition of independence of trials implies that the result of a trial must not be affected by the result of previous trials. The result would be an estimate of the probability of the drug producing an improvement in the single patient tested. Under these conditions the proportion of successes in repeated sets of n trials should follow the binomial distribution.

At first sight it might be thought, because it is doubtless true that the probability of a success outcome B, will differ from patient to patient, that if the n trials were conducted on n *different patients*, the proportion of successes in repeated sets of n trials would *not* follow the binomial distribution. This would quite probably be so if, for example, each set of n patients was selected in a different part of the country. However, if the sets of n patients were selected *strictly at random* (see § 2.3) from a *large* population of patients, then the proportion of patients in the population who will show outcome B (i.e. the probability, given random sampling, of outcome B) would not change between the selection of one patient and the next, or between the selection of one sample of n patients and the next. Therefore the conditions of constant probability and independence would be met in spite of the fact that patients differ in their reactions to drugs. Notice the critical importance of strictly random selection of samples, already emphasized in § 2.3.

From the rules of probability discussed in § 2.4 it is easy to find the probability of any specified result (number of successes out of n trials) if $\mathscr{P}(\text{B})$, the *true* (population) proportion of cases in which the patient improves, is known. This is a deductive, rather than inductive, procedure. A true probability is given and the probability of a particular result calculated. The reverse process, the inference of the population proportion from a sample, is discussed later in § 7.7 and exemplified by the assay of purity in heart by the Oakley method, described in § 7.8.

Two different drugs will be considered.

1. Suppose that drug X is completely inactive, but nevertheless 50 per cent of patients, in the long run, improve spontaneously, i.e.

$$\mathscr{P}(B) = 0.5 \text{ and } \mathscr{P}(W) = 1 - \mathscr{P}(B) = 0.5. \qquad (3.2.1)$$

2. Suppose that drug Y is effective, and that the percentage of patients improving in the long run is increased to 90 per cent. Thus

$$\mathscr{P}(B) = 0.9 \text{ and } \mathscr{P}(W) = 1 - \mathscr{P}(B) = 0.1. \qquad (3.2.2)$$

In both cases, because the outcomes B and W are mutually exclusive, the special case of the addition rule (2.4.2) gives $\mathscr{P}(B \text{ or } W) = \mathscr{P}(B) + \mathscr{P}(W)$, and because B and W are exhaustive (the only possible outcomes) $\mathscr{P}(B \text{ or } W) = 1$.

Two trial administrations of the drug ($n = 2$)

Out of two trials 0, 1, or 2 successes might be observed. The possible outcomes of the two trials are shown in Table 3.2.1 and from these probabilities, $P(r)$, of observing r successes ($r = 0$, 1, or 2), are calculated using the multiplication rule, (2.4.6), and the addition rule (2.4.2).

TABLE 3.2.1

r	1st trial	2nd trial	Prob. of outcome, $\mathscr{P}(r)$	$P(r)$ when $\mathscr{P}(B) = 0.5$	$P(r)$ when $\mathscr{P}(B) = 0.9$
0	W	W	$\mathscr{P}(W) \times \mathscr{P}(W)$	0·25	0·01
1	W	B	$\mathscr{P}(W) \times \mathscr{P}(B)$	0·25⎫	0·09⎫
1	B	W	$\mathscr{P}(B) \times \mathscr{P}(W)$	0·25⎬	0·09⎬ 0·18
2	B	B	$\mathscr{P}(B) \times \mathscr{P}(B)$	0·25	0·81
			Total	1·0	1·0

It can be seen that $\sum_{r=0}^{r=n} P(r) = 1.0$ in each case, as it should by the addition rule, because it is certain that r will take some value between 0 and n.

It is also clear from the table that the calculations are affected by the number of ways in which a given result can occur. One success out of two can occur in two ways, either at the first trial or at the second, so the probability of one success out of two trials, if the order in which they occur is immaterial, is 0·5 when $\mathscr{P}(B) = 0·5$ (drug X), and 0·18 when $\mathscr{P}(B) = 0·9$ (drug Y). This follows from the addition rule, being the probability of either (B at first trial **and** W at second) **or** (W at first trial **and** B at second).

The mean number (*expectation*, see Appendix 1) of successes out of n trials is $n\mathscr{P}$, i.e. 1 success out of 2 trials when $\mathscr{P}(B) = 0·5$ (drug X),

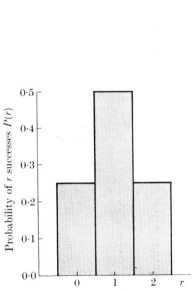 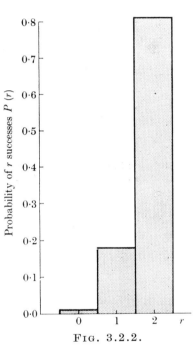

FIG. 3.2.1. Binomial distribution of r, the number of successes out of $n = 2$ trials of an event, the probability of success at each trial being $\mathscr{P} = 0·5$.

FIG. 3.2.2.
As in Fig. 3.2.1, but $\mathscr{P} = 0·9$.

and 1·8 successes out of two trials when $\mathscr{P}(B) = 0·9$ (drug Y). The results in the table are plotted in Figs. 3.2.1 and 3.2.2.

Three trial administrations of the drug $(n = 3)$

Calculations, exactly similar to those above are shown in Table 3.2.2 for the case when three trial administrations of the drug are performed. In this case there are three possible orders in which one success may occur in three trials, and the same number for two successes in three trials. Check the figures in the table to make sure you have got the idea.

These distributions are plotted in Figs. 3.2.3 and 3.2.4.

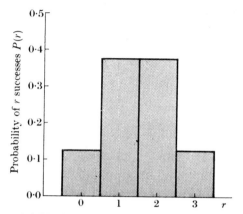

FIG. 3.2.3.   Binomial distribution of r, the number of successes out of $n = 3$ trials of an event, the probability of success at each trial being $\mathscr{P} = 0.5$.

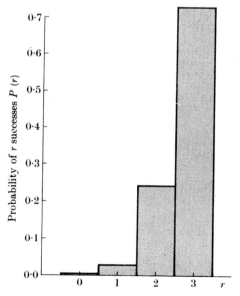

FIG. 3.2.4.   As in Fig. 3.2.3 but $\mathscr{P} = 0.9$.

TABLE 3.2.2

r	1st trial	2nd trial	3rd trial	$P(r)$ when $\mathscr{P}(B) = 0\cdot5$ (Drug X)	$P(r)$ when $\mathscr{P}(B) = 0\cdot9$ (Drug Y)
0	W	W	W	0·125	0·001
1	W	W	B	0·125 ⎫	0·009 ⎫
1	W	B	W	0·125 ⎬ 0·375	0·009 ⎬ 0·027
1	B	W	W	0·125 ⎭	0·009 ⎭
2	B	B	W	0·125 ⎫	0·081 ⎫
2	B	W	B	0·125 ⎬ 0·375	0·081 ⎬ 0·243
2	W	B	B	0·125 ⎭	0·081 ⎭
3	B	B	B	0·125	0·729
		Total		1·000	1·000

3.3. Illustration of the danger of drawing conclusions from small samples

Suppose that it is wished to compare the treatments, X and Y, used in the previous section (see (3.2.1) and (3.2.2)). An experiment is performed by testing three subjects with treatment X and three subjects with treatment Y, the subjects being randomly selected from the population of subjects, and randomly allocated to X or Y using random number tables (see § 2.3). The probabilities of obtaining r successes in each set of 3 trials have already been given in Table 3.2.2 and are reproduced in Table 3.3.1 together with the products which, by the multiplication rule, give the probabilities of observing both (r successes with X) **and** (r successes with Y).

TABLE 3.3.1

r	$P(r)$ when $\mathscr{P}(B) = 0\cdot5$ (treatment X)	$P(r)$ when $\mathscr{P}(B) = 0\cdot9$ (treatment Y)	product
0	0·125	0·001	0·000125
1	0·375	0·027	0·010125
2	0·375	0·243	0·091125
3	0·125	0·729	0·018225
Totals	1·000	1·000	0·1196

The sum of the products, 0·1196, gives, by the addition rule, the probability of obtaining *either* (0 successes with both drugs) **or** (1 success with both) **or** (2 successes with both) **or** (3 successes with both). Thus in 11·96 per cent of experiments in the long run, treatment X will appear to be equi-effective with treatment Y, though in fact the latter is considerably better.

Furthermore, in some experiments X will actually produce a *better* result than

Y. By enumerating the ways in which this can happen, and applying the addition and multiplication rules, the probability of this outcome is seen to be

$$(0\cdot375 \times 0\cdot001) + 0\cdot375(0\cdot027 + 0\cdot001) + 0\cdot125(0\cdot243 + 0\cdot027 + 0\cdot001) = 0\cdot04475.$$

For example, the second term is the probability of obtaining both (2 successes with X) **and** (either 0 **or** 1 successes with Y). The treatments will be placed in the *wrong* order of effectiveness in 4·475 per cent of trials in the long run.

The result of these calculations is the prediction that in the long run X will appear to be as good as, or even better than Y in $11\cdot96 + 4\cdot475 = 16\cdot4$ per cent of experiments. It would thus be quite likely that a good new treatment would remain undetected if an experiment were conducted with samples as small as those in this illustration. The hazards of small samples are dealt with further in § 7.7 and in § 7.8, which describes the use of the binomial for the assay of purity in heart.

3.4. The general expression for the binomial distribution and for its mean and variance

The probability, $P(r)$, of observing r successes out of n trials when the probability of a success is $\mathscr{P}$, and the probability of a failure is therefore $1 - \mathscr{P}$ from (2.4.3)), can be inferred by generalization of the deductions in § 3.2. It is

$$P(r) = \mathscr{P}^r(1 - \mathscr{P})^{n-r} \tag{3.4.1}$$

if the order in which the successes occur is specified. Commonly the order is of no interest, and therefore, by the addition rule, this must be multiplied by the number of ways in which r successes can occur in n trials namely

$$\frac{n!}{r!(n-r)!}, \tag{3.4.2}$$

which is the number of possible combinations of r objects selected from n.† Thus, when the order of the successes ignored,

$$P(r) = \frac{n!}{r!(n-r)!}\mathscr{P}^r(1 - \mathscr{P})^{n-r}. \tag{3.4.3}$$

The proof that the sum of these probabilities, for all possible values of r from 0 to n, is 1 follows from the fact that (3.4.3) is a term in the

† This quantity is often denoted by the symbol $\binom{n}{r}$, or by ^nC_r. It is the number of possible ways of dividing n objects into two groups containing r and $n-r$ objects ('successes' and 'failures' in the present case). The n objects can be arranged in $n!$ different orders (permutations), and in each case the first r selected for one group, the remaining $n-r$ for the other. However the $r!$ permutations of the objects *within* the first group, and the $(n-r)!$ permutations within the second group, all result into the same division into two groups, hence the denominator of (3.4.2).

expansion $(\mathcal{Q}+\mathcal{P})^n$, where $\mathcal{Q} = 1-\mathcal{P}$, by the binomial theorem. Thus

$$\sum_{r=0}^{r=n} P(r) = (\mathcal{Q}+\mathcal{P})^n = 1^n = 1.$$

Example 1. If $n = 3$ and $\mathcal{P} = 0\cdot5$, then the probability of one success ($r = 1$) out of three trials is, using (3.4.3),

$$P(1) = \frac{3!}{1!2!}0\cdot5^1 0\cdot5^2 = 3\times0\cdot125 = 0\cdot375$$

as found already in Table 3.2.2.

Example 2. If $n = 3$ and $\mathcal{P} = 0\cdot9$, then the probability of three trials all being successful ($r = 3$) is , similarly,

$$P(3) = \frac{3!}{3!0!}0\cdot9^3 \ 0\cdot1^0 = 1\times0\cdot729 = 0\cdot729$$

as found in Table 3.2.2 (because $0! = 1$, see § 2.1, p. 10).

Estimation of the mean and variance of the binomial distribution

When it is required to estimate the probability of a success from experimental results the obvious method is to use the observed proportion of successes in the sample, r/n, as an estimate of $\mathcal{P}$. Conversely, the average number of successes in the long run will be $n\mathcal{P}$ as exemplified in § 3.2 (this can be found more rigorously using appendix equation (A1.1.1)).

If many samples of n were taken it would be found that the number of successes, r, varied from sample to sample (see § 3.2). Given a number of values of r this scatter could be measured by estimating their variance in the usual way, using (2.6.2). However, in the case of the binomial distribution (unlike the Gaussian distribution) it can be shown (see eqn (A1.2.7)) that the variance that would be found in this way can be predicted even from a single value of r, using the formula

$$var(r) = n\mathcal{P}(1-\mathcal{P}) \tag{3.4.4}$$

into which the experimental estimate of $\mathcal{P}$, viz. r/n, can be substituted.

The meaning of this equation can be illustrated numerically. Take the case of $n = 2$ trials when $\mathcal{P} = 0\cdot5$, which was illustrated in § 3.2. The mean number of successes in 2 trials (long run mean of r) will be $\mu = n\mathcal{P} = 2\times0\cdot5 = 1$. Suppose that a sample of 4 sets of 2 trials were performed, and that the results were $r = 0$, $r = 1$, $r = 1$, and $r = 2$ successes out of 2 trials (that is, by good luck, the results were exactly typical of the population, each of these values of r being equiprobable, see Table 3.2.1). The variance of r could now be estimated using (2.6.3). $N = 4$ is used in the denominator, not $N-1$, because the population mean, μ, not the sample mean, is being used—see § 2.6). It would be

$$\text{var}(r) = \frac{\Sigma(r-\mu)^2}{N} = \frac{(0-1)^2+(1-1)^2+(1-1)^2+(2-1)^2}{4} = 0.5,$$

which is exactly the result found using (2.4.4); thus

$$var(r) = n\mathscr{P}(1-\mathscr{P}) = 2\times0.5(1-0.5) = 0.5.$$

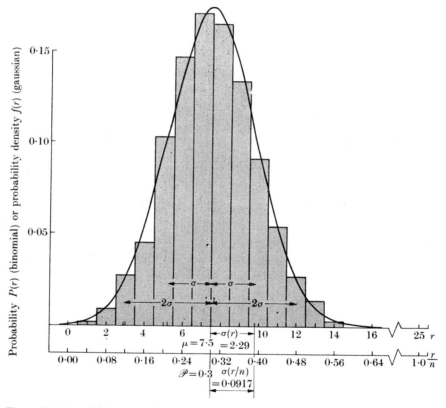

FIG. 3.4.1. *Histogram.* Binomial distribution with $n = 25$ and $\mathscr{P} = 0.3$. Mean number of successes out of 25 trials $= n\mathscr{P} = 7.5$. Variance of r, $var(r)$ $= n\mathscr{P}(1-\mathscr{P}) = 5.25$; $\sigma(r) = \sqrt{(5.25)} = 2.29$; $var(r/n) = \mathscr{P}(1-\mathscr{P})/n = 0.0084$; $\sigma(r/n) = \sqrt{0.0084} = 0.0917$. *Continuous distribution.* Calculated Gaussian ('normal') distribution with $\mu = 7.5$ and $\sigma = 2.29$.

The agreement is only *exact* because the sample happened to be *perfectly* representative of the population. If the calculations are based on small samples the estimate of variance obtained from (3.4.4) will agree approximately, but not exactly, with the estimate from (2.6.2). A similar situation arises in the case of the Poisson distribution and a numerical example is given in § 3.7.

Results are often expressed as the proportion (r/n), rather than the number (r), of successes out of n trials. The variance of the proportion of successes follows directly from the rule (2.7.5) for the effect of multiplying a variable (r in this case) by a constant ($1/n$ in this case). Thus, from (3.4.4),

$$var(r/n) = \frac{var(r)}{n^2} = \frac{\mathscr{P}(1-\mathscr{P})}{n}. \tag{3.4.5}$$

The use of these expressions is illustrated in Fig. 3.4.1 in which the abscissa is given in terms both of r and of r/n. As might be supposed from Fig. 3.2.1–3.2.4, the binomial distribution is only symmetrical if $\mathscr{P} = 0·5$. However, Fig. 3.4.1 shows that as n increases the distribution becomes more nearly symmetrical even when $\mathscr{P} \neq 0·5$. The binomial distribution in Fig. 3.4.1 is seen to be quite closely approximated by the superimposed continuous and symmetrical Gaussian distribution (see Chapter 4), which has been constructed to have the same mean and variance as the binomial.

3.5. Random events. The Poisson distribution

Genesis of the distribution. Relationship to the binomial

The Poisson distribution describes the occurrence of purely random events in a continuum of space or time. The sort of events that *may* be described by the distribution (it is a matter for experimental observation) are the number of cells visible per square of haemocytometer, the number of isotope disintegrations in unit time, or the number of quanta of acetylcholine released at a nerve-ending in response to a stimulus. The Poisson distribution is used as a criterion of the randomness of events of this sort (see § 3.6 for examples). It can be derived in two ways.

First, it can be derived directly by considering random events, when (3.5.1) follows (using the multiplication rule for independent events, (2.4.6)) from the assumption that events occurring in non-overlapping intervals of time or space are *independent*. This derivation is given in § A2.2 (Chapter 5 should be read first). The independence of time intervals is part of the definition of random events (see Chapter 5 and Appendix 2).

Secondly, the Poisson distribution can be derived from the binomial distribution (§ 3.4). In the examples cited the number of 'successes' (e.g. disintegrations per second) can be counted, but it does not obviously make sense to talk about the 'number of trials of the event'. Consider an interval of time Δt seconds long (or an interval of space) divided into n small intervals. If the true (or population, or long-term) average number of events in Δt seconds is called m, then the probability of *one* event occurring ('success') in a small interval of length $\Delta t/n$ is $\mathscr{P} = m/n$.† Because of the independence of time intervals the n intervals are like n *independent* trials with a *constant* probability $\mathscr{P} = m/n$ of success at each trial, just like n tosses of a coin. These properties of independence and constancy *define* (plausibly enough) what is meant by 'random'. If n is finite, the number of successes in n *trials* is therefore given by the binomial distribution, (3.4.3), with $\mathscr{P} = m/n$. In order to consider very short time intervals let $n \to \infty$ (and thus $\mathscr{P} \to 0$) in eqn (3.4.3), so that $m = n\mathscr{P}$ remains fixed. The result is (3.5.1), a limiting form of the binomial distribution in which neither n nor $\mathscr{P}$, but only m appears. The derivation is discussed by Feller (1957, p. 146), Mood and Graybill (1963, p. 70), and Lindley (1965, p. 73). It is easy to follow if it is remembered that as $n \to \infty$, $\lim (1 - m/n)^n = e^{-m}$. See Thompson (1965, Chapter 14) if it is not remembered.

The distribution gives the true probability of observing r events per unit of time (or space) as

$$P(r) = \frac{m^r}{r!}e^{-m}, \tag{3.5.1}$$

where m is the true *mean* number of events per unit of time or space. (It is shown in Appendix 1, (A1.1.7), that m is the population mean value of r.) This is a discontinuous distribution because r must be an

---

† You may object that m could be bigger than n, giving a probability bigger than 1! But the argument only applies to *very short* intervals so that $m \ll n$ and the chance of *more than one* event occurring in a short interval (length $\Delta t/n$) is negligible. For example, if $\Delta t = 1$ hour (3600 s) and $m = 36$ events/h, then if $n = 3600$ it follows that $\mathscr{P} = 36/3600 = 0.01$. On average, 99 out of 100 1s intervals contain no event ('failure'), 1 in 100 contains 1 event ('success') and a negligible proportion contains more than one event. The 'negligible proportion' is dealt with more rigorously in Appendix 2. It becomes zero if the intervals are made infinitely short, which is why we let $n \to \infty$ in the derivation.

integer. It has the basic property of all probability distributions that it should be certain ($P = 1$) that one or other of the possible outcomes ($r = 0$, $r = 1$,...) will be observed. From the addition rule (2.4.2), this means that $P[r = 0 \text{ or } r = 1 \text{ or}...\infty]$ is the sum of the separate probabilities i.e., from (3.5.1),

$$\sum_{r=0}^{\infty} P(r) = e^{-m} \sum_{r=0}^{\infty} \frac{m^r}{r!} = e^{-m}(1+m+\frac{m^2}{2!}+...) = e^{-m}e^{m} = 1. \qquad (3.5.2)$$

(See Thompson (1965, p. 118) if you do not recognize the expansion of e^{m}.)

The variance of the Poisson distribution

According to (3.4.4) the variance of the number of 'successes', r, for the binomial distribution, is $var(r) = n\mathscr{P}(1-\mathscr{P})$. Because $m = n\mathscr{P}$ this can be written $m(1-\mathscr{P})$, and because, as discussed above, the Poisson distribution can be derived from the binomial by letting $\mathscr{P} \to 0$, the variance of the Poisson becomes simply

$$var(r) = m, \qquad (3.5.3)$$

the same as the mean. As in the case of the binomial distribution, but not the normal distribution, this allows an estimate of variance to be made with even a single observation of r (a single estimate of m), as well as by the conventional method of estimation. This is illustrated numerically in § 3.7.

3.6. Some biological applications of the Poisson distribution

Cell distribution

If the number of cells per unit area of a counting chamber were observed to be Poisson-distributed this would imply that the cells were independent and randomly distributed, for example that they have no tendency to clump.

Thus, if the number of red cells present in the volume represented by one small square of a haemocytometer is r, and the number of squares observed to contain r cells is f, then, using the observations in Table 3.6.1, the estimated mean number of cells per square is the total number of cells divided by the total number of squares, i.e.

$$\bar{r} = \frac{\Sigma fr}{\Sigma f} = \frac{531}{80} = 6 \cdot 625.$$

The Poisson distribution (3.5.1) gives the probability of a square containing r cells as $P(r) = e^{-m}m^r/r!$, where m, the mean number of cells per square, is estimated by $\bar{r}$. For example, the probability of a square containing 3 cells is predicted to be

$$P(3) = e^{-6\cdot625}\frac{(6\cdot625)^3}{3!} \simeq 0\cdot064.$$

Multiplying this *probability* by the number of squares counted (80) gives the predicted *frequency* (f_{calc}) of squares containing 3 cells, viz. $80 \times 0\cdot064 = 5\cdot1$, i.e. about 5 squares. The rest of the values are given in Table 3.6.1. The observed distribution is slightly more clumped than the calculated Poisson distribution. In § 8.6, p. [133], a test is carried out to see whether this tendency can be reasonably be attributed to random errors. For this purpose some categories are pooled as indicated by the brackets in Table 3.6.1.

TABLE 3.6.1

r	obs. freq. (f)	calc. freq.	fr
0	0⎫	0⎫	0
1	0⎪	1⎪	0
2	1⎬4	2⎬8	2
3	3⎭	5⎭	9
4	5	9	20
5	10	11	50
6	15	13	90
7	20	12	140
8	17	10	136
9	6	7	54
10	3	5	30
11	0⎫0	3⎫5	0
12 or more	0⎭	2⎭	0
Totals	80	80	531

Bacterial dilutions

If samples of a dilute suspension of bacteria are subcultured into several replicate tubes then bacterial growth will result in those tubes in which the added sample contained one or more viable bacteria. The proportion of tubes showing growth is therefore an estimate of the probability that a sample contains one or more organisms, $P(r \geqslant 1)$. *If the bacteria in the sample suspensions were randomly and independently distributed* throughout the suspending medium the number of bacteria in unit volume of solution (r) would follow the Poisson distribution; this enables an estimate of the mean number of cells per sample (m) to be made from the observed proportion of subcultures showing growth ($P(r \geqslant 1) = p$, say).

From (3.5.1) the probability of the sample being sterile ($r = 0$) is $P(0) = e^{-m}$ and therefore, by (2.4.3)(cf. (3.5.2)),

$$p = P(r \geqslant 1) = 1 - P(0) = 1 - e^{-m}.$$

By solving this for m the mean number of viable organisms per sample is estimated to be

$$m = -\log_e(1 - p)$$

(remember $\log_e e^x \equiv x$). For example, if 40 per cent of cultures are non-sterile, p is 0·4 and $m = -\log_e(1 - 0·4) = 0·51$ organisms per sample. The error of this estimate depends on the number of subcultures on which the estimate of p is based and is usually quite large.

The quantal release of acetylcholine at nerve terminals

In a low-calcium, high-magnesium medium the muscle end-plate (or post-synaptic) potential elicited by nerve stimulation is reduced in size because the number of quanta of acetylcholine released is reduced. A certain proportion of stimuli produce no response at all ('failures').

The number of quanta of acetylcholine released per stimulus has been found to be Poisson distributed (see Martin 1966). In other words, the proportion of stimuli causing release of r quanta, $P(r)$, is observed to be predicted well by (3.5.1). This is illustrated by an example given by Katz (1966). The mean response to a single quantum (mean of 78 spontaneous miniature end-plate potentials) was 0·4 mV. The mean of the responses to 198 nerve impulses was 0·933 mV, the individual responses tending to be either zero ('failures', $r = 0$) or integer multiples of 0·4 mV corresponding to the release of an integral number (r) of quanta. Assuming that the response (mV) is proportional to the number of quanta released, the mean number released is estimated to be $m = 0·933/0·4 = 2·33$ quanta per stimulus. The proportion of stimuli releasing r quanta is therefore predicted, from (3.5.1), to be $2·33^r e^{-2·33}/r$! The predicted *number* of impulses out of 198 releasing r quanta is simply 198 times this proportion. The results in Table 3.6.2 show that the Poisson prediction agrees well with observations.

TABLE 3.6.2

Comparison of observed and Poisson distributions of the number of quanta
of acetyl choline released per stimulus (Katz, 1966; based on Boyd and
Martin, 1956)

r number of quanta	predicted frequency $198\, m^r e^{-m}/r!$	observed frequency
0	19	18
1	44	44
2	52	55
3	40	36
4	24	25
5	11	12
6	5	5
7	2	2
8	1	1
9	0	0
.	.	.
.	.	.
.	.	.
Total	198	198

The predicted frequencies are only approximate because the observed
mean m has been substituted for the population mean, m, in (3.5.1).

The observed frequencies also are only approximate because the response to a
single quantum is itself quite variable (standard deviation 0·086 mV, coefficient
of variation $100 \times 0\cdot086/0\cdot4 = 21\cdot5$ per cent, so the responses (in mV) to 0, 1, 2
. . . quanta overlap somewhat. Also, if the response is large, the depolarization
(in mV) is no longer directly proportional to the number of quanta. The details
are discussed by Martin (1966) and Katz (1966). When corrections are made for
these factors the observed distribution of responses (in mV) is fitted closely by
the calculated distribution.

Furthermore, assuming a Poisson distribution of r, m can be estimated
from the observed number of failures, viz. 18 (from Table 3.6.2), because
$P(0) = \mathrm{e}^{-\mathit{m}}$ (from 3.5.1)). Thus $m = -\log_e P(0) = \log_e 1/P(0) = \log_e$
$198/18 = 2\cdot4$ quanta per stimulus, agreeing quite well with the
independent estimate 2·33 quanta, which was found above without
assuming a Poisson distribution.

Estimation of the quantal content, m, by the 'coefficients of variation method'

If the depolarization produced by a single quantum (miniature end plate
potential) is denoted z, and the quantal content is m as above, then the end plate

potential can be represented (when S is small enough for the z_i to be additive) by

$$S = \sum_{i=1}^{m} z_i, \tag{3.6.1}$$

which is the sum of a variable number (m) of random variables (z). It is stated in (2.7.19), and proved in § A1.4, that if the miniature end-plate potentials are independent of each other (which is probably so), and if the end-plate potential, S, is produced by a random sample (of variable size, m) from the population of single quanta (which is less certain), then the square of coefficient of variation of the end-plate potential size is given by

$$\mathscr{C}^2(S) = \frac{\mathscr{C}^2(z)}{m} + \mathscr{C}^2(m), \tag{3.6.2}$$

where $\mathscr{C}(z)$ and $\mathscr{C}(m)$ are the population coefficients of variation of z and m defined in (2.6.4). This result does not depend on assuming any particular distribution for either z or m (see § A1.4).

Suppose, for example, that m is binomially distributed, which might be expected if the nerve impulse caused there to be a constant probability $\mathscr{P}$ of releasing each of a population of N quanta, so the true mean number of quanta released is $m = N\mathscr{P}$ as in § 3.4, and, on average, a proportion $\mathscr{P}$ of the population is released. According to (3.4.4), $var(m) = N\mathscr{P}(1-\mathscr{P}) = m(1-\mathscr{P})$ and therefore $\mathscr{C}^2(m) \equiv var(m)/m^2 = (1-\mathscr{P})/m$. Substituting this into (3.6.2) gives

$$\mathscr{C}^2(S) = \frac{\mathscr{C}^2(z)}{m} + \frac{(1-\mathscr{P})}{m}. \tag{3.6.3}$$

Solving for m, gives, in this case of binomial distribution of m,

$$m = \frac{\mathscr{C}^2(z) + 1 - \mathscr{P}}{\mathscr{C}^2(S)}. \tag{3.6.4}$$

The case where m is Poisson-distributed is obtained when $\mathscr{P}$ tends to zero (see § 3.5), or directly from (3.6.2) using $var(m) = m$ from (3.5.3), i.e. $\mathscr{C}^2(m) = 1/m$, giving

$$m = \frac{\mathscr{C}^2(z) + 1}{\mathscr{C}^2(S)}. \tag{3.6.5}$$

This, and the other results in this section, are discussed in the review by Martin (1966). An estimate of m is obtained by substituting the experimental estimates of $\mathscr{C}(z)$ and $\mathscr{C}(S)$ into (3.6.5).

Equations (3.6.4) and (3.6.5) do not entirely account for the experimenta observations and it was pointed out by del Castillo and Katz (see Martin 1966) that if we drop the rather unreasonable assumption that all the quanta have the same probability of release, then $\mathscr{C}^2(m)$ will be less than the binomial value $(1-\mathscr{P})/m$, which in turn is less than the Poisson value, $1/m$. It can be shown (e.g. Kendall and Stuart (1963, p. 127)) that if each quantum has a different probability $(\mathscr{P}_j)$ of release, and that if these probabilities are constant from one nerve impulse to the next, then $\mathscr{C}^2(m) = (1-\bar{\mathscr{P}}-var(\mathscr{P})/\bar{\mathscr{P}})/m$ where $\bar{\mathscr{P}}$ is the mean probability of a quantum being released (i.e. $\Sigma\mathscr{P}_j/N$) and $var(\mathscr{P})$ is the variance of the $\mathscr{P}_j$ values (zero in the binomial case when all the $\mathscr{P}$ are identical). If this is substituted in (3.6.2), solving for m gives

$$m = \frac{\mathscr{C}^2(z) + 1 - \overline{\mathscr{P}} - var(\mathscr{P})/\overline{\mathscr{P}}}{\mathscr{C}^2(S)}, \tag{3.6.6}$$

which is smaller than is given by either (3.6.4) or (3.6.5). In the case where N is very large (3.6.6), like (3.6.4), tends to the Poisson form (3.6.5) despite the variability of $\mathscr{P}$.

As an example consider the observations discussed above. The observed values for the response to one quantum was $\bar{z} = 0.4$ mV with standard deviation 0.086 mV, i.e. coefficient of variation $C(z) = 0.086/0.4 = 0.215$. The observed mean end-plate potential was $\overline{S} = 0.933$ mV with a standard deviation of 0.634 mV (this value is taken for the purposes of illustration, the original figure not being available) and hence $C(S) = 0.634/0.933 = 0.680$. If m were Poisson-distributed its mean value could be estimated from (3.6.5) as

$$\bar{m} = \frac{0.215^2 + 1}{0.680^2} = \frac{1.046}{0.462} = 2.26,$$

which agrees quite well with estimate (viz. 2.4) from the proportion of failures, which also assumes a Poisson distribution, and the direct estimate $0.933/0.4 = 2.33$ which does not.

The number of spontaneous miniature end-plate potentials in unit time

The number of single quanta released in unit time is observed to follow the Poisson distribution, i.e. quanta appear to be released spontaneously in a random fashion. This phenomenon is discussed in Chapter 5, after continuous distributions have been dealt with, so the continuous distribution of intervals between random events can be discussed.

3.7. Theoretical and observed variances: a numerical example concerning random radioisotope disintegration

The number of unstable nuclei disintegrating in unit time is observed to be Poisson-distributed over periods during which decay is negligible (see Appendix A2.5), and disintegration is therefore a random process in time.

Since the variance of the Poisson distribution (3.5.3) is estimated by the mean number of disintegrations per unit time, the uncertainty of a count depends only on the number of disintegrations counted and not on how long they took to count, or on whether one long count or several shorter counts were done. The example is based on one given by Taylor (1957). The values of x listed are $n = 10$ replicate counts, each over a period of 5 min, of a radioactive sample. The decay over the period of the experiment is assumed to be negligible. The x values are

10536	10636	10398	10393	10586
10381	10479	10401	10262	10403

The total number of counts is $\Sigma x = 104475$ counts/50 min, mean count $\bar{x} = \Sigma x/n = 10447\cdot5$ counts/5 min, and count-rate $= 10447\cdot5/5 = 2089\cdot5$ counts/min. What is the uncertainty in the count-rate? Its variance can be calculated in two ways.

(a) *Theoretical Poisson variance*

The number of counts observed in a 50-min unit of time was 104475 so *if* the number of counts in unit were Poisson-distributed the estimate of the variance of the variable 'number of counts in 50 minutes' would be 104475 (from (3.5.3)). In this case the total number of counts was the sum of ten 5-min counts. In general, according to (2.7.4), $var(\Sigma x) = n\ var(x)$, so if x is the number of counts in 5 min, the variance of a single 5-min count is estimated to be

$$var(x) = \frac{var(\Sigma x)}{n} = \frac{104475}{10} = 10447\cdot5.$$

If there had only been one 5-min count, say the first one, its variance would have been estimated as 10536, a similar figure.

However, what is really wanted is the variance of the count-rate per minute, determined from 50 min of counting in the experiment, not the variance of a 5-min count. The count-rate is $\Sigma x/50$ counts/min. In general, from (2.7.5), $var(ax) = a^2 var(x)$, where a is a constant (1/50 in this case), therefore

$$var\left(\frac{\Sigma x}{50}\right) = \frac{var(\Sigma x)}{50^2} = \frac{104475}{50^2} = 41\cdot79.$$

The standard deviation of the mean count–rate (2089·5 counts/min) is therefore $\sqrt{(41\cdot79)} = 6\cdot46$ counts/min.

If there had been only a single 5-min count, say 10536, the mean count-rate would have been $10536/5 = 2107\cdot2$ counts/min, and, by a similar argument, its estimated standard deviation would have been $\sqrt{(10536/5^2)} = 20\cdot6$ counts/min. Thus when the number of observations is reduced tenfold, the standard deviation of the mean goes up by $\sqrt{(10)}$, as expected from (2.7.9) ($6\cdot46 \times \sqrt{(10)} = 20\cdot4$).

It can be seen that the uncertainty in the count depends only on the total number of counts. If it is known that the count-rate has a Poisson distribution (as it will have *if* the counter is functioning correctly) its uncertainty can be estimated without having to do replicate observations.

Observed variance

In this particular case there *are* replicate counts so the variance of an observation (a 5-min count) can be estimated in the usual way using (2.6.2),

$$\text{var}(x) = \frac{\Sigma(x-\bar{x})^2}{n-1} = \frac{111938}{9} = 12437 \cdot 5.$$

This is quite close to the estimate of 10447·5 found above. Because there are ten 5-min counts the estimate of count-rate will be based on the mean of these, the variance of which is estimated, using (2.7.8), to be

$$\text{var}(\bar{x}) = \frac{\text{var}(x)}{n} = \frac{12437 \cdot 5}{10} = 1243 \cdot 75.$$

And the variance of the mean count-rate per minute will be, from (2.7.5),

$$\text{var}\left(\frac{\bar{x}}{5}\right) = \frac{\text{var}(\bar{x})}{5^2} = \frac{1243 \cdot 75}{25} = 49 \cdot 75.$$

By using the scatter of replicate counts, the standard deviation of the count-rate (2089·5 counts/min) is therefore estimated to be $\sqrt{49 \cdot 75} = 7 \cdot 05$ counts/min. This estimate, which has *not* involved any assumption about the distribution of the observations, agrees well with the estimate (6·46 counts/min) calculated assuming that the count-rate was Poisson-distributed. This suggests that the assumption was not far wrong. With either estimate the coefficient of variation of the count-rate, by (2.6.4), comes out to about 0·3 per cent.

The effect of allowing for background count-rate

Counting equipment registers a background rate even when there is no sample in it and this must be subtracted from the sample count-rate. There is uncertainty in the background count as well as the sample count and this must be allowed for.

To illustrate what happens when the sample count-rate is not much above the background rate suppose that 20 000 background counts were recorded in 10 min. The net count is thus $2089 \cdot 5 - 2000$ $= 89 \cdot 5$ counts/min.

By arguments similar to those above:

estimated variance of background count/min = var(count/10) = var $(\text{count})/10^2 = \text{count}/10^2 = 20000/100 = 200$.

The estimated variance of the net count-rate (sample minus background) is required. Because the counts are independent this is, by (2.7.3), the sum of the variances of the two count-rates

$$\begin{aligned} \text{var}(\text{sample} - \text{background}) &= \text{var}(\text{sample}) + \text{var}(\text{background}) \\ &= 49 \cdot 75 + 200 = 249 \cdot 75, \end{aligned}$$

and the estimated standard deviation of the net count-rate ($89 \cdot 5$ counts/min) is therefore $\sqrt{(249 \cdot 75)} = 15 \cdot 8$ counts/min. The coefficient of variation of the net count, by (2.6.4), is now quite large ($17 \cdot 7$ per cent), and if the net count had been much smaller the difference between sample and background would have been completely swamped by the counting error (for a fuller discussion see Taylor (1957)).

6

4. Theoretical distributions. The Gaussian (or normal) and other continuous distributions

4.1. The representation of continuous distributions in general

So far only discontinuous variables have been dealt with. In many cases it is more convenient, though, because only a few significant figures are retained, not strictly correct, to treat the experimental variables as continuous. For example, changes in blood pressure, muscle tension, daily urinary excretion, etc. are regarded as potentially able to have any value. The difficulties involved in dealing with this situation have already been mentioned in § 3.1 and will now be elucidated.

The discontinuous distributions so far discussed have been represented by histograms in which the height (along the ordinate) of the blocks was a measure of the probability or frequency of observing a particular value of the variable (along the abscissa). However, if one asks 'What is the probability (or frequency) of observing a muscle tension of *exactly* 2·0000 . . . g ?', the answer must be that this probability is infinitesimally small, and cannot therefore be plotted on the ordinate. What *can* sensibly be asked is 'What is the probability (or frequency) of observing a muscle tension between, say, 1·5 and 2·5 g ?'. This frequency will be finite, and if many observations of tension are made a histogram can be plotted using the frequency (along the ordinate) of making observations between 0 and 0·5 g, 0·5 and 1·0 g, etc., as shown in Fig. 4.1.1. If there were enough observations it would be better to reduce the width of the classes from 0·5 g to, say, 0·1 g as shown in Fig. 4.1.2. This gives a smoother-looking histogram, but because there are more classes the number (or probability) of observations falling in a particular class will be reduced. The blocks will also be drawn narrower though it will usually be convenient to keep them about the

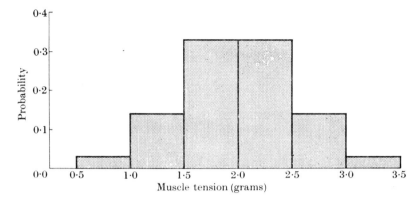

FIG. 4.1.1. Histogram of muscle tensions; 'observations' grouped into classes 0·5 g wide and proportion of observations in each class plotted as ordinate. Total *height* of all blocks = 1·0.

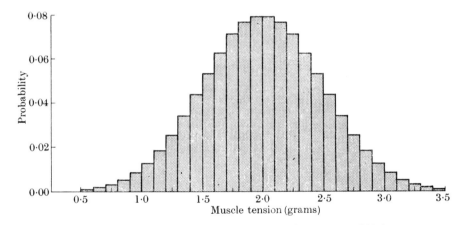

FIG. 4.1.2. Same 'observations' as Fig. 4.1.1 but grouped into narrower classes (0·1 g), showing shape of distribution more clearly. Total *height* of all blocks = 1·0 as before.

same height, as shown. This suggests that it might be more convenient to represent the probability of an observation falling in a particular class by the area of the block rather than its height. If the width of the block (class interval) is constant then the area of the block is proportional to its height so in ordinary histograms the area is in fact proportional to the probability. When the class width is reduced, the reduction in width of the blocks will reduce their area, and hence the probability they represent, without having to reduce their height. An example of a histogram with unequal block widths occurs in §§ 14.1–14.3. Fig. 14.2.3(a) shows it drawn with height representing probability and Fig. 14.2.3(b) shows the preferable representation of the histogram with area representing probability.

Using the convention that the probability is represented by the area of the blocks rather than their height, the condition that the sum of all the probabilities must be 1·0 is expressed by defining the total area of the blocks as 1·0 (see (3.5.2) for example).

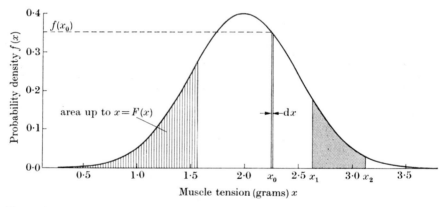

FIG. 4.1.3.   Continuous distribution of muscle tensions. Ordinate is probability density, $f(x)$, i.e. a function such that the area under the curve represents probability. The total *area* under the curve is 1·0. The probability of observing a value equal to or less than x is denoted p, or $F(x)$ and is the area under the curve up to x.

If the class intervals are made narrower and narrower, the probability attached to each becomes smaller. The probability of observing a muscle tension (x) between 1·999 and 2·001 g is small, and a very large number of observations would be necessary to use intervals as narrow as this. When the class interval becomes infinitesimally narrow then the probability represented by each block (i.e. the probability that x will lie within the interval dx) is also infinitesimal, say dP, and the graph becomes continuous instead of being made up of finite blocks, as shown in Fig. 4.1.3. It now represents an infinite population and it can never be observed exactly. It is a mathematical idealization.

The area of a block, i.e. the probability of x falling in the interval of width dx (between x and $x+dx$), must now be written

$$dP = f(x)dx, \tag{4.1.1}$$

where the function $f(x)$ is the ordinate of the curve shown in Fig. 4.1.3 (i.e. the height of the block), and x is the continuous variable (e.g. blood pressure or muscle tension) the distribution of which is being defined (see, e.g., Thompson, 1965, if the notation of (4.1.1) is not understood). The function $f(x)$ is known as the probability density function (or simply density) of x. A value of this function is called the probability density of a particular value of x. It is *not* the probability of that value of x, but merely a function that defines a curve such that the *area under the curve* represents probability. For example, the uniformly shaded area in Fig. 4.1.3, as a proportion of the whole area under the curve, is the probability that a value of x will lie between two specified values, x_1 and x_2. The summation of the infinitesimal blocks of which this area is made up is handled mathematically by integration so this area can be written as the integrated form of (4.1.1),

$$P(x_1 \leqslant x \leqslant x_2) = \int_{x_1}^{x_2} f(x)dx. \tag{4.1.2}$$

Similarly, the probability that a value of x greater than x_2 will be observed is equal to the area above the point x_2. How far along the x-axis the distribution curve extends depends on the particular distribution under consideration. The curve may reach the axis at some finite minimum or maximum value of x, implying that observations less or greater than this value are impossible; or the curve may, like the Gaussian (or normal) distribution, be asymptotic to the x-axis so that *any* value of x is allowed, through the probability of observing values far removed from the mean soon becomes small. In the latter case the probability of observing a value of x equal to or less than x_1 (the area under the distribution curve below x_1) would be written

$$P(x \leqslant x_1) = \int_{-\infty}^{x_1} f(x)dx. \tag{4.1.3}$$

This area is said, in statistical jargon, to be the *lower tail* of the distribution. It can be called p, or $F(x_1)$, and is vertically shaded in Fig. 4.1.3. It depends, of course, on the value of x_1 chosen, i.e. it is a function of x_1.

A more satisfactory way of writing the same thing is to use a special symbol, say $\tilde{x}$, to distinguish x considered as a random variable, from a particular value of the random variable, denoted simply x. The probability of observing a value of the variable (e.g. muscle tension) equal to or less than some specified value x (e.g. 2·0 g) as in (4.1.3), is written in this notation as†

$$P(\tilde{x} \leqslant x) = \int_{-\infty}^{x} f(x)\mathrm{d}x \equiv F(x), \text{ or } p. \qquad (4.1.4)$$

This is referred to as the *distribution function* of x, or as the *cumulative distribution*. The area below x in Fig. 4.1.3, $F(x)$, is plotted against x

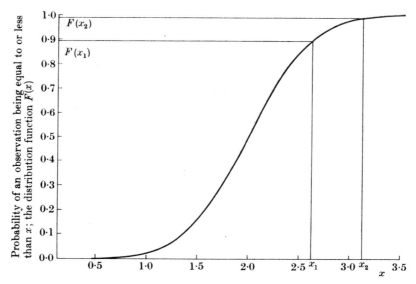

FIG. 4.1.4.   Distribution function, $F(x)$, for the distribution shown in Fig. 4.1.3. The probability of observing a value of x or less is plotted against x. The area between x_1 and x_2 in Fig. 4.1.3 is $F(x_2) - F(x_1) = 0·988 - 0·894 = 0·094$, the probability of an observation falling between x_1 and x_2.

in Fig. 4.1.4. Examples of cumulative distributions occur in §§ 5.1 and 14.2. The area, $F(x)$, approaches 1·0 as x becomes very large, i.e. it is almost certain that the variable (e.g. muscle tension) will be less than

† Another, mathematically better, way of writing exactly the same thing

$$P(\tilde{x} \leqslant x) = \int_{-\infty}^{x} f(v)\mathrm{d}v.$$

The variable v does not appear in the final answer.

a specified very large value (e.g. 100 kg). Differentiating (4.1.4) shows that the distribution function is related to the probability density, as suggested by (4.1.1), thus

$$\frac{\mathrm{d}F(x)}{\mathrm{d}x} = f(x). \tag{4.1.5}$$

4.2. The Gaussian, or normal, distribution. A case of wishful thinking?

The assumption that the errors in real experimental observations can be described by the normal distribution (4.2.1) has dominated large areas of statistics for many years. The assumption is virtually always untested, and the extent to which it is mere wishful thinking will be discussed in this section, after the distribution has been defined, and in § 6.2, where the merits of methods not involving the assumption of Gaussian errors are considered.

Definition of the distribution

The Gaussian distribution, often, but inappropriately, known as the normal distribution, is defined by the probability density function (see § 4.1)

$$f(x) = \frac{1}{\sigma\sqrt{(2\pi)}}\exp[-(x-\mu)^2/2\sigma^2], \tag{4.2.1}$$

where π has its usual value, and μ and σ are constants. The factor $1/\sigma\sqrt{(2\pi)}$ is a constant such that the total area under the curve (from $x = -\infty$ to $x = +\infty$) is 1·0. The notation $\exp(x)$ is used to stand for e^x when the exponent, x, is a long expression that would be inconvenient to write as a superscript. If $f(x)$ is plotted against x the graph comes out as shown in Fig. 4.2.1.

It is a symmetrical bell-shaped curve asymptotic to, i.e. never quite reaching, the x-axis. Being continuous it represents an infinite population (see § 4.1). The constant μ is the population mean† and also the population median and mode because the distribution is symmetrical and unimodal; see §§ 2.5 and 4.5. The constant σ measures the width‡ of

† This is proved in § A1.1.
‡ The distance from μ to the point of inflection (maximum slope) on each side of the mean. Differentiating (4.2.1) twice with respect to x and equating to zero gives $x = \mu \pm \sigma$. The population variance is defined in § A1.2.

the curve as shown in Fig. 4.2.1, i.e. it is a measure of the scatter of the values of x, and is the population standard deviation of x. An estimate of σ could be made from a sample of observations, taken from the population represented by Fig. 4.2.1, using (2.6.2). The distribution is completely defined by the two parameters μ and σ.

Is the widespread use of the normal distribution justified?

'Everybody firmly believes in it [the normal distribution] because the mathematicians imagine that it is a fact of observation, and observers that it is a theory of mathematics' (quoted by Poincaré 1892).

From the point of view of someone trying to interpret real observations (and who else is statistics for?) the only possible justification for the common assumption of normality would be the *experimental*

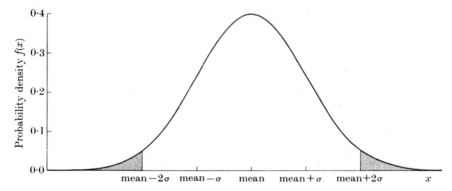

FIG. 4.2.1.   Gaussian ('normal') distribution. 4·6 per cent of the observations in the population are more than two population standard deviations from the mean (the shaded area is 4·6 per cent of the total area). The value 4·6 does not apply to *samples* or, in general, to distributions other than the Gaussian (see §§ 4.4 and 4.5).

demonstration that the methods based on this assumption give results that are correct, or at least sufficiently nearly correct for the purpose in hand.

The truth is that no such demonstration exists. The many textbooks, elementary and not so elementary, describing methods that mostly depend on this assumption evade this awkward fact in a variety of ways. The more advanced books usually say something like 'If x were normally distributed then . . . would follow', which is true but not very helpful in real life. In more elementary books one often finds (to quote two) remarks such as 'It is not infrequently found that a population represented in this way [i.e. by a Gaussian curve] is sufficiently accurately specified for the purpose of the inquiry', or 'Many of the frequency functions applicable to observed distribution do have a normal form'. Such remarks are, at least as far as most laboratory investigations are concerned, just wishful thinking. Anyone with experience of doing experiments must know that it is rare for the distribution of the observations to be investigated. The number of observations *from a single population* needed to get an idea of the form of the distribution is quite large—a hundred or two at least—so this is not surprising. In the vast majority of cases the form of the distribution is simply not known: and, in an even more overwhelming majority of cases there is no substantial evidence regarding whether or not the Gaussian curve, is a sufficiently good approximation for the purposes of the inquiry. It is simply not known how often the assumption of normality is seriously misleading. See § 4.6 for tests of normality.

That most eminent amateur statistician, W. S. Gosset ('Student', see § 4.4), wrote, in a letter dated June 1929 to R. A. Fisher, the great mathematical statistician, '. . . although when you think about it you agree that "exactness" or even appropriate use depends on normality, in practice you don't consider the question at all when you apply your tables to your examples: *not one word*.'

For these reasons some methods have been developed that do not rely on the assumption of normality. They are discussed in § 6.2. However, many problems can still be tackled only by methods that involve the normality assumption, and when such a problem is encountered there is a strong temptation to forget that it is not known how nearly true the assumption is. A possible reason for using the Gaussian method in the absence of evidence one way or the other about the form of the distribution, is that an important use of statistical methods is to prevent the experimenter from making a fool of himself (see Chapters 1, 6, and 7). It would be a rash experimenter who presented results that would not pass a Gaussian test, unless the distribution was definitely known to be *not* Gaussian.

It is commonly said that if the distribution of a variable is not normal, the variable may be transformed to make the distribution normal (for example, by taking the logarithms of the observations, see § 4.5). As pointed out above, there are hardly ever enough observations to find out whether the distribution is normal or not, so this approach can rarely be used. Transformations are discussed again in §§ 4.6, 11.2 (p. 176) and § 12.2 (p. 221).

Various other reasons are often given for using Gaussian methods. One is that *some* Gaussian methods have been shown to be fairly immune to *some* sorts of deviations from normality, if the samples are not too small. Many methods involve the estimation of means and there is an ingenious bit of mathematics known as the *central limit theorem* that states that the distribution of the means of samples of observations will tend more and more nearly to the Gaussian form as the sample size increases whatever (almost) the form of the distribution of the observations themselves (even if it is skew or discontinuous). These remarks suggest that when one is dealing with reasonably large samples, Gaussian methods may be used as an approximation. The snag is that it is impossible to say, in any particular case, what is a 'reasonable' number of observations, or how approximate the approximation will be.

Further discussion of the assumptions made in statistical calculations will be found particularly in §§ 6.2 and 11.2.

4.3. The standard normal distribution

Applications of the normal distribution often involve finding the proportion of the total area under the normal curve that lies between particular values of the abscissa x. This area must be obtained by evaluating the integral (4.1.2), with the normal probability density function (4.2.1) substituted for $f(x)$. The integral cannot be explicitly solved. The answer comes out as the sum of an infinite number of terms (obtained by expanding the exponential). In practice the only convenient method of obtaining areas is from tables. For example, the Biometrika Tables, Pearson and Hartley (1966, Table 1), give the area under the standard normal distribution (defined below) below u (or the area above $-u$ which is the same), i.e. the area between $-\infty$ and u (see below). In this table u and the area are denoted X and $P(X)$ respectively. Fisher and Yates (1963, Table II_1, p. 45) give the area above u ($=$ area below $-u$), the value of u being denoted x in this table.†

If tables had to be constructed for a wide enough range of values of μ and σ to be useful they would be very voluminous. Fortunately this is not necessary since it is found that the area lying within any given number of standard deviations on either side of the mean is the

† Tables of Student's t (see § 4.4) give, on the line for infinite degrees of freedom, the area below $-u$ plus the area above $+u$, i.e. the area in *both* tails of the distribution of u.

same whatever the values of the mean and standard deviation. For example it is found that:

(1) 68·3 per cent of the area under the curve lies within one standard deviation on either side of the mean. That is, in the long run, 68·3 per cent of random observations from a *Gaussian population* would be found to differ from the population mean by not more than one population standard deviation.

(2) 95·4 per cent of the area lies within two standard deviations (or 95·0 per cent within ±1·96σ). The 4·6 per cent of the area outside ±2σ is shaded in Fig. 4.2.1.

(3) 99·7 per cent of the area lies within three standard deviations. Only 0·3 per cent of random observations from a Gaussian population are expected to differ from the mean by more than three standard deviations.

It follows that all normal distributions can be reduced to a single distribution if the abscissa is measured in terms of deviations from the mean, expressed as a number of population standard deviations. In other words, instead of considering the distribution of x itself it is simpler to consider the distribution of

$$u = \frac{x - \mu}{\sigma}. \tag{4.3.1}$$

The distribution of u is called the *standard normal distribution*. It is still a normal distribution because u is linearly related to the normally distributed x (μ and σ being constants for any particular distribution), but it necessarily† always has a mean of zero and a standard deviation of 1·0. The numerator, $x - \mu$, is a normally distributed variable with a population mean of zero (because the long run average value of x is μ) and variance σ^2. To illustrate this consider a normally distributed variable x with population mean $\mu = 6$ and population standard deviation $\sigma = 3$. It can be seen from Fig. 4.3.1 that the distribution of $(x - \mu)$, i.e. of $(x - 6)$, has a mean of zero but a standard deviation unchanged at 3 (cf. (2.7.7)), and that when this quantity is divided by σ the standard normal distribution (mean = 0 standard deviation = 1) results.

† See §§ A1.1 and A1.2. The standard form of a distribution is defined in § A1.2.

In terms of the standard normal distribution the areas (obtainable from the tables referred to above) become

(1) 68·3 per cent of the area lies between $u = -1$ and $u = +1$ (and thus 15·85 per cent lies below -1, and 15·85 per cent above $+1$),

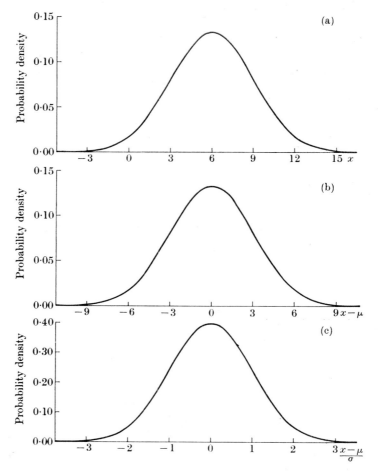

Fig. 4.3.1.   Relation of normal distribution to standard normal distribution. (a) x is normally distributed with population mean $\mu = 6$ and population standard deviation $\sigma = 3$. (b) $(x-\mu)$ is normally distributed with population mean $= 0$ and population standard deviation $= 3$. (c) $u = (x-\mu)/\sigma = \dfrac{x-6}{3}$ in this case is normally distributed with population mean $= 0$ and population standard deviation $= 1$.

(2) 95 per cent of the area lies between $u = -1.96$ and $u = +1.96$,

(3) 99·7 per cent of the area lies between $u = -3$ and $u = +3$.

In order to convert an observation x into a value of $u = (x-\mu)/\sigma$, it is, of course, necessary to know the values of μ and σ. In real life the values of μ and σ will not generally be known, only more or less accurate *estimates* of them, viz. $\bar{x}$ and s, will be available. If the normal distribution is to be used for induction as well as deduction this fact must be allowed for, and the method of doing this is discussed in the next section.

4.4. The distribution of *t* (Student's distribution)

The variable t is defined as

$$t = \frac{x-\mu}{s(x)}, \tag{4.4.1}$$

where x is any normally distributed variable and $s(x)$ is an estimate of the standard deviation of x from a sample of observations of x (see § 2.6). Tables of its distribution are referred to at the end of the section. It is the same as u defined in (4.3.1), except that the deviation of a value of x from the true mean (μ) is expressed in units of the *estimated* or *sample* standard deviation of x, $s(x)$ (eqn (2.6.2)), rather than the *population* standard deviation $\sigma(x)$. As in § 4.3 the numerator, $(x-\mu)$, is a normally distributed variable with population mean zero (because the long run average value of x is μ, see Appendix, eqn (A1.1.8), and estimated standard deviation $s(x)$.

The 'distribution of t' means, as usual, a formula, too complicated to derive here, for calculating the frequency with which the value of t would be expected to fall between any specified limits; see example below. The distribution of t was found by W. S. Gosset, who wrote many papers on statistical subjects under the pseudonym 'Student', in a classical paper called 'The probable error of a mean' which was published in 1908.

Gosset was not a professional mathematician. After studying chemistry and mathematics he went to work in 1899 as a brewer at the Guinness brewery in Dublin, and he worked for this firm for the rest of his life. His interest in statistics had strong practical motives. The majority of statistical work being done at the beginning of the century involved large samples and the drawing of conclusions from small samples was regarded as a very dubious process. Gosset realized that the methods used for dealing with large samples would need modification if the results were to be applicable to the small samples he had to work in the laboratory.

Gosset spent a year (1906–7) away from the brewery, mostly working in the Biometric Laboratory of University College London with Karl Pearson, and in 1908 published a paper on the distribution of t.

As an example, suppose that the normally distributed variable of interest is $\bar{x}$, the mean of a sample of 4 observations selected randomly from a population of normally distributed values of x with population mean μ and population standard deviation $\sigma(x)$. The population standard deviation of $\bar{x}$ (or 'standard error', see § 2.7) will be $\sigma(\bar{x})$

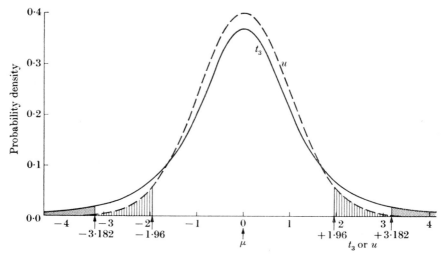

FIG. 4.4.1.   *Continuous line.* Distribution of Student's t with 3 degrees of freedom. Ninety-five per cent of values lie between $-3{\cdot}182$ and $+3{\cdot}182$ (see text for example). The 5 per cent of the area outside these values is shaded. *Broken line.* Standard Gaussian (normal) distribution. 95 per cent of u values lie between $u = -1{\cdot}96$ and $u = +1{\cdot}96$. The 5 per cent of the area outside these values is shaded vertically. As the sample size (degrees of freedom) become very large, the t distribution becomes identical with the standard normal distribution.

$= \sigma(x)/\sqrt{4}$ (by (2.7.9)) and the population mean of $\bar{x}$ will be μ, the same as for x. (See Appendix 1, (A1.2.3).) Therefore if a very large number of samples of 4 were taken, and if for each $u = (\bar{x} - \mu)/\sigma(\bar{x})$ (from the definition (4.3.1)) were calculated, it would be found that in the long run 95 per cent of the values of u would lie between $u = -1{\cdot}96$ and $u = +1{\cdot}96$, as discussed in § 4.3 and illustrated in Fig. 4.4.1.

However, if $\sigma(x)$ were not known, an estimate of it, $s(x)$, could be calculated from each sample of 4 observations using (2.6.2) as in the example in § 2.7, and from each sample $s(\bar{x}) = s(x)/\sqrt{4}$ obtained by (2.7.9). For each sample $t = (\bar{x}-\mu)/s(\bar{x})$ (from the definition (4.4.1)) could be now calculated. The values of $\bar{x}$ would be the same as those used for calculating u, but the value of $s(\bar{x})$ would differ from sample to sample, whereas the same population value, $\sigma(\bar{x})$, would be used in calculating every value of u). The extra variability introduced by variability of $s(\bar{x})$ from sample to sample means that t varies over a wider range than u, and it can be found from the tables referred to below, that it would be expected that, in the long run, 95 per cent of the values of t would lie between $-3\cdot182$ and $+3\cdot182$, as illustrated in Fig. 4.4.1.

Notice that both the distributions in Fig. 4.4.1 are based on observations from the normal distribution with population standard deviation σ. The distribution of t, unlike that of u, is not normal, though it *is* based on the assumption that x is normally distributed.

Although the definition of t (4.4.1) takes account of the uncertainty of the estimate of $\sigma(x)$, it still involves knowledge of the true mean μ and it might be thought at first that this is a big disadvantage. It will be found when tests of significance and confidence limits are discussed that, on the contrary, everything necessary can be done by giving μ a hypothetical value.

The use of tables of the distribution of t

The extent to which the distribution of t differs from that of u will clearly depend on the size of the sample used to estimate $s(x)$. The appropriate measure of sample size, as discussed in § 2.6, is the number of degrees of freedom associated with $s(x)$. If $s(x)$ is calculated from a sample of N observations the number of degrees of freedom associated with $s(x)$ is $N-1$ as in § 2.6. Clearly, t with an infinite number of degrees of freedom is the same as u, because in this case the estimate $s(x)$ is very accurate and becomes the same as $\sigma(x)$.

Fisher and Yates (1963, Table 3, p. 46, 'The distribution of t') denote the number of degrees of freedom n and tabulate values such that t has the specified probability of falling above the tabulated value or below minus the tabulated value. Looking in the table for $n = 4-1 = 3$ and $P = 0\cdot05$ gives $t = 3\cdot182$ as discussed in the example above, and illustrated in Fig. 4.4.1 in which the 5 per cent of the area outside $t = \pm3\cdot182$ is shaded.

The Biometrika Tables of Pearson and Hartley (1966, Table 12, p. 146, 'Percentage points of the *t* distribution') give the same sort of table. The number of degrees of freedom is denoted ν and the probability $2Q$, Q being the shaded area in *one* tail of Fig. 4.4.1.

4.5. Skew distributions and the lognormal distribution

In § 4.2 it was stressed that the normal distribution is a mathematical convenience that cannot be supposed to represent real life adequately, and that it is very rare in experimental work for the distribution of observation to be known. In those cases where the distribution has been investigated it has often been found to be non-normal. Distributions may be more flat-topped or more sharp-topped than the normal distribution, and they may be *unsymmetrical*. Unsymmetrical distributions may have *positive skew* as in Fig. 4.5.1 (an even more extreme case is the exponential distribution Fig. 5.1.2), or *negative skew*, as in the mirror image of Fig. 4.5.1.

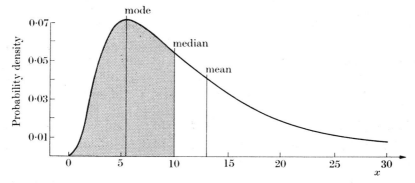

Fig. 4.5.1. The lognormal distribution; a positively skewed probability distribution. The mean value of x is greater than the median, and the mode is less than the median. The 50 per cent of the area that lies (by definition) below the median is shaded. For the lognormal distribution, in general, mode = antilog$_{10}$ $(\mu - 2 \cdot 3026\sigma^2)$ (= 5·81 in this example), median = antilog$_{10}\mu$ (= 10·0 in this example), mean = antilog$_{10}(\mu + 1 \cdot 1513\sigma^2)$ (= 13·1 in this example), where μ and σ^2 are mean and variance of the (normal) distribution of the log$_{10}x$ shown in Fig. 4.5.2. Reproduced from *Documenta Geigy scientific tables*, 6th edn, by permission of J. R. Geigy S.A., Basle, Switzerland.

In the case of symmetrical distributions (such as the normal) the population mean, median, and mode (see § 2.5) are all the same, but this is not so for unsymmetrical distributions. For example, when the distribution of x has a positive skew, as in Fig. 4.5.1, the population mean is greater than the population median which is in turn larger than the population mode. There is no particular reason to prefer the mean to the median or mode as a measure of the 'average' value of the variable in a case like this. A reason for preferring the median is mentioned below (see also Chapter 14). The distribution of personal incomes has a positive skew so the most frequent income (the mode) is less than the mean income, and more people earn less than the mean income than earn more than the mean income, because incomes above the mean are,

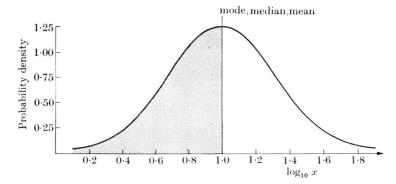

FIG. 4.5.2.   The distribution of $\log_{10}x$, when x is lognormally distributed as shown in Fig. 4.5.1. This distribution is normal (by definition of the lognormal distribution). In this example the mean (= median = mode) of $\log_{10}x$ is $\mu = 1.0$, and the standard deviation of $\log_{10}x$ is $\sigma = 0.32$. See text and Chapter 14. Reproduced from *Documenta Geigy scientific tables*, 6th edn by permission of J. R. Geigy S.A., Basle, Switzerland.

on the whole, further from the mean than incomes below it, i.e. more than 50 per cent of the area under the curve is below the mean, as shown by the shading in Fig. 4.5.1.

It is usually recommended that non-normal distributions be converted to normal distributions by transforming the scale of x (see §§ 4.2, 11.2, and 12.2). This should be done when possible, but in most experimental investigations there is not enough information to allow the correct transformation to be ascertained. In Chapter 14 an example is given of a variable (individual effective dose of drug) with a positively skewed distribution (Fig. 14.2.1). In this particular example the logarithm of the variable is found to be approximately normally distributed (Fig. 14.2.3). In general, a variable is said to follow the *lognormal distribution*, which looks like Fig. 4.5.1, if the logarithm of the variable is normally distributed, as in Fig. 4.5.2.

In Chapter 14 the median value of the variable (rather than the mean) is estimated. The median is unchanged by transformation, i.e. the population median of the (lognormal) distribution of x is the antilog of the population median ($=$ mean $=$ mode) of the (normal) distribution of log x, whereas the population mode of x is smaller, and the population mean of x greater than this quantity (cf. (2.5.4)). For example, in Fig. 4.5.2 the median $=$ mean $=$ mode of the distribution of $\log_{10} x$ is 1·0, and the median of the distribution of x in Fig. 4.5.1 is $\mathrm{antilog}_{10}\ 1 = 10$, whereas the mode is less than 10 and the mean larger than 10.

Because of the rarity of knowledge about the distribution of observations in real life these theoretical distributions will not be discussed further here, but they occur often in theoretical work and good accounts of them will be found in Bliss (1967, Chapters 5–7) and Kendall and Stuart (1963, p. 168; 1966, p. 93).

4.6. Testing for Gaussian distribution. Rankits and probits

If there are enough observations to be plotted as a histogram, like Figs. 14.2.1 and 14.2.2, the probit plot described in §§ 14.2 and 14.3 can be used to test whether variables (e.g. z and log z in § 14.2) follow the normal distribution. For smaller samples the rankit method described, for example, by Bliss (1967, pp. 108, 232, 337) can be used. For two way classifications and Latin squares (see Chapter 11) there is no practicable test. It must be remembered that a small sample gives very little information about the distribution; but *consistent* non-linearity of the rankit plot over a series of samples would suggest a non-normal distribution. The N observations are ranked in ascending order, the rankit corresponding to each rank looked up in tables (see Appendix, Table A9). Each observation (or any transformation of the observation that is to be tested for normality) is then plotted against its rankit.

The rankit corresponding to the smallest (next to smallest, etc.) observation, is defined as the long run average (expectation, see § A1.1) of the smallest (next to smallest, etc.) value in a random sample of N standard normal deviates (values of u, see § 4.3). Thus, if the observations (or their transformations) are normally distributed, the observations and rankits should differ only in scale, and by random sampling, so the plot should, on average, be straight.

5. Random processes. The exponential distribution and the waiting time paradox

5.1. The exponential distribution of random intervals

DYNAMIC processes involving probability theory such as queues, Brownian motion, and birth and deaths are called *stochastic processes*. This subject is discussed further in Appendix 2. An example of interest in physiology is the apparently random occurrence of miniature post-junctional potentials at many synaptic junctions (reviewed by Martin (1966)). It has been found that when the observed number (n) of the time intervals between events (miniature end-plate potentials), of duration equal to or less that t seconds is plotted against t, the curve has the form shown in Fig. 5.1.1. Similar results would be obtained with the intervals between radiosotope disintegrations; see § A2.5.

The observations are found to be fitted by an exponential curve,

$$n = N(1-e^{-t/T}), \qquad (5.1.1)$$

where N = total number of intervals observed and T = mean duration of all N intervals (an estimate of the population mean interval, $\mathscr{T}$).

If the events were occurring randomly it would be expected that the number of events in unit time would follow the Poisson distribution, as described in § § 3.5 and A2.2. How would the intervals between events be expected to vary if this were so?

The true mean number of events in t seconds (called m in § 3·5) is $t/\mathscr{T}$, which may be written as λt, where $\lambda = 1/\mathscr{T}$ is the true mean number of events in 1 second. Thus $\mathscr{T} = 1/\lambda$ is the mean number of seconds per event, i.e. the mean interval between events (see (A1.1.11)). According to the Poisson distribution (3.5.1) the probability that no event ($r = 0$) occurs within time t from any specified starting point, i.e. the probability that the interval before the first event is greater than t, is $P(0) = e^{-m} = e^{-\lambda t}$. The first event must occur either at a time greater than, or at a time equal to or less than t. Because these cannot both happen it follows from the addition rule, (2.4.3), that

$$P[\text{interval} > t] + P[\text{interval} \leqslant t] = 1$$

and thus

$$P[\text{interval} \leqslant t] \equiv F(t) = 1 - e^{-\lambda t} \quad (\text{for } t \geqslant 0). \qquad (5.1.2)$$

(The distribution function, F, was defined in (4.1.4).

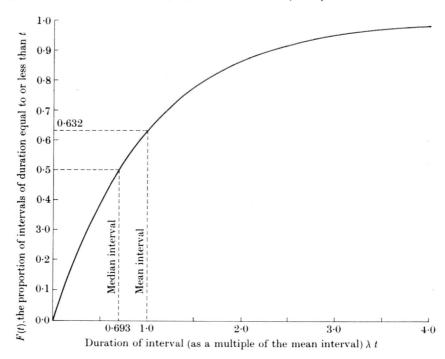

FIG. 5.1.1.   The cumulative exponential distribution (eqn. (5.1.2)). The intervals between random events are observed to fall on a curve like this. The abscissa is the interval duration expressed as a multiple of the population mean interval, i.e. it is $t/\mathscr{T} \equiv \lambda t$. For example, if the population mean were $\mathscr{T} = 10$ s (i.e. $\lambda = 0.1$ s^{-1}), the graph shows that 63.21 per cent of intervals would be 10 s or shorter, and that 50 per cent of intervals (by definition of the median) would be equal to or less than 6.93 s, the population median.

Multiplying this *probability* by N predicts the *number* of intervals shorter than t as $N(1 - e^{-\lambda t})$, as observed (see (5.1.1)).

This implies that the *exponential distribution* is the distribution of the interval between any specified point of time and the point at

which the next event occurs. And, in particular, it is the distribution of
the time interval between successive events (see § 5.2 and Appendix 2).
Because the intervals can be of *any* length this is a continuous distribution, unlike the Poisson, and it has probability density (see § 4.1), using
(5.1.2) and (4.1.5),

$$f(t) = \frac{\mathrm{d}F(t)}{\mathrm{d}t} = \frac{\mathrm{d}}{\mathrm{d}t}(1 - e^{-\lambda t}) = \lambda e^{-\lambda t} \text{ (for } t \geqslant 0\text{)}, \qquad (5.1.3)$$

$$= 0 \qquad\qquad\qquad \text{(for } t < 0\text{)}.$$

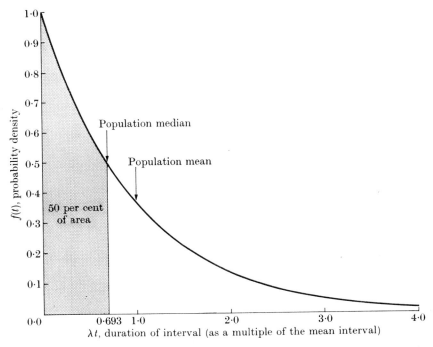

FIG. 5.1.2. The exponential distribution (an extreme case of the positive
skew illustrated in Fig. 4.5.1). Fifty per cent of the area under the curve lies
below the median. The area up to t is plotted against t in Fig. 5.1.1. The abscissa
is plotted in the same way as in Fig. 5.1.1. (If the abscissa is multiplied by $\mathscr{T}$
$\equiv \lambda^{-1}$ to convert it to time units, the probability density would be divided by $\mathscr{T}$,
so the area under the curve remained 1·0.)

This exponential distribution of the lengths of random intervals is
plotted in Fig. 5.1.2. It is an extreme form of positively skewed distribution (see § 4.5), the mode being zero, the mean $1/\lambda = \mathscr{T}$, and the
median $0·693\mathscr{T}$ (this is proved in Appendix 1, (A1.1.11) and (A1.1.14)).

Fig. 5.1.1 is the cumulative form, $F(t)$ (see (4.1.4)), of the exponential distribution (cf. Fig. 4.1.4, which is the cumulative form of the normal distribution in Fig. 4.1.3). To obtain Fig. 5.1.1 from Fig. 5.1.2 notice that the probability of observing an interval $\leqslant t$ is given by the area under the distribution curve (Fig. 5.1.2) below t, i.e. between 0 and t (see § 4.1). This, using (4.1.4), is

$$P[0 \leqslant \text{interval} \leqslant t] \equiv F(t) = \int_0^t \lambda e^{-\lambda t} dt = 1 - e^{-\lambda t}, \quad (5.1.4)$$

which is (5.1.2) again. Further discussion will be found in Appendix 2.

A more complete discussion of the Poisson process would require consideration of the distribution of the sum of n intervals. When this is done it is seen that the observation of an exponential distribution does not necessarily imply a Poisson distribution of events in unit time unless the intervals are independent of each other. Independence has been checked experimentally by Burnstock and Holman (1962). This independence is one of the defining properties of the Poisson process (see § 3.5 and Appendix 2).

5.2. The waiting time paradox

It was implied in § 5.1 that, for completely random events, the average length of time from a randomly selected arbitrary point of time (midday, for example) until the next event is the same (viz. $\mathscr{T}$) as the average length of the interval between two events (both intervals have the same exponential distribution). This is proved in § A2.6. (An *arbitrary* point, in this context, means a point of time chosen by any method that is independent of the occurrence of events.) It must be so since the events in non-overlapping time intervals are supposed independent, i.e. the process has no 'memory' of what has gone before†. Yet it seems 'obvious' that, since the arbitrarily selected time is equally likely to fall anywhere in the interval between two events, the average waiting time from the selected time to the next event must be $\frac{1}{2}\mathscr{T}$.

For example, if buses were to arrive at a bus stop at *random* intervals, with a mean interval of $\mathscr{T} = 10$ min, then a person arriving at the bus stop at an arbitrary time might be supposed, on the average, to have to wait 5 min for the next bus.‡ In fact, the true average waiting time would be 10 min.

---

† See §§ 3.5, A2.1 and A2.2 for details.
‡ 5 min would be the right answer if the buses arrived regularly not randomly, so that *all* intervals were exactly 10 min.

The subtle flaw in the argument for a waiting time of $\frac{1}{2}\mathscr{T}$ lies in the implicit assumption that the interval in which an arbitrarily selected time falls is a random selection from all intervals. In fact, longer intervals have a better chance of covering the selected time than shorter ones, and it can be shown that the average length of the interval in which an arbitrarily selected time falls is not the same as the average length of *all* intervals, $\mathscr{T}$, but is actually $2\mathscr{T}$ (see § A2.7). Since the selected time may fall anywhere in this interval, the average waiting time is half of $2\mathscr{T}$, i.e. it is $\mathscr{T}$, the average length of all intervals, as originally supposed. The paradox is resolved. In the bus example this means that a person arriving at the bus stop at an arbitrary time would, on average, arrive in a 20-min interval. On average, the previous bus would have passed 10 min before his arrival (as long as this was not too near the time when buses started running) and, on average, it would be another 10 min until the next bus.

These assertions, which surprise most people at first, are discussed (with examples of biological importance), and proved, in Appendix 2.

6. Can your results be believed? Tests of significance and the analysis of variance

'. . . before anything was known of Lydgate's skill, the judgements on it had naturally been divided, depending on a sense of likelihood, situated perhaps in the pit of the stomach, or in the pineal gland, and differing in its verdicts, but not less valuable as a guide in the total deficit of evidence.'

GEORGE ELIOT
(*Middlemarch*, Chap. 45)

6.1. The interpretation of tests of significance

THIS has already been discussed in Chapter 1. It was pointed out that the function of significance tests is to prevent you from making a fool of yourself, and not to make unpublishable results publishable. Some rather more technical points can now be discussed.

(1) *Aids to judgement*

Tests of significance are only aids to judgement. The responsibility for interpreting the results and making decisions always lies with the experimenter whatever statistical calculations have been done.

The result of a test of significance is always a probability and should always be given as such, along with enough information for the reader to understand what method was used to obtain the result. Terms such as 'significant' and 'very significant' should never be used. If the reader is unlikely to understand the result of a significance test then either explain it fully or omit reference to it altogether.

(2) *Assumptions*

Assumptions about, for example, the distribution of errors, must always be made before a significance test can be done. Sometimes some of the assumptions are tested but usually none of them are (see §§ 4.2 and 11.2). This means that the uncertainty indicated by the test can be taken as only a minimum value (see §§ 1.1 and 7.2). The assumptions of tests involving the Gaussian (normal) distribution are discussed in §§ 11.2 and 12.2. Other assumptions are discussed when the methods are described.

Some tests (*nonparametric* tests), which make fewer assumptions than those based on a specified, for example normal, distribution (*parametric* tests such as the *t* test and analysis of variance), are described in the following sections. Their relative merits are discussed in § 6.2. Note, however, that whatever test is used, it remains true that if the test indicates that there is *no* evidence that, for example, an experimental group differs from a control group then the experimenter cannot reasonably suppose, on the basis of the experiment, that a real difference exists.

(3) *The basis and the results of tests*

No statements of *inverse probability* (see § 1.3) are, or at any rate need be, made as a result of significance tests. The result, P, is always the probability that certain observations would be made given a particular hypothesis, i.e. if that hypothesis were true. It is *not* the probability that a particular hypothesis is true given the observations.

It is often convenient to start from the hypothesis that the effect for which one is looking does not exist.† This is called a *null hypothesis*. For example, if one wanted to compare two means (e.g. the mean response of a group of patients to drug A with the mean response of another group, randomly selected from the same population, to drug B) the variable of interest would be the difference between the two means. The null hypothesis would be that the *true* value of the difference was zero. The amount of scatter that would be expected in the difference between means if the experiment were repeated many times can be predicted from the experimental observations (see § 2.7 for a full discussion of this process), and a distribution constructed with this amount of scatter and with the hypothetical mean value of zero, as illustrated in Fig. 6.1.1. From this it can be predicted what *would* happen *if* the null hypothesis that the true difference is zero were true. In practice it will be necessary to allow for the inexactness of the experimental estimate of error by considering, for example, the distribution of Student's *t*, see §§ 4.4 and 9.4, rather than the distribution of the difference between means itself. If the differences are supposed to have a continuous distribution, as in Fig. 6.1.1, it is clearly not possible to calculate the probability of seeing *exactly* the observed difference (see § 4.1); but it is possible to calculate the probability of seeing a difference equal to or *larger than* the observed value. In the example illustrated this is $P = 0.04$ (the vertically shaded area) and

† See p. 93 for a more critical discussion.

this figure is described as the result of a *one-tail significance test*. Its interpretation is discussed in (4) below. It is the figure that would be used to test the null hypothesis against the alternative hypothesis that the *true* difference is positive. When the alternative hypothesis is that true difference is *positive,* the result of a one-tail test for the difference between two means always has the following form.

> *If* there were no difference between the true (population) means *then* the probability of observing, because of random sampling error, a difference between sample means equal to or greater than that observed in the experiment would be P (assuming the assumptions made in carrying out the test to be true).

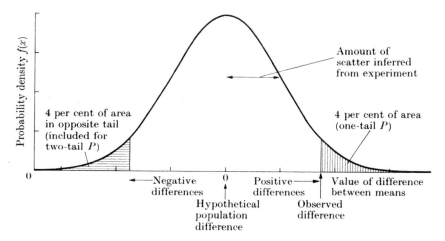

FIG. 6.1.1.   Basis of significance tests. See text for explanation.

If the only possible alternative to the null hypothesis is that the true difference is negative, then the interpretation is the same, except that it is the probability (on the null hypothesis) of a difference being equal to or *less than* the observed one that is of interest.

In practice, in research problems at least, the alternative to the null hypothesis is usually *not* that the true difference is positive (or that it is negative) but simply that it differs from zero† (in either direction), because it is usually not reasonable to say in advance that only positive (or negative) differences are possible (or that only positive differences are of interest so the test is not required to detect negative differences).

† See also p. 93.

If the alternative to the null hypothesis is the hypothesis that the true difference between means is, say, positive, this implies that however large a negative difference was *observed* it would be attributed to chance rather than a *true* (population) negative difference (or at least that it would be considered of no interest if real).

Suppose now that it cannot be specified beforehand whether the *true* difference between means is positive, zero, or negative. In the example above there would be probability of 0·04 of seeing a difference at least as large as the positive difference observed in the experiment *if* the null hypothesis were true. But there would also be a probability of 0·04 (the horizontally shaded area) of seeing a deviation from the null hypothesis at least as extreme as that actually observed but in the opposite direction. The total probability of observing a deviation from the null hypothesis (in either direction) at least as extreme as that actually observed would be $P = 0·04 + 0·04 = 0·08$ *if* the null hypothesis were true. This is the appropriate probability because, if it were resolved to reject the null hypothesis as false every time an experiment gave a difference between means as large as, or larger than that observed in this experiment, then, *if* the null hypothesis were actually true it would be rejected (wrongly) not in 4 per cent of repeated experiments, but in 8 per cent. This is because negative observed differences in the lower tail of Fig. 6.1.1, which would also lead to wrong rejection of the null hypothesis, would be just as common, in the long run, as positive differences. The probability is chosen so as to control the frequency of this sort of error. This is discussed in more detail in subsection (6) below.

The value $P = 0·08$ is described as the result of a *two-tail test of significance*. Its interpretation is discussed in subsection (4) below. The value of P is usually† twice that for a one-tail test. The result of a two-tail test always has the following form.

> *If* the null hypothesis were actually true *then* the probability of a sample showing a deviation from it, in either direction, as extreme,† or more extreme, than that observed in the experiment would be P (assuming the assumptions made in carrying out the test to be true).

† In the case of the normal distribution (§ 4.2), or any other distribution that is symmetrical, whether continuous or discontinuous, for example the binomial distribution with $\mathscr{P} = 0·5$ (§§ 3·2 and 3·4) or Student's distribution, (§ 4.4), one could say here ' . . . a deviation from it, in either direction, *as large as, or larger than*, that observed in the

Notice that P is not the probability that the null hypothesis *is* true but the probability that certain observations would be made *if it were*.

Perhaps the best popular interpretation of P is that it is the 'probability of the results occurring by chance'. Although this is inaccurate and vague, and should therefore be avoided, it is not too misleading.

(4) *Interpretation of the results*

If P is very small the conclusion drawn is that either

(a)   an unlikely event has taken place, the null hypothesis being true. As Fisher (1951) said: ' . . . no isolated experiment, however significant in itself, can suffice for the experimental demonstration of any natural phenomenon; for the "one chance in a million" will undoubtedly occur, with no less and no more than its appropriate frequency, however surprised we may be that it should occur to *us*,' or

(b)   the assumptions on which the test was based were faulty, for example the samples were not drawn randomly, or

(c)   the null hypothesis is not true, for example the true (population) means in the above example are different, so that the drugs do in fact differ in their effects on patients (see also subsection (7), below).

Whether (b) can be ruled out, and what level of improbability is enough to make one favour explanation (c) rather than (a), are

---

experiment . . . ' In general, this simpler statement is not possible, however. Two other cases must be considered. (1) The sampling distribution (e.g. Fig. 6.1.1) is continuous but unsymmetrical (see § 4.5). In this case different sized positive and negative deviations will be needed to cut off equal areas in the upper and lower tails (respectively) of the distribution. It is the *extremeness* (i.e. *rarity*) of the deviation measured by the area it cuts off in the tail of the distribution (rather than its *size*) that matters. The two-tail probability is still twice the one-tail probability, however. (2) The sampling distribution is both unsymmetrical and discontinuous (as often happens in the very important sort of tests known as randomization tests, see §§ 8.2, 9.2, 9.3, and 10.2–10.4). A greater difficulty arises in this case because the most extreme observations in the opposite tail of the distribution (that not containing the observation) will not generally cut off an area exactly the same as that cut off by the observation in its own tail so P for the two-tail test cannot be exactly twice that for the one-tail test. There is no definite rule about what to do in this case. Most commonly a deviation is chosen in the opposite direction to that observed that cuts cuts off an area in the opposite tail *not greater than* the value found in the one-tail test, so the two-tail P is not greater than twice the one-tail P. However, it may be decided to choose a deviation that cuts off an area in the opposite tail that is *as near as possible* to that of the one-tail test. This is exemplified at the end of § 8.2 where the deviations of a from the null hypothetical value are stated, to show exactly what has been done. With small unequal samples the most extreme possible observation in the opposite tail may cut off an area far greater than that in the one tail test. This problem is discussed in § 8.2.

entirely matters for personal judgement. The calculations throw no light whatsoever on these problems. It is often found in the biomedical literature that $P = 0.05$ is taken as evidence for a 'significant difference'. However 1 in 20 is not a level of odds at which most people would want to stake their reputations as an experimenters and, if there is no other evidence, it would be wiser to demand a much smaller value before choosing explanation (c).

A twofold change in the value of P given by a test should make little difference to the inference made in practice. For example, $P = 0.03$ and $P = 0.06$ mean much the same sort of thing, although one is below and the other above the conventional 'significance level' of 0.05. They both suggest that the null hypothesis may not be true without being small enough for this conclusion to be reached with any great confidence.

In any case, as mentioned above, no single test is ever enough. To quote Fisher (1951) again: 'In relation to the test of significance, we may say that a phenomenon is experimentally demonstrable when we know how to conduct an experiment which will rarely fail to give us a statistically significant result'.

(5) *Generalization of the result*

Whatever the interpretation of the statistical calculations it is tempting to generalize the conclusion from the experimental sample to other samples (e.g. to other patients); in fact this is usually the purpose of the experiment. To do this it is necessary to assume that the new samples are drawn randomly from the same population as that from which the experimental samples were drawn. However, because of differences of, for example, time or place this must usually remain an untested assumption which will introduce an unknown amount of bias into the generalization (see §§ 1.1 and 2.3).

(6) *Types of error and the power of tests*

If the null hypothesis is not rejected on the basis of the experimental results (see subsection (7), below) this does *not* mean that it can be accepted. It is only possible to say that the difference between two means is *not demonstrable*, or that a biological assay is *not demonstrably invalid*. The converse, that the means are identical or that the assay is valid, can never be shown. If it could it would always be possible to find that there was, for example, 'no difference between two means' but doing such a bad experiment that even a large real difference was not apparent. Although this may seem gloomy, it is only common sense. To show that two population means are identical *exactly*, the whole population, usually infinite, is obviously needed.

An example. The supposition that a large P value constitutes evidence in favour of the null hypothesis is, perhaps, one of the most frequent abuses of 'significance' tests. A nice example appears in a paper just received. The essence of it is as follows. Differences between membrane potentials before and after applying three drugs were measured. The mean differences ($\bar{d}$) are shown in Table 6.1.1.

TABLE 6.1.1

d stands for the difference between the membrane potentials (millivolts) in the presence and absence of the specified drug. The mean of n such differences is $\bar{d}$, and the observed standard deviation of d is $s(d)$. The standard deviation of the mean difference is $s(\bar{d}) = s(d)/\sqrt{n}$ and values of Student's t are calculated as in § 10.6.

	$\bar{d}$	$s(d)$	n	$s(\bar{d})$	t	P (approx)
Noradrenaline	2·7	10·1	40	1·60	1·7	0·1
Adrenaline	3·4	12·2	80	1·36	2·5	<0·02
Isoprenaline	3·9	10·8	60	1·39	2·8	<0·01

The potentials were about 90 mV so the percentage change is small, but by doing many ($n = 40$–80) pairs of measurements, evidence was found against the null hypothesis that adrenaline has no effect, using the paired t test (see § 10.6). Similarly it was inferred that isoprenaline increases membrane potential. These inferences are reasonable, though the order in which treatments were applied was not randomized. In contrast, the P value for noradrenaline was 0·1 and the authors therefore inferred that 'noradrenaline had no effect on membrane potential', i.e. that the null hypothesis was true. This is completely unjustified. The apparent effect of noradrenaline, 2·7 mV, was not much smaller than that for other drugs, and, although the significance test shows that we cannot be sure that repeating the measurements would give a similar result, it certainly does not show that we *would not* get similar results. Suppose, perfectly plausibly, that 80 experiments had been done with noradrenaline (as with adrenaline) instead of 40. And suppose the mean difference was 2·7 mV and the standard deviation of the differences was 10·1. In this case $t = 2\cdot7/(10\cdot1/\sqrt{80}) = 2\cdot4$ giving $P < 0\cdot02$ a 'significant' result. The size of the difference $\bar{d} = 2\cdot7$ mV, and the scatter of the observations $s(d) = 10\cdot1$, is just the same as in

Table 6.1.1, but despite this the authors would presumably have come to the opposite conclusion. This is clearly absurd. But if the original experiment with $n = 40$ differences had been interpreted as 'no evidence for a real effect of noradrenaline' or 'effect, if any, masked by experimental error' there would have been no trouble. It is reasonable that the larger experiment should be capable of detecting differences that escape detection in the smaller experiments.

These ideas can be formalized by considering the *power* of a significance test which is defined as the probability that the test will reject the null hypothesis (e.g. that two population means are equal), this probability being considered as a function of the *true* difference between the means. For example, if the null hypothesis was always rejected whenever a test gave $P \leqslant 0.05$ then, if the null hypothesis really were true it would be rejected (*wrongly*) in 5 per cent of trials, as explained in subsection (3) above (see subsection (7), below). The wrong rejection of a correct hypothesis is *called an error of the first kind*, and, in this case, the probability (α) of an error of the first kind would be $\alpha = 0.05$. If in fact there was a difference between true population means, and this real difference was, for example, equal in size to the true standard deviation of the difference between means (see §§ 2.7 and 9.4) (i.e. the difference, although real, is similar in size to the experimental errors), then it can be shown that a two-tail normal deviate test† would reject the null hypothesis (this time correctly) in 17 per cent of experiments. However, if the null hypothesis was accepted as true every time it was not rejected then it would be *wrongly* accepted in 83 per cent of experiments. The wrong acceptance of a false hypothesis is called *an error of the second kind*, and, in this case, the probability (β) of this sort of error is $\beta = 0.83$.

The power curve for a two-tailed normal deviate test for the difference between two means is shown in Fig. 6.1.2 and compared with the power curve for the (non-existent) ideal test that would always accept true hypotheses and reject false ones. The power of even the best tests to detect real differences that are similar in size to the experimental error is quite small.

(7) *Some more subtle points about significance tests*

The critical reader will, no doubt, have some objections to the arguments presented in this section. It is difficult to give a consensus of informed opinion

† A t test (see § 9.4) in which the standard deviation is accurately known (e.g. because the samples are large) so the standard normal deviate, u (see § 4.3), can be used in place of t (see §4.4).

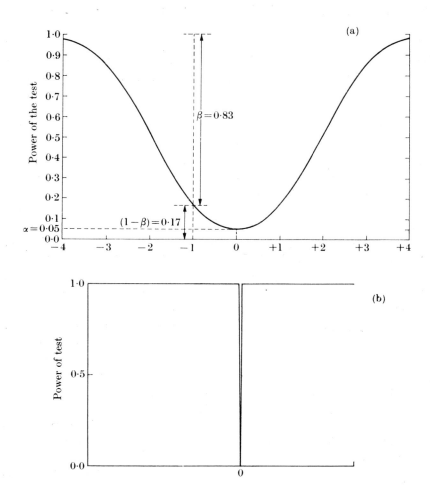

FIG. 6.1.2. In both figures the abscissa gives the difference between the *population* means (expressed as a multiple of the standard deviation of the difference between means: see § 9.4). (a) The power curve for a two-tail normal deviate test for difference between two means (see text) when $\alpha = 0.05$, i.e. the null hypothesis is rejected whenever $P < 0.05$, so if it were actually true it would be wrongly rejected in 5 per cent of repeated experiments. If the null hypothesis were false, i.e. there *is* a difference between the *population* means (in this example, a difference equal in size to one standard deviation of the difference between means: see § 9.4) the null hypothesis would be rejected (correctly) in 17 per cent of experiments and not rejected (wrongly) in $\beta = 83$ per cent of experiments. (b) Power curve for the (non-existent) ideal test that always rejects a hypothesis (population means equal) when it is false, and never rejects it when it is true.

because, although there is much informed opinion, there is rather little consensus. A personal view follows.

The first point concerns the role of the null hypothesis and the role of prior knowledge, i.e. knowledge available before the experiment was done. It is widely advocated nowadays (particularly by Bayesians, see §§ 1.3 and 2.4) that prior information should be used in making statistical decisions. There is no doubt that this is desirable. All relevant information should be taken into account in the search for truth, and in some fields there are reasonable ways of doing this. But in this book the view is taken that attention must be restricted to the information that can be provided by the experiment itself. This is forced on us because, in the sort of small-scale laboratory or clinical experiment with which we are mostly concerned, no one has yet devised a way that is acceptable to the scientist, as opposed to the mathematician, of putting prior information in a quantitative form.

Now it has been mentioned already that in most real experiments it is unrealistic to suppose that the null hypothesis† could ever be true, that two treatments could be *exactly* equi-effective. So is it reasonable to construct an experiment to test a null hypothesis? The answer is that it is a perfectly reasonable way of approaching our aim of preventing the experimenter from making a fool of himself if, as recommended above, we say only that 'the *experiment* provides evidence against the null hypothesis' (if P is small enough), or that 'the *experiment* does not provide evidence against the null hypothesis' (if P is large enough). The fact that there may be prior evidence, *not* from the experiment, against the null hypothesis does not make it unreasonable to say that the experiment itself provides no evidence against it, in those cases where the observations in the experiment (or more extreme ones) would not have been unusual in the (admittedly improbable) event that the null hypothesis was exactly true.

And, because it has been stressed that if there is no evidence against the null hypothesis it does not imply that the null hypothesis is true, the inference from a large P value does not contradict the prior ideas about the null hypothesis. We may still be convinced on prior grounds that there is a real difference of some sort, but as it is apparently not large enough, relative to the experimental error and method of analysis, to be detected in the experiment, we have no idea of its size *or direction*. So the prior knowledge is of no practical importance.

Another point concerns the discussion of power. It has been recommended that the result of significance test should be given as a value of P. It would be silly to reject the null hypothesis automatically whenever P fell below arbitrary level (0·05 say). Each case must be judged on its merits. So what is the justification for discussing in subsections (3) and (6) above, what would happen 'if the null hypothesis were always rejected when $P \leqslant 0.05$'? As usual, the aim is to prevent the experimenter making a fool of himself. Suppose, in a particular case, that a significance test gave $P = 0.007$, and the experimenter decided that, all things considered, this should be interpreted as meaning that the experiment provided evidence against the null hypothesis, then it is certainly of interest to the experimenter to known what would be the consequences of acting consistently in this way, in a series of imaginary repetitions of the experiment in question. This does not in any way imply that given a different experiment, under different circumstances, the experimenter should behave in the same way, i.e. use $P = 0.007$ as a critical level.

---

† This remark applies to point hypotheses, i.e. those stating that means, populations, etc., are *identical*. All the null hypotheses used in this book are of this sort.

8

6.2. Which sort of test should be used, parametric or nonparametric?

Parametric tests, such as the *t* test and the analysis of variance are those based on an assumed form of distribution, usually the normal distribution, for the population from which the experimental samples are drawn. Nonparametric tests are those that, although they involve some assumptions, do not assume a particular distribution. A discussion of the relative 'advantages' of the tests is ludicrous. If the distribution is known (not assumed, but *known;* see § 4.6 for tests of normality), then use the appropriate parametric test. Otherwise do not. Nevertheless the following observations are relevant.

Characteristics of nonparametric methods

(1) Fewer untested assumptions are needed for nonparametric methods. This is the main advantage, because, as emphasized in § 4.2, there is rarely any substantial evidence that observations follow a normal, or any other, distribution. The assumptions involved in parametric methods are discussed in § 11.2. Nonparametric methods do involve some assumptions (e.g. that two distributions are of the same, but unspecified, form), and these are mentioned in connection with individual methods.

(2) Nonparametric methods can be used for classification (Chapter 8) or rank (Chapters 9–11) measurements. Parametric methods cannot.

(3) Nonparametric methods are usually easier to understand and use.

Characteristics of parametric methods

(1) Parametric methods are available for analysing for more sorts of experimental results. For example there are, at the moment, no widely available nonparametric methods for the more complex sort of analysis of variance or curve fitting problems. This is not relevant when choosing which method to use, because there is only a choice if a nonparametric method *is* available.

(2) Many problems involving the estimation of population parameters from a sample of observations have so far only been dealt with by parametric methods.

(3) It is sometimes listed as an advantage of parametric methods that *if* the assumptions they involve (see § 11.2) are true, they are more powerful (see § 6.1, para. (6)), i.e. more sensitive detectors of real differences, than nonparametric. However, if the assumptions are not true, which is normally not known, the nonparametric methods may

well be more powerful, so this cannot really be considered an advantage. In any case, even when the assumptions of parametric methods *are* fulfilled the nonparametric methods are often only slightly less powerful. In fact the randomization tests described in §§ 9.2 and 10.3 are as powerful as parametric tests even when the assumptions of the latter are true, at least for large samples.

There is a considerable volume of knowledge about the *asymptotic relative efficiencies* of various tests. These results refer to infinite sample sizes and are therefore of no interest to the experimenter. There is less knowledge about the relative efficiencies of tests in small samples. In any case, it is always necessary to specify, among other things, the distribution of the observations before the relative efficiencies of tests can be deduced; and because it is part of the problem that nothing is known about this distribution, even the results for small samples are not of much practical help. Of the alternative tests to be described, each can, for certain sorts of distribution, be more efficient than the others.

There is, however, one rather distressing consequence of lack of knowledge of the distribution of error, which is, of course, not abolished by assuming the distribution known when it is not.

As an example of the problem, consider the comparison of the effects of two treatments, A and B. The experimenter will be very pleased if a large and consistent difference between the effects of A and B is observed, and will feel, reasonably, that not many observations are necessary. But it turns out that with very small samples it is impossible to find evidence against the hypothesis that A and B are equi-effective, however large, and however consistent, the difference observed between their effects, unless something is known about the distributions of the observations. Suppose, for the sake of argument, that the experimenter is prepared to accept $P = 1/20$ (two tail) as small enough to constitute evidence against the hypothesis of equi-effectiveness (see § 6.1). If the experiment is conducted on two independent samples, each sample must contain at least 4 observations (for all the nonparametric tests described in Chapter 9, q.v., the minimum possible two-tail P value with samples of 3 and 4 would be $2.3!4!/7! = 1/17\frac{1}{2}$, however large and consistent the difference between the samples). Similarly, if the observations are paired, at least 6 pairs of observations are needed; with 5 pairs of observations the observations on the nonparametric methods described in Chapter 10, q.v., can *never* give a two-tail P less than $2.(\frac{1}{2})^5 = 1/16$. (See also the discussion in §§ 10.5 and 11.9.)

In contrast, the parametric methods can give a very low P with the smallest samples if the difference between A and B is sufficiently large and consistent. *Nevertheless, these facts mean that it is a disadvantage not to know the distribution of the observations.* They do not constitute a disadvantage of nonparametric tests. The problem is less acute with samples larger than the minimum sizes mentioned.

In view of these remarks it may be wondered why parametric tests are used at all when there are nonparametric alternatives. In fact they are still widely used even now. This is partly because of familiarity. The t test and analysis of variance were in use for many years before most nonparametric methods were developed. It probably also results from the sacrifice of relevance to the real world for the sake of mathematical elegance. Methods based on the assumption of a normal distribution have been developed to cover a wide range of problems within a single, admittedly elegant, mathematical framework.

It is not uncommon for those who are dubious about the assumptions necessary for parametric tests to be told something along the lines 'experience has shown that the t test (for example) will not mislead us'. Unfortunately, as Mainland (1963) has pointed out, this is just wishful thinking. There is no knowledge at all of the number of times people have been misled by using the t test when they would not have been misled by a nonparametric test (see §§ 4.2 and 4.6).

A plausible reason for using tests based on the normal distribution is that some of them have been shown to be fairly insensitive to some sorts of deviations from the assumptions on which they are based if the samples are reasonably big. The tests are said to be fairly *robust*. But this knowledge can usually be used only by intuition. One is never sure how large is large enough for the purposes in hand. When the nature and extent of deviations from the assumptions is unknown, the amount of error resulting from assuming them true is also unknown. It is much simpler to avoid as many as possible of the assumptions.

> If a nonparametric test is available it should be used in preference to the parametric test, unless there is *experimental* evidence about the distribution of errors.

In spite of what has just been said parametric methods are discussed in the following chapters, even when nonparametric methods exist. This is necessary as an approach to the more complex experimental designs, curve-fitting problems, and biological assay for which there are still hardly any nonparametric methods available, so parametric tests or nothing must be used. Whichever test is used, it should be interpreted as suggested in §§ 1.1, 1.2, 6.1, and 7.2, the uncertainty indicated by the test being taken as the minimum uncertainty that it is reasonable to feel.

6.3. Randomization tests

The principle of *randomization tests*, also known as *permutation* tests, is of great importance because these tests are among the most powerful of nonparametric tests (see § 6.1 and 6.2). Moreover, they are easier to understand, at the present level, than almost all other sorts of test and they make very clear the fundamental importance of randomization. Examples are encountered in §§ 8.2, 8.3, 9.2, 9.3, 10.2, 10.3, 10.4, 11.5, 11.7, and 11.9.

6.4. Types of sample and types of measurement

When comparing two groups the groups may be related or independent. For example, to compare drugs A and B two groups could be selected *randomly* (see § 2.3) from the population of patients, and one group given A, the other B. The two samples are independent. Independent samples are discussed in Chapters 8 and 9, and in §§ 11.4, 11.5, and 11.9. On the other hand, the two drugs might both be given, in *random* order, to the same patient, or to a patient randomly selected from a pair of patients who had been matched in some way (e.g. by age, sex, or prognosis). The samples of observations on drug A and drug B are said to be related in this case. This is usually a preferable arrangement if it is possible; but it may not be possible because, for example, the effects of treatments are too long-lasting, or because of ignorance of what characteristics to match. Related samples are discussed in Chapter 10 and in §§ 8.6, 11.6, 11.7, and 11.9.

The method of analysis will also depend on what sort of measurements are made. The three basic types of measurement are (1) classification (the nominal scale), (2) ranking (the ordinal scale), and (3) numerical measurements (the interval and ratio scales). For further details see, for example, Siegel (1956a, pp. 21–30). If the best that can be done is *classification* as, for example, improved or not improved, worse or no change or better, passed or failed, above or below median, then the methods of analysis in Chapter 8 are appropriate. If the measurements cannot be interpreted in a quantitative numerical way but can be *arranged (ranked) in order of magnitude* (as, for example, with arbitrary scores such as those used for subjective measurements of the intensity of pain) then the rank methods described in §§ 9.3, 10.4, 10.5, 11.5, 11.7, and 11.9 should be used. For *quantitative numerical measurements* the methods described in the remaining sections of Chapters 9–11 are appropriate.

Methods for dealing with a single sample are discussed in Chapter 7 and those for more than two samples in Chapter 11.

7. One sample of observations. The calculation and interpretation of confidence limits

'Eine Hauptursache der Armut in den Wissenschaften ist meist eingebildeter Reichtum. Es is nicht ihr Ziel, der unendlichen Weisheit eine Tür zu öffnen, sondern eine Grenze zu setzen dem unendlichen Irrtum.'†

† 'One of the chief causes of poverty in science is usually imaginary wealth. The aim of science is not to open a door to infinite wisdom, but to set a limit to infinite error'.

GALILEO in Brecht's *Leben des Galilei*

7.1. The representative value: mean or median?

IT is second nature to calculate the arithmetic mean of a sample of observations as the representative central value (see §2.5). In fact this is an arbitrary procedure. *If* the distribution of the observations were normal it would be a reasonable thing to do since the sample mean would be an estimate of the same quantity (the population mean = population median) as the sample median (§§ 2.5 and 4.5), and it would be a more precise estimate than the median. However, the distribution will usually *not* be known, so there is usually no reason to prefer the mean to the median. For more discussion of the estimation of 'best' values see §§ 12.2 and 12.8 and Appendix 1.

7.2. Precision of inferences. Can estimates of error be trusted?

The answer is that they cannot be trusted. The reasons why will now be discussed. Having calculated an estimate of a population median or mean, or other quantity of interest, it is necessary to give some sort of indication of how precise the estimate is likely to be. Again it is second nature to calculate the standard deviation of the mean—the so-called 'standard error'—see § 2.7. This is far from ideal because there is no simple way of interpreting the standard deviation unless the distribution of observations is known. *If* it were normal then the confidence limits, sometimes called confidence intervals, based on the t distribution (§ 7.4) would be the ideal way of specifying precision since it allows for the fact that the sample standard deviation is itself only a more or less inaccurate estimate of the population value (see § 4.4).

As usual it must be emphasized that the distribution is hardly ever known, so it will usually be preferable to use the nonparametric confidence intervals for the median (§ 7.3), which do not assume a normal distribution.

No sort of confidence interval, nonparametric or otherwise, can make allowance for samples not having been taken in a strictly random fashion (see §§ 1.1 and 2.3), or for systematic (non-random) errors. For example, if a measuring instrument were wrongly calibrated so that every reading was 20 per cent below its correct value, this error would not be detectable and would not be allowed for by any sort of confidence limits.

Therefore in the words of Mainland (1967*a*), confidence limits 'provide a kind of minimum estimate of error, because they show how little a particular sample would tell us about its population, even if it were a strictly random sample'. It seems then that estimates cannot be trusted very far. To quote Mainland (1967 *b*) again,

'Any hesitation that I may have had about questioning error estimates in biology disappeared when I recently learned more about error estimates in that sanctuary of scientific precision—physics.

'One of the most disturbing things about scientific work is the failure of an investigator to confirm results reported by an earlier worker. For example in the period 1895 to 1961, some 15 observations were reported on the magnitude of the astronomical unit (the mean distance from the earth to the sun). You will find these summarized in a table . . . which lists the value obtained by each worker and his estimates of plus or minus limits for the error of the estimate. It is both entertaining and shocking to note that, in every case, a worker's estimate is *outside* the limits set by his immediate predecessor. Clearly there is an unresolved problem here, namely, that experimenters are apparently unable to arrive at realistic estimates of experimental errors in their work" (Youden 1963).

If we add to the problems of the physicist the variability of biological and human material, and the nonrandomness of our samples from it, we may well marvel at the confidence with which "confidence intervals" are presented.'

Confidence limits purport to predict from the results of one experiment what will happen when the experiment is repeated under the same (as nearly as possible) conditions (see § 7.9). But the experimentalist will not need much persuading that the only way to find out what will happen is actually to repeat the experiment and see. And on the few occasions when this has been done in the biological field the results have been no more encouraging than those just quoted. For example, Dews and Berkson (1954) found that the internal estimates of error calculated in individual biological assays were mostly considerably lower than the true error found by actual repetition of the assay. As Dews and Berkson point out, if the assays were performed at different times or in different laboratories it would probably be said that there were 'inter-time' or 'inter-laboratory' differences; and if there were no such 'obvious' reasons for the interval error estimates being too low, then probably 'the animals would be stigmatized as "heterogeneous", with more than a hint that there had been too little incestuous activity among them'. The moral is once again that confidence limits, or other estimates of error calculated from the internal evidence of an experiment, must be interpreted as lower bounds for the real error.

Nevertheless, on the grounds that a minimum estimate of error is better than none at all, examples follow. Their interpretation is discussed further in § 7.9.

7.3. Nonparametric confidence limits for the median

Limits can be found very simply indeed, without any calculation at all, using the table of Nair (1940) which is reproduced as Table A1.

Consider, for example, determinations of the glomerular filtration rate (ml/min) from nine randomly selected dogs:

$$135 \quad 133 \quad 154 \quad 124 \quad 153 \quad 142 \quad 140 \quad 134 \quad 138.$$

The observations will be denoted in the usual way (§ 2.1), $y_i (i = 1, 2, ..., n)$ and $n = 9$. Now rank the observations in ascending order; 124, 133, 134, 135, 138, 140, 142, 153, 154. These observations will be denoted $y_{(i)} (i = 1, 2,...9)$, the parenthesized subscript being used to indicate that the observations have been ranked, i.e. y_1 simply denotes the first observation written down, whereas $y_{(1)}$ indicates the smallest of the observations. The sample estimate of the population median is $y_{(5)} = 138$ ml/min (using (2.5.5)). Reference to Table A1, for the approximately 95 per cent confidence limits, and for a sample size $n = 9$, gives a value $r = 2$. This means that the second (i.e. the rth) observation from each end, viz. 133 ml/min ($= y_{(2)}$) and 153 ml/min ($= y_{(8)}$), are to be taken as the confidence limits for the estimated median, 138 ml/min. The table also gives, in the next column after r, the figure 96·1, which indicates that these are actually 96·1 per cent confidence limits. The fact that r has to be a whole number makes it impossible to get exactly 95 per cent limits. There is a probability of 0·961 that the population median is between $y_{(2)}$ and $y_{(8)}$ in the sense explained in § 7.9.

The reasoning behind the construction of Table A1 is roughly as follows (see Nair (1940) and Mood and Graybill (1963, p. 407)). Let m denote the population (true) median. By definition of the median (§ 2.5) the probability is $1/2$ that an observation selected *at random* from the population, which is assumed to follow *any* continuous distribution, will be less than m. The probability that i observations out of n fall below m follows directly from the binomial distribution (3.4.3) with $\mathscr{P} = \frac{1}{2}$, i.e.

$$\frac{n!}{i!(n-i)!}\left(\frac{1}{2}\right)^n. \tag{7.3.1}$$

To find from this the probability that the rth ranked observation, $y_{(r)}$, in a sample of n observations, will be greater than the population median, note that this will be the case if the sample contains either $i = 0$ **or** 1 **or**...**or** $(r-1)$ observations below the median, so, by using the addition rule (2.4.2),

$$P(y_{(r)} > m) = \sum_{i=0}^{i=r-1} \frac{n!}{i!(n-i)!}\left(\frac{1}{2}\right)^n. \tag{7.3.2}$$

If a 95 per cent confidence limit is required, r is now chosen so as to make this expression as near as possible to $0 \cdot 025$ ($2 \cdot 5$ per cent). In the above example this means taking $r = 2$ giving

$$P(y_{(2)} > m) = \sum_{i=0}^{i=1} \frac{9!}{i!(9-i)!}\left(\frac{1}{2}\right)^9 = \frac{1}{512} + \frac{9}{512} = 0 \cdot 0195,$$

i.e. it is unlikely that $y_{(2)}$ will be *above* the population median. Because of the symmetry of the binomial distribution when $\mathscr{P} = \frac{1}{2}$ (§ 3.4) this is also the probability that $y_{(8)} < m$; it is equally unlikely that $y_{(8)}$ is *less than* the population median. Thus, in general, (7.3.2) also gives $P(y_{(n-r+1)} < m)$. So the probability of the event that either $m < y_{(2)}$ **or** $m > y_{(8)}$ is, again by the addition rule (2.4.2), $0 \cdot 0195 + 0 \cdot 0195 = 0 \cdot 039$. If this event does not occur then it must† be that

† If you find this argument takes you by surprise, in spite of its mathematical impeccability, you may be relieved to find that this view is shared by some of the most eminent mathematical statisticians. For example, Lindley (1969) says 'The procedure which transfers a distribution on x to one on θ through a pivotal quantity such as $x - \theta$ has always seemed to me to be reminiscent of a conjuring trick: it all looks very plausible, but you cannot see how it is done . . . . As a young man I remember asking E. C. Fieller to suggest a really difficult problem. His answer was beautifully simple: "The probability that an observation is less than the median is 1/2: explain why this means that the probability that the median is greater than the observation is also 1/2." I could offer no really sound explanation then, and I still cannot.' You may also be relieved to find that, in spite of the difficulties, virtually all statisticians, faced with experimental results such as those in this section, would reach a conclusion that differed little, if at all, from that presented here.

$y_{(2)} \leqslant m \leqslant y_{(8)}$, and the probability of this must be $1 - 0 \cdot 039 = 0 \cdot 961$, as discovered above from Table A1. The general result is

$$P[y_{(r)} \leqslant m \leqslant y_{(n-r+1)}] = 1 - 2 \sum_{i=0}^{i=r-1} \frac{n!}{i!(n-i)!} \left(\frac{1}{2}\right)^n. \quad (7.3.3)$$

and r is chosen so that this is as near as possible, given that r must be a whole number, to $0 \cdot 95$, or whatever other confidence probability is required. A very similar sort of statement is found for the mean in the next section.

The method assumes that the distribution of the observations is continuous (see § 4.1) so it is not possible for two observations to be *exactly* the same. In practice there may be ties because of rounding errors but this does not matter even though very occasionally a sample could give the same, say 95 and 99 per cent limits. If the distribution is really discontinuous then the method is not appropriate.

7.4. Confidence limits for the mean of a normally distributed variable†

Confidence limits for the population mean

In the improbable event that the glomerular filtration rate of dogs was known to follow the normal distribution it would be possible to calculate confidence limits for the mean of the nine observations given in § 7.3. The sample mean is $\Sigma y/n = 1253/9 = 139 \cdot 2$ ml/min, compared with the sample median of 138 ml/min. The sum of squared deviations is given by (2.6.5) as

$$\sum_{i=1}^{i=n} (y_i - \bar{y})^2 = \sum_{i=1}^{n} y_i^2 - \frac{(\Sigma y_i)^2}{n} = 175179 - \frac{(1253)^2}{9} = 733 \cdot 56.$$

Therefore the variance of y is estimated to be $s^2(y) = 733 \cdot 56/(9-1) = 91 \cdot 69$; the variance of the mean is $s^2(\bar{y}) = 91 \cdot 69/9 = 10 \cdot 19$ by eqn. (2.7.8), and the estimated standard deviation of the sample mean glomerular filtration rate is $s(\bar{y}) = \sqrt{(10 \cdot 19)}$ ml/min $= 3 \cdot 192$ ml/min. These estimates have $n-1 = 8$ degrees of freedom (§ 2.6). From this estimate of scatter, and the assumption that y (and therefore $\bar{y}$) is normally distributed, limits can be calculated within which the mean of the population from which the observations were drawn (which may

---

† The assumption of normality could be tested as in § 4.6 if there were more observations, but with one sample of 9 no useful test can be made.

or not be the population in which the investigator is really interested) is likely to lie.

The limits must be based on Student's t distribution (§ 4.4) because only the estimated standard deviation is available. Reference to tables (see § 4.4) shows that, in the long run, 95 per cent of values of t (with 8 d.f.) will fall between $t = -2 \cdot 306$ and $t = +2 \cdot 306$. The definition of t (eqn. (4.4.1)) is $(x-\mu)/s(x)$ where x is normally distributed. In the present example the (assumed) normally distributed variable of interest is the sample mean, $\bar{y}$, so t is defined as $(\bar{y}-\mu)/s(\bar{y})$.

It follows that in 95 per cent of experiments $t = (\bar{y}-\mu)/s(\bar{y})$ is expected to lie between $-2 \cdot 306$ and $+2 \cdot 306$, i.e.

$$P[-2 \cdot 306 \leqslant (\bar{y}-\mu)/s(\bar{y}) \leqslant +2 \cdot 306] = 0 \cdot 95,$$
$$\therefore P[-2 \cdot 306.s(\bar{y}) \leqslant (\bar{y}-\mu) \leqslant +2 \cdot 306.s(\bar{y})] = 0 \cdot 95,$$
$$\therefore P[\bar{y}-2 \cdot 306.s(\bar{y}) \leqslant \mu \leqslant \bar{y}+2 \cdot 306.s(\bar{y})] = 0 \cdot 95.\dagger$$

This statement, which is analogous to (7.3.3), indicates our confidence that the population mean, μ, lies between the $P = 0 \cdot 95$ confidence limits, viz. $\bar{y}-2 \cdot 306 s(\bar{y}) = 139 \cdot 2 -(2 \cdot 306 \times 3 \cdot 192) = 131 \cdot 8$ ml/min and $\bar{y}+2 \cdot 306 s(\bar{y}) = 139 \cdot 2 +(2 \cdot 306 \times 3 \cdot 192) = 146 \cdot 6$ ml/min. Compare the mean $139 \cdot 2$ ml/min and its $P = 0 \cdot 95$ Gaussian confidence limits, $131 \cdot 8$ to $146 \cdot 6$ ml/min, with the median and its confidence limits found, with fewer assumptions, in § 7.3.

Condensing the above argument into one formula, the Gaussian confidence limits for μ (given an estimate of it, $\bar{y}$, the mean of a sample of n normally distributed observations) are

$$\bar{y} \pm t \sqrt{\left(\frac{s^2(y)}{n} \right)}. \qquad (7.4.1)$$

In general, the confidence limits for any normally distributed variable, x, are

$$x \pm ts(x), \qquad (7.4.2)$$

where the value of Student's t is taken from tables (see § 4.4) for the probability required and for the number of degrees of freedom associated with $s(x)$.

To be more sure that the limits will include μ (the population value of $\bar{y}$; see § A1.1), they must be made wider. For example, the value of t for $P = 0 \cdot 99$ with 8 d.f. is, from tables, $3 \cdot 355$. That is $0 \cdot 5$

$\dagger$ If this argument shakes you, see the footnote in § 7.3 (p. 104), reading y for x and μ for θ.

per cent (0·005) of the area under the curve for the distribution of t with 8 d.f. lies below $-3·355$ and another 0·5 per cent above $+3·355$, and 99 per cent lies between these figures. The 99 per cent Gaussian confidence limits are then $\bar{y} \pm ts(\bar{y})$, i.e. 128·5 to 149·9 ml/min.

Confidence limits for new observations

The limits just found were those expected to contain, μ, the population mean value of y (and also of $\bar{y}_n$ and $\bar{y}_m$ see below). If limits are required within which a new *observation* from the same population is expected to lie the result is rather different. Suppose, as above, that n observations are made of a normally distributed variable, y. The sample mean is $\bar{y}_n$ and the sample deviation $s(y)$, say. If a further m independent observations were to be made on the same population, within what limits would their mean, $\bar{y}_m$, be expected to lie? The variable $\bar{y}_n - \bar{y}_m$ will be normally distributed with a population value $\mu - \mu = 0$, so $t = (\bar{y}_n - \bar{y}_m)/s[\bar{y}_n - \bar{y}_m]$, (see § 4.4). Using (2.7.3) and (2.7.8) the estimated variance is $s^2[\bar{y}_n - \bar{y}_m] = s^2(\bar{y}_n) + s^2(\bar{y}_m) = s^2(y)/n + s^2(y)/m = s^2(y).(1/n + 1/m)$. The best prediction of the new observation, $\bar{y}_m$, will, of course, be the observed mean, $\bar{y}_n$. This is the same as the estimate of μ, but the confidence limits must be wider because of the error of the new observations. As above, $P[-t < (\bar{y}_n - \bar{y}_m)/s[\bar{y}_n - \bar{y}_m] < +t] = 0·95$ so, by rearranging this as before, the confidence limits for $\bar{y}_m$ are found to be

$$\bar{y}_n \pm t \sqrt{\left[s^2(y).\left(\frac{1}{n} + \frac{1}{m} \right) \right]}. \tag{7.4.3}$$

For example, a single new observation ($m = 1$) of glomerular filtration rate would have a 95 per cent chance (in the sense explained in § 7.9) of lying within the limits calculated from (7.4.3), viz. $139·2 \pm 2·306\sqrt{[91·69(1/9 + 1)]}$; that is, from 115·9 ml/min to 162·5 ml/min. These limits are far wider than those for μ.

Where m is very large, (7.4.3) reduces to the Gaussian limits for μ, eqn (7.4.1) (as expected, because in this case $\bar{y}_m$ becomes the same thing as μ).

It is important to notice the condition that the m new observations are from the same Gaussian population as the original n. As they are probably made later in time there may have been a change that invalidates this assumption.

7.5. Confidence limits for the ratio of two normally distributed observations

If a and b are normally distributed variables, their ratio $m = a/b$ will not be normally distributed, so if the approximate variance of the ratio is obtained from (2.7.16), it is only correct to calculate limits by the methods of § 7.4 if the denominator is large compared with its standard deviation (i.e. if g is small, see § 13.5). This problem is quite a common one because it is often the ratio of two observations that is of interest rather than, say, their difference.

If a and b were *lognormally* distributed (see § 4.5) then log a and log b would be normally distributed, and so log $m = \log a - \log b$ would be normally distributed with $\mathrm{var}(\log m) = \mathrm{var}(\log a) + \mathrm{var}(\log b)$ from (2.7.3) (given independence). Thus confidence limits for $\log m$ could be calculated as in § 7.4, $\log m \pm t\sqrt{[\mathrm{var}(\log m)]}$, and the anti-logarithms found. See § 14.1 for a discussion of this procedure.

When a and b are *normally* distributed the exact solution is not difficult. But, because it looks more complicated at first sight, it will be postponed until § 13.5 (see also § 14.1), nearer to the numerical examples of its use in §§ 13.11–13.15.

7.6. Another way of looking at confidence limits

A more general method of arriving at confidence intervals will be needed in §§ 7.7 and 7.8. The ingenious argument, showing that limits found in the following way can be interpreted as described in § 7.9, is discussed clearly by, for example, Brownlee (1965, pp. 121–32). It will be enough here to show that the results in § 7.4 can be obtained by a rather different approach.

For simplicity it will be supposed at first that the population standard deviation, σ, is known. It is expected that in the long run 95 per cent of randomly selected observations of a normally distributed variable y, with population mean μ and population standard deviation σ, will fall within $\mu \pm 1.96\sigma$ (see § 4.2). In § 7.4 the normally distributed variable of interest was $\bar{y}$, the mean of n observations, and similarly, in the long run, 95 per cent of such means would be expected to fall within $\mu \pm 1.96\sigma(\bar{y})$, where $\sigma(\bar{y}) = \sigma/\sqrt{n}$ (by (2.7.9)). The problem is to find limits that are likely to include the unknown value of μ.

Now consider various possible values of μ. It seems reasonable to take as a lower limit, μ_{L} say, a value which, *if* it were the true value, would make the the observation of a mean as large as that actually observed ($\bar{y}_{\mathrm{obs}}$) or larger a rare event—an event that would only occur in 2·5 per cent of repeated trials in the long run, for example. In Fig. 7.6.1(a) the normal distribution of $\bar{y}$ is shown with the known standard deviation $\sigma(\bar{y})$, and the hypothetical mean μ_{L} chosen in the way just described.

Similarly, the highest reasonable value for μ, say μ_{H}, could be chosen so that, *if* it were the true value, the observation of a mean equal to or less than $\bar{y}_{\mathrm{obs}}$ would be a rare event (again $P = 0.025$, say). This is shown in Fig. 7.6.1 (b). It is clear from the graphs that $\bar{y}_{\mathrm{obs}} = \mu_{\mathrm{L}} + 1.96\sigma(\bar{y}) = \mu_{\mathrm{H}} - 1.96\sigma(\bar{y})$. Rearranging this gives $\mu_{\mathrm{L}} = \bar{y}_{\mathrm{obs}} - 1.96\sigma(\bar{y})$, and $\mu_{\mathrm{H}} = \bar{y}_{\mathrm{obs}} + 1.96\sigma(\bar{y})$. If σ is not known but has to be estimated from the observations, then $\sigma(\bar{y})$ must be replaced by

$s(\bar{y})$, and so 1·96 must be replaced by the appropriate value of Student's t (for example 2·306 for $P = 0.95$ limits in the example in § 7.4; see also § 4.4). When this is done μ_L and μ_H are the limits previously found using (7.4.1).

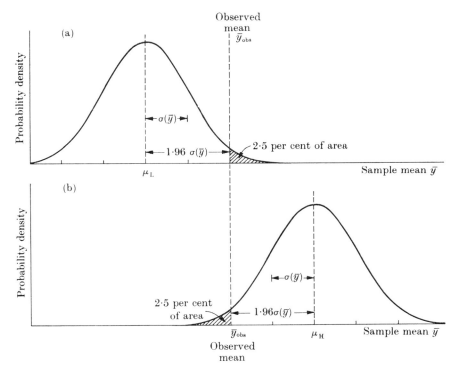

FIG. 7.6.1.   One way of looking at confidence limits. See text.

7.7. What is the probability of 'success'? Confidence limits for the binomial probability

In §§ 3.2–3.5 it was described how the number of successes (r) out of n trials of an event would be expected to vary in repeated sets of n trials when the probability of 'success' was $\mathscr{P}$ at each trial and the probability of 'failure' was $1-\mathscr{P}$. Usually, of course, the problem is reversed. $\mathscr{P}$ is unknown and must be estimated from the experimental results. For example, if a drug were observed to cause improvement in $r = 7$ out of $n = 8$ patients, who had been selected strictly randomly (see § 2.3) from some population of patients, then the best *estimate* of the proportion ($\mathscr{P}$) of patients in the population that would improve when given the drug is r/n (as in § 3.4), i.e. $7/8 = 0.875$ or 87.5 per cent. What is the error of this estimate? Would it be unreasonable, for example, to suppose that the population contained only 50 per cent of 'improvers'? The answer can be found without any calculation at all using Table A2, which is based on the following reasoning.

The approach described in § 7.6 can be used to find confidence limits for the population value of $\mathscr{P}$. For concreteness suppose that 95 per cent (or $P = 0.95$) confidence limits are required for the population value of $\mathscr{P}$ when r_{obs} 'successes' have been observed out of n trials. The highest reasonable value of $\mathscr{P}$, $\mathscr{P}_H$ say, will be taken as the value that, *if* it were the true value, would make the observation of r_{obs} or fewer successes a rare event (an event occurring in only 2.5 per cent of repeated sets of n trials). Now the probability of r successes $P(r)$, is given by (3.4.3), and $r \leqslant r_{obs}$ if $r = 0$ or 1 or...or r_{obs}, so, using (3.4.3) and the addition rule, (2.4.2), it is required that

$$P[r \leqslant r_{obs}] = \sum_{r=0}^{r=r_{obs}} \frac{n!}{r!(n-r)!}\mathscr{P}_H^r(1-\mathscr{P}_H)^{n-r} = 0.025. \qquad (7.7.1)$$

The only unknown in this equation is $\mathscr{P}_H$, the upper confidence limit for the population proportion, so it can be solved for $\mathscr{P}_H$. There is no simple way of rearranging the equation to get $\mathscr{P}_H$ however, so tables are provided (Table A2) giving the solution. Similarly, the lowest reasonable value, $\mathscr{P}_L$, for the population $\mathscr{P}$ (the lower confidence limit for $\mathscr{P}$) is taken as the value that, *if* it were the true value, would make the observation of r_{obs} successes or more (i.e. $r = r_{obs}$ or $r_{obs}+1$ or...or n) a rare event. Thus $\mathscr{P}_L$ is found by solving

$$\mathscr{P}[r \geqslant r_{obs}] = \sum_{r=r_{obs}}^{r=n} \frac{n!}{r!(n-r)!}\mathscr{P}_L^r(1-\mathscr{P}_L)^{n-r} = 0.025. \qquad (7.7.2)$$

Again the solution is tabulated in Table A2.

The use of Table A2

Confidence limits (95 and 99 per cent) for the population value of $100\mathscr{P}$ are tabulated for any observed r, and sample sizes from $n = 2$ to $n = 30$, and also some values for $n = 1000$ for comparison. Other sample sizes are tabulated in the *Documenta Geigy Scientific Tables* (1962, pp. 85–103). In the example at the beginning of this section $r = 7$ out of $n = 8$ patients improved ($100r/n = 87.5$ per cent improvement). Consulting Table A2 with $n = 8$ and $r = 7$ shows that the $P = 0.95$ confidence limits ($100\mathscr{P}_L$ to $100\mathscr{P}_H$ from (7.7.1)) and (7.7.2) are 47.35 to 99.68 per cent. In other words, if repeated samples of 8 were taken from a population that actually contained 47.35 per cent of improvers, 2.5 per cent of the samples would contain 7 or more (i.e. 7 or 8) improvers. And if the population actually contained 99.68 per cent of improvers then, in the long run, 2.5 per cent of samples would contain 7 or fewer improvers. Thus, if the drug were tested on an infinite sample (rather than only 8) it would not be surprising (see § 7.9 for a more precise interpretation) to find any proportion of patients improving between $\mathscr{P}_L = 0.4735$ and $\mathscr{P}_H = 0.9968$. The observation is compatible with any *hypothetical* population $\mathscr{P}$ that lies between the confidence limits (see § 9.4) so the observation of 7 improving out of 8 cannot be considered incompatible with a true improvement rate of 50 per cent ($\mathscr{P} = 0.5$) at the $P = 0.95$ level of significance. For greater certainty the $P = 0.99$ confidence limits would be found from the tables. They are, of course, even wider, 36.85 to 99.94 per cent. A sample of 8 gives surprisingly little information about the population it was drawn from, even when all the assumptions of randomness and simple sampling (see § 3.2) are fulfilled.

The comparison of two *observed* binomial proportions is a different problem. It is discussed in Chapter 8.

7.8. The black magical assay of purity in heart as an example of binomial sampling

In a sadly neglected paper, Oakley (1943) proposed an assay method for purity in heart. Oakley points out that lack of statistical knowledge may vitiate a worth-while experiment the apparent failure of which may deter others from repeating it, and that this fate seems to have overtaken an experiment carried out many years ago in Germany. The only known source (Anon 1932) describes the experiment thus:

'The legend of the Brocken (the famous peak in the Harz Mountains noted for its "spectre" and as the haunt of witches on Walpurgis Night), according to which a "virgin he-goat" can be converted into "a youth of surpassing beauty" by spells performed in a magic circle at midnight, was tested on June 17th by British and German scientists and investigators, including Professor Joad and Mr. Harry Price of the National Institute of Psychical Research. The object was to expose the fallacy of Black Magic and also to pay a tribute to Goethe, who used the legend in "Faust". Some wore evening dress. The goat was anointed with the prescribed compound of scrappings from church bells, bats' blood, soot and honey The necessary "maiden pure in heart" who removed the white sheet from the goat at the critical moment, was Fräulein Urta Bohn, daughter of one of the German professors taking part in the test. Her mother was a Scotswoman (formerly Miss Gordon). The scene was floodlit and filmed. As our photographs show, the goat remained a goat, and the legend of the Brocken was dispelled!'

The main variables are the virgin he-goat and the maiden pure in heart. Virginity may for the present be regarded as an absolute character, but purity in heart no doubt varies from person to person.

Oakley therefore supposed that it might be possible to estimate the purity in heart index (PHI) of a maiden by observing how many of a group of he-goats are converted into young men. The original experimenters were clearly guilty of a grave scientific error in using only one he-goat.

We shall assume, as Oakley did, that the conversion of he-goats into young men is an all-or-nothing process; either complete conversion or nothing occurs. Oakley supposed, on this basis, that a comparison could be made between, on one hand, the percentage of he-goats converted by maidens of various degrees of purity in heart, and, on the other hand, the sort of pharmacological experiment that involves the measurement of the percentage of individuals showing a specified effect in response to various doses of a drug. In conformity with the common pharmacological practice he supposed that a plot of percentage he-goat conversion against log purity in heart index (log PHI) would have the sigmoid form shown in Fig. 14.2.4. As explained in Chapter 14, this implies that log PHI required to convert individual he-goats is a normally distributed variable. Furthermore it means that infinity purity in heart is required to produce a population he-goat conversion rate (HGCR) of 100 per cent.

Although there is a lack of experimental evidence on this point, the present author feels that the assumption of a normal distribution is, as so often happens, without foundation (see § 4.2). The implication of the normality assumption, that there exist he-goats so resistant to conversion that infinite purity in heart is needed to affect them, has not been (and cannot be) experimentally verified. Furthermore the very idea of infinite purity in heart seems likely to cause despondency in most people, and should therefore be avoided until such time as its necessity may be demonstrated experimentally. Oakley's treatment of the problem requires, in addition, that PHI be treated as an independent variable (in the regression sense, see Chapter 12), which raises problems because there is no known method of measuring PHI other than he-goat conversion.

In the light of these remarks it appears to the present author desirable that the purity in heart index should be redefined simply as the population percentage of he-goats converted.† This simple operational definition means that the PHI of all maidens will fall between 0 and 100, and confidence limits for the true PHI can be found easily from

---

† i.e., in the more rigorous notation of Appendix 1, PHI = E[HGCR].

the observed conversion rate (which should be binomially distributed, see §§ 3.2–3.5) using Table A2, as explained in § 7.7.

For example, if it were observed that a particular maiden caused conversion of $r = 2$ out of $n = 4$ he-goats, the estimated PHI would be $100 \times 2/4 = 50$ per cent, and from Table A2 the confidence limits ($P = 0.95$) for the PHI are 6·76–93·24 per cent. Clearly the information to be gained from a sample of only four he-goats is so imprecise that it is difficult to conceive what use it could be put to. Oakley recommended that for *preliminary* experiments at least $n = 10$ he-goats should be used. If $r = 5$ (50 per cent) of these were observed to be converted Table A2 would give the confidence limits ($P = 0.95$) for the true PHI as 18·71–81·29 per cent. While the most extreme forms of vice and of virtue appear to be ruled out by this result, there is still considerable uncertainty about the PHI. If a greater degree of confidence were required, as might happen, for example, if a potential husband demanded a certain minimum (or, alternatively, a certain maximum) PHI before committing himself, the $P = 0.99$ confidence limits could be found from Table A2. They are 12·83–87·17 per cent. The most tolerant suitor might be forgiven for requiring a larger sample.

These calculations show that the assay is subject to considerable experimental error; and the problem of measuring very high or very low PHIs is even more difficult (because percentage responses around 50 per cent are the most accurately determined).† If the practical difficulties involved in using samples of $n = 1000$ he-goats could be overcome, PHIs not too far from 50 per cent could be determined with reasonable accuracy. For $r = 500$ converted, the confidence limits ($P = 0.95$) from Table A2 are 46·85–53·15 per cent. If only $r = 10$ he-goats were converted out of 1000 (1 per cent) the confidence limits ($P = 0.95$) should be 0·48–1·84 per cent. Although the *relative* error is a good deal bigger than for conversion rates near 50 per cent, this is likely to be precise enough for practical purposes.

A more precise and economical assay is clearly needed, but until more experimental work is done the present method will have to do. However, as Oakley points out, 'All thoughtful persons must regard the indiscriminate conversion of he-goats into young men with concern, for there is no knowledge of what education or social, political or

† This depends on what is meant by accuracy. It is true if one is interested in the *relative* error of the proportion converted (or not converted, whichever is the smaller). It is also true if, as in Chapter 14, one is interested in the error of the dose producing a specified proportion converted in quantal experiments.

economic views such young men might have, and it might well be that their behaviour would bring scientific experiment into disrepute. This is, however, a problem for necromancers rather than statisticians.'

7.9. Interpretation of confidence limits

The logical basis and interpretation of confidence limits are, even now, a matter of controversy. However, few people would contest the statement that if $P = 0.95$ (say) limits were calculated according to the above rules in each of a large number of experiments then, in the

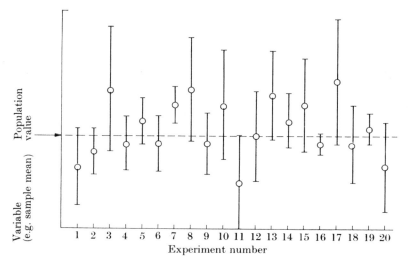

Fɪɢ. 7.9.1. Interpretation of confidence limits. Repeated estimates (e.g. sample mean) of a parameter (e.g. population mean), and their 95 per cent confidence limits. In this (ideal) case one experiment (number 7) out of twenty gave confidence limits that do not include the population value. One in twenty is the predicted long-run frequency.

long run, 95 per cent of the intervals so calculated would include the population mean (§ 7.4) or median (§ 7.3), μ, *if* the assumptions made in the calculation were true. *The limits must be regarded as optimistic as explained in* § 7.2.

In any particular experiment a single confidence interval is calculated which obviously either does or does not include μ. It might therefore be thought that it could be said *a priori* that the probability that the interval includes μ is either 0 or 1, but not some intermediate value. However, in a series of identically conducted experiments, somewhat different values of the sample median or mean, and of the sample scatter, for example of $s(\bar{y})$, will, in general, be found in every experiment. The confidence limits will therefore be different from experiment to experiment. The prediction is that, in the long run 95 per cent (19 out of 20) of such limits will include μ as illustrated in Fig. 7.9.1. It is *not* predicted that in 95 per cent of experiments the true mean will fall within the particular set of limits calculated in the one actual experiment.

Thus, if one were willing to consider that the actual experiment was a random sample from the population of experiments that might have been done, i.e. that 'nature has done the shuffling' one could go further and say that there was a 95 per cent chance of having done an experiment in which the calculated limits include the true mean, μ.

Another interpretation of confidence intervals will be mentioned later during the discussion of significance tests.

8. Classification measurements

8.1. Two independent samples. Relationship between various methods

CLASSIFICATION measurements and independent samples were discussed in § 6.4. Before starting any analysis § 8.7 should be read to
make sure that the results are not actually the incorrectly presented
results of an experiment with related samples. The fundamental
method of analysis for the 2×2 table (§ 8.2) is the randomization
method (see § 6.3), which is known as the Fisher exact test (see § 8.2).
There is an approximate method that gives similar results to the
exact test with sufficiently large samples. This method can be written
in two ways, as the normal approximation described in § 8.4 or as the
chi-squared test described in § 8.5. The exact test (§ 8.2) should be
used when the total number of observations, N, is up to 40. Published
tables, which only deal with $N \leqslant 40$, make this easy. When $N > 40$
the exact test should be calculated directly from (8.2.1) if the frequency
in any cell of the table is very small. When the smallest expected value,
x_e (see § 8.5), is 5 or more there is reason to believe that the chi-
squared test (§ 8.5) corrected for continuity will be a good approximation to the exact test (Cochran 1952).

8.2. Two independent samples. The randomization method and the Fisher test

Randomization tests were introduced in § 6.3. As an example of the result of classification measurements (see § 6.4), consider the clinical comparison of two drugs, X and Y, on seven patients. It is fundamental to any analysis that the allocation of drug X to four of the patients, and of Y to the other three be done in a strictly random way using random number tables (see §§ 2.3 and 8.3). It is noted, by a suitable blind method, whether each patient is improved (I) or not improved (N). The result is shown in Table 8.2.1 (b).

TABLE 8.2.1

Possible results of the trial. Result (b) *was actually observed*

	I	N	Total	I	N	Total	I	N	Total	I	N	Total
Drug X	4	0	4	3	1	4	2	2	4	1	3	4
Drug Y	0	3	3	1	2	3	2	1	3	3	0	3
Total	4	3	7	4	3	7	4	3	7	4	3	7
	(a)			*(b)*			*(c)*			*(d)*		

With drug X 75 per cent improve (3 out of 4), and with drug Y only 33 1/3 per cent improve. Would this result be likely to occur if X and Y were really equi-effective? If the drugs are equi-effective then it follows that whether an improvement is seen or not cannot depend on which drug is given. In other words, each of the patients would have given the same result even if he had been given the other drug, so the observed difference in 'percentage improved' would be merely a result of the particular way the random numbers in the table came up when the drugs were being allocated to patients.† For example, for the experiment in Table 8.2.1 the null hypothesis postulates that of the 7 patients, 4 would improve and 3 would not, quite independently of which drug was given. If this were so, would it be reasonable to suppose that the random numbers came up so as to put 3 of the 4 improvers, but only 1 of the 3 non-improvers, in the drug X group (as observed, Table 8.2.1(b))? Or would an allocation giving a result that appeared to

---

† Of course, if a subject who received treatment X during the trial were given an equi-effective treatment Y at a later time, the response of the second occasion would not be exactly the same as during the trial. But it is being postulated that *if* X and Y are equi-effective then if one of them is given to a given subject at a given moment in time, the response would have been exactly the same if the other had been given to the same subject at the same moment.

favour drug X by as much as (or more than) this, be such a rare happening as to make one suspect the premise of equi-effectiveness ?

Now if the selection was really random, every possible allocation of drugs to patients should have been equally probable. It is therefore simply a matter of counting permutations (possible allocations) to find out whether it is improbable that a random allocation will come up that will give such a large difference between X and Y groups as that observed (or a larger difference). Notice that attention is restricted to the actual 7 patients tested without reference to a larger population (see also § 8.4). Of the 7 patients, 4 improved and 3 did not.

Three ways of arriving at the answer will be described.

(a) *Physical randomization.* On four cards write 'improved' and on three write 'not improved'. Then rearrange the cards in random order using random number tables (or, less reliably, shuffle them), mimicking exactly the method used in the actual experiment. Call the top four cards drug X and the bottom three drug Y, and note whether or not the difference between drugs resulting from this allocation of drugs to patients is as large as, or larger than, that in the experiment. Repeat this say, 1000 times and count the proportion of randomizations that result in a difference between drugs as large as or larger than that in the experiment. This proportion is P, the result of the (one-tail) significance test. If it is small it means that the observed result is unlikely to have arisen solely because of the random allocation that happened to come up in the real experiment, so the premise (null hypothesis) that the drugs are equi-effective may have to be abandoned (see § 6.2). This method would be tedious by hand, though not on a computer, but fortunately there are easier ways of reaching the same results. The two-tail test is discussed below.

(b) *Counting permutations.* As each possible allocation of drugs to patients is equally probable, if the randomization was properly done, the results of the procedure just described can be predicted in much the same way that the results of coin tossing were predicted in § 3.2. If the seven patients are distinguished by numbers, the four who improve can be numbered 1, 2, 3, and 4, and those who do not can be numbered 5, 6, and 7. According to the null hypothesis each patient would have given the same response whichever drug had been given. How many way can the 7 be divided into groups of 3 and 4? The answer is given, by (3.4.2), as $7!/(4!3!) = 35$ ways. It is not necessary to write out about both groups since once the number improved has been found in one group (say the smaller group, drug Y, for convenience),

TABLE 8.2.2

Enumeration of all 35 *possible ways of selecting a group of* 3 *patients from* 7 *to be given drug Y. Patients* 1, 2, 3, *and* 4 *improved and patients* 5, 6, *and* 7 *did not. Number of subjects improving with* $Y = b$ (see Table 8.2.3(a))

Patients given drug Y			Result	Patients given drug Y			Result
5	6	7	$b = 0$ improve. 1 way giving Table 8.2.1(a). $P = 1/35 = 0{\cdot}029$	1	2	5	
				1	2	6	
				1	2	7	
1	5	6		1	3	5	
1	5	7		1	3	6	
1	6	7		1	3	7	$b = 2$ improve. 18 ways all
2	5	6		1	4	5	giving
2	5	7	$b = 1$ improve.	1	4	6	Table 8.2.1(c).
2	6	7	12 ways all giving	1	4	7	$P = 18/35 = 0{\cdot}514$
3	5	6	Table 8.2.1(b).	2	3	5	
3	5	7	$P = 12/35 = 0{\cdot}343$	2	3	6	
3	6	7		2	3	7	
4	5	6		2	4	5	
4	5	7		2	4	6	
4	6	7		2	4	7	
				3	4	5	
				3	4	6	
				3	4	7	
				1	2	3	b = 3 improve.
				1	2	4	4 ways all
				1	3	4	giving
				2	3	4	Table 8.2.1(d). $P = 4/35 = 0{\cdot}114$

the number improved in the other group follows from the fact that the total number improved is necessarily 3. All 35 ways in which the drug Y group could have been constituted are listed systematically in Table 8.2.2. If the randomization was done properly each way should have had an equal chance of being used in the experiment. Notice that *proper randomization in conducting the experiment is crucial for the analysis of the results*. It is seen that 12 out of the 35 result in one improved, two not improved in the drug Y group, as was actually observed. Furthermore, 1 out of 35 shows an even more extreme result, no patient at all improving on drug Y group, as shown in Table 8.2.1(a).

Thus $P = 12/35 + 1/35 = 0.343 + 0.029 = 0.372$ for a one-tail test† (see § 6.1). This is the probability (the long-run proportion of repeated experiments) that a random allocation of drugs to patients would be picked that would give the results in Table 8.2.1(a) or 8.2.1(b), i.e. that would give results in which X would appear as superior to Y as in the actual experiment (Table 8.2.1(b)), or even more superior (Table 8.2.1(a)), *if* X and Y were, in fact, equi-effective. This probability is not low enough to suggest that X is really better than Y. Usually a two-tail test will be more appropriate than this one-tail test, and this is discussed below.

Using the results in Table 8.2.2, the sampling distribution under the null hypothesis, which was assumed in constructing Table 8.2.2, is plotted in Fig. 8.2.1. This is the form of Fig. 6.1.1 that it is appropriate to consider when using the randomization approach. The variable on the abscissa is the number of patients improved on drug Y, i.e. b in the notation of Table 8.2.3(a). Given this figure the rest of the table can be filled in, using the marginal totals, so each value of b corresponds to a particular difference in percentage improvement between drugs X and Y. Fig. 8.2.1 is described as the *randomization* (or *permutation*) *distribution* of b, and hence of the difference between samples, given the null hypothesis. The result of a one-tail test of significance when the experimentally observed value is $b = 1$ (Table 8.2.1(b)), is the shaded area (as explained in § 6.1), i.e. $P = 0.372$ as calculated above.

The two-tail test. Suppose now that the result in Table 8.2.1(a) had been found in the experiment ($b = 0$). A one-tail test would give $P = 1/35 = 0.029$, and this is low enough for the premise of equi-effectiveness of the drugs to be suspect *if* it is known beforehand that Y cannot possibly be better than X (the opposite result to that observed). As this is usually not known a two-tail test is needed (see § 6.1). However, the most extreme result in favour of drug Y ($b = 3$ as in Table

---

† This is a one-tail test of the null hypothesis that X and Y are equi-effective, when the alternative hypothesis is that X is better than Y. If the alternative to the null hypothesis had been that Y was better than X (the alternative hypothesis must, of course, be chosen before the experiment) then the one-tail P would have been $12/35 + 18/35 + 4/35 = 0.971$, the probability of result as favourable to Y as that observed, or *more* favourable, when X and Y are really equi-effective.

8.2.1(d)) is seen to have P = 0·114. It is therefore impossible that a verdict in favour of drug Y could have been obtained with *these* patients. *If* the drugs were really equi-effective then, if the hypothesis of equi-effectiveness were rejected every time $b = 0$ or $b = 3$ (the two most extreme results), it would be (wrongly) rejected in 2·9+11·4 = 14·3 per cent of trials—far too high a level for the probability of an error of the first kind (see § 6.1, para. 7). A two-tail test is therefore not possible with such a small sample. This difficulty, which can only occur with very small samples—it does not happen in the next example, has been discussed in a footnote in § 6.1 (p. 89).

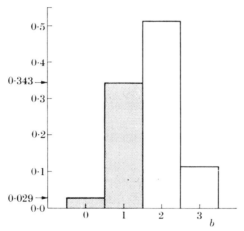

FIG. 8.2.1. Randomization distribution of b (the number of patients improving on drug Y), when X and Y are equi-effective (i.e. null hypothesis true).

(c) *Direct calculation. The Fisher test.* It would not be feasible to write out all permutations for larger samples. For two samples of ten there are $20!/(10!10!) = 184756$ permutations. Fortunately it is not necessary. If a general 2×2 table is symbolized as in Table 8.2.3(a)

TABLE 8.2.3

	success	failure	total	success	failure	total
treatment X	a	$A-a$	A	8	7	15
treatment Y	b	$B-b$	B	1	11	12
total	C	D	N	9	18	27
	(a)			(b)		

then Fisher has shown that the proportion of permutations giving rise to the table is

$$P = \frac{A\,!\,B\,!\,C\,!\,D\,!}{N\,!\,a\,!\,(A-a)\,!\,b\,!\,(B-b)\,!}. \tag{8.2.1}$$

For example, for Table 8.2.1(b), $P = 4\,!\,3\,!\,4\,!\,3\,!/(7\,!\,3\,!\,1\,!\,1\,!\,2\,!) = 12/35 = 0\cdot343$ as already found. With larger figures (8.2.1) is most conveniently evaluated using tables of logarithms of factorials (e.g. Fisher and Yates, 1963).

In fact no calculation at all is necessary as tables have been published (Finney, Latscha, Bennett, and Hsu, 1963) for testing any 2×2 table with A and B, or C and D, both not more than 40. Unfortunately, to keep the tables a reasonable size it is not possible to find the exact P value for all 2×2 tables, but it is given for those 2×2 tables with marginal totals up to 30 for which $P \leqslant 0\cdot05$ (one tail). The published tables are for $B \leqslant A$ and $b < a$ only, to avoid duplication. If the table to be tested does not comply with this, rows and/or columns must be interchanged until it does. As an example, the table in Table 8.2.3(b), which is from the introduction to the table of Finney *et al.* (1963), is tested using the appropriate part of their table, which has been reproduced in Table 8.2.4.

TABLE 8.2.4

Exact test for the 2×2 table (Extract from tables of Finney *et al.* (1963))

	a	0·05	0·025	0·01	0·005
			Probability (nominal)		
$A = 15$ $B = 12$	15	8 0·028	7 0·010⁻	7 0·010⁻	6 0·003
	14	7 0·043	6 0·016	5 0·006	4 0·002
	13	6 0·049	5 0·019	4 0·007	3 0·002
	.	.	.	.	.
	.	.	.	.	.
	.	.	.	.	.
	9	2 0·028	1 0·007	1 0·007	0 0·001
	8	1 0·018	1 0·018	0 0·003	0 0·003
	7	1 0·038	0 0·007	0 0·007	———
	.	.	.	.	.
	.	.	.	.	.
	.	.	.	.	.

The observed Table 8.2.3(b) has $A = 15$, $B = 12$, and $a = 8$. Entering Table 8.2.4 with these values shows under each nominal probability a figure in bold type which is the largest value of b that is just 'significant in a one-tail test at the 5 per cent (or 2·5, 1, or 0·5 per cent) level', i.e. for which the *one-tail*† $P \leqslant 0\cdot05$, (or $0\cdot025$, $0\cdot01$, or $0\cdot005$). The exact value of P is given in smaller type. It is the nearest value, given that b must be a whole number, that is not greater than the nominal value. In this example the one-tail P corresponding to the observed $b = 1$ is $0\cdot018$. This is the sum of the P values calculated from (8.2.1) for the observed table ($a = 8$, $b = 1$, $P = 0\cdot017$), and the only possible more extreme one with the same marginal totals ($a = 9$, $b = 0$, $P = 0\cdot001$). To find the two-tail P value (see § 6.1 and above) consider the distribution of b analogous to Fig. 8.2.1. In this case b can vary from 0 to 9 and if the null hypothesis were true it would be 4 on the average (see § 8.5). The one-tail P found is the tail of the distribution for $b \leqslant 1$. It is required to cut off an area as near as possible to this in the other tail of the distribution ($b > 4$), as in Fig. 6.1.1. No value of b cuts off exactly $P = 0\cdot018$ but $b = 7$ cuts off an area of $P = 0\cdot019$ that is near enough (see footnote, § 6.1, p. 89). This is the sum of the probabilities of $b = 7$ and all the more extreme ($b = 8$ and $b = 9$) results. It can be found from the tables of Finney *et al.* by the method described in their introduction. The table has $a = 2$, $b = 7$, $A-a = 13$, $B-b = 5$ so columns are interchanged, as mentioned above, and the table entered with 13 and 5 rather than 2 and 7, as marked in Table 8.2.4. Therefore if it were resolved to reject the null hypothesis whenever $b \leqslant 1$ (as observed) or when $b \geqslant 7$ (opposite tail) then, if the null hypothesis was in fact true, the probability that it would be rejected (wrongly)—an error of the first kind—would be $P = 0\cdot018 + 0\cdot019 = 0\cdot037$. This result for the two-tail test is small enough to make one question the null hypothesis, i.e. to suspect a real difference between the treatments, (see § 6.1).

In practice, if the samples are not too small, it would be adequate, and much simpler, to double the one-tail P from the table to get the required two-tail P.

8.3. The problem of unacceptable randomizations

Sometimes it will be found that when two samples are selected at random one sample contains, for example, all the men and the other all the women. In fact if this does *not* happen sometimes, the selection cannot be random. It seems silly to carry out an experiment in which

---

† For the case when it is decided, before the experiment, that the only alternative to the null hypothesis is a difference between X and Y in the observed direction.

treatment X is given only to men and treatment Y only to women. Yet the logical basis of significance tests will be destroyed if the experimenter rejects randomizations producing results he does not like. Often this will be preferable to the alternative of doing an experiment that is, on scientific grounds, silly. But it should be realized that the choice must be made.

There is a way round the problem if randomization tests are used. If it is decided *beforehand* that any randomization that produces two samples differing to more than a specified extent in sex composition— or weight, age, prognosis, or any other criterion—is unacceptable then, if such a randomization comes up when the experiment is being designed, it can be legitimately rejected *if*, in the analysis of the results, those of the possible randomizations that differ excessively, according to the previously specified criteria, are also rejected, in exactly the same way as when the real experiment was done. So, in the case of Table 8.2.2, the number of possible allocations of drugs to the 7 patients could be reduced to less than 35. This can only be done when using the method of physical randomization, or a computer simulation of this process, or writing out the permutations as in Table 8.2.2. The shorter methods using calculation (e.g. from (8.2.1), or published tables (e.g. for the Fisher exact test, § 8.2, or the Wilcoxon tests, §§ 9.3 and 10.4), cannot be modified to allow for rejection of randomizations.

8.4. Two independent samples. Use of the normal approximation

Although the reasoning in § 8.2 is perfectly logical, and although there is a great deal to be said for restricting attention to the observations actually made since it is usually impossible to ensure that any further observations will come from the same population (see §§ 1.1 and 7.2), the exact test has nevertheless given rise to some controversy among statisticians. It is possible to look at the problem differently. If, in the example in Table 8.2.1, the 7 patients were thought of as being selected from a larger population of patients then another sample of 7 would not, in general, contain 4 who improved and 3 who did not. This is considered explicitly in the approximate method described in this section. However there is reason to suppose that the exact test of § 8.2 is best, even for 2×2 tables in which the marginal totals are not fixed (Kendall and Stuart 1961, p. 554).

Consider Table 8.2.3 again but this time imagine two infinite populations (e.g. X-treated and Y-treated) with true probabilities of success (e.g. improved) $\mathscr{P}_1$ and $\mathscr{P}_2$ respectively. From the first population a sample of A individuals is drawn at random and is observed to contain a successes (e.g. improved patients). Similarly b successes out of B are observed in the sample from the second population. The experimental estimates of $\mathscr{P}_1$ and $\mathscr{P}_2$ are, as in § 3.4, $p_1 = a/A$ and $p_2 = b/B$, the observed proportions of successes in the samples from the two populations. In repeated trials a and b should vary as predicted by the binomial distribution (see § 3.4).

Use of the normal approximation to the binomial

It is required to test the null hypothesis that $\mathscr{P}_1 = \mathscr{P}_2$, both being $\mathscr{P}$, say. If this were so then *on the average* the observed proportions would be the same too, so $p_1 - p_2$ would be distributed about a mean value of zero (cf. Fig. 6.1.1). It was mentioned in § 3.4 and illustrated in Fig. 3.4.1) that if n is reasonably large the discontinuous binomial distribution of p is quite well approximated by a continuous normal distribution. It will therefore be supposed, as an approximation, that p_1 and p_2 are both normally distributed. This implies that the difference between them $(p_1 - p_2)$ will be normally distributed with, according to the null hypothesis, a population mean (μ) of zero. The standard deviation of this distribution can now be found by using (3.4.5) to find the true variances of p_1 and p_2 which, given the null hypothesis, are

$$var(p_1) = \frac{\mathscr{P}(1-\mathscr{P})}{A}, \qquad var(p_2) = \frac{\mathscr{P}(1-\mathscr{P})}{B}. \qquad (8.4.1)$$

If p_1 and p_2 are independent, as they will be if the samples are independent as assumed, (cf. § 6.4), the variance of their difference will be, using (2.7.3),

$$var(p_1 - p_2) = var(p_1) + var(p_2) = \mathscr{P}(1-\mathscr{P})\left(\frac{1}{A} + \frac{1}{B}\right). \qquad (8.4.2)$$

The true value, $\mathscr{P}$, is, of course, unknown, and it must be estimated from the experimental results. No allowance is made for this, which is another reason why the method is only approximate. The natural estimate of $\mathscr{P}$, under the null hypothesis, is to pool the two samples and divide the total number of successes by the total number of trials (e.g. total number improved by total number of patients), i.e. $p = (a+b)/(A+B)$. Thus, taking $x = (p_1 - p_2)$ as the normal variable, with, according to the null hypothesis, $\mu = 0$, an approximate normal deviate (see § 4.3) can be calculated, using (4.3.1) and (8.4.2). This value of u can then be referred to tables of the standard normal distribution (see § 4.3).

$$u = \frac{x-\mu}{\sigma(x)} \simeq \frac{(p_1-p_2)}{\sqrt{[p(1-p)(1/A+1/B)]}}. \qquad (8.4.3)$$

Applying this method to the results in Table 8.2.3 gives $p_1 = a/A = 8/15$, $p_2 = b/B = 1/12$, $p = (a+b)/(A+B) = 9/27$ and so, using (8.4.3), the approximate normal deviate is

$$u \simeq \frac{8/15 - 1/12}{\sqrt{\left[\frac{9}{27}\left(1-\frac{9}{27}\right)\left(\frac{1}{15}+\frac{1}{12}\right)\right]}} = 2 \cdot 4648.$$

According to Table 1 of the Biometrika tables† about 1·4 per cent of the area of the standard normal distribution lies outside $u \pm 2 \cdot 4648$ (0·7 per cent in each tail). The result of the test, $P = 0 \cdot 014$, is seen to be a poor approximation to the exact result, $P = 0 \cdot 037$, found at the end of § 8.2. A better approximation can be found by using a 'correction for continuity' and this should always be done.

Yates' correction for continuity

Say $p = r/n$ in general. It can be shown (e.g. Brownlee (1965, pp. 139, 152)) that the approximation of the continuous normal distribution to the discontinuous binomial is improved if 0·5 is added to or subtracted from r (or $0 \cdot 5/n$ is added to or subtracted from p), so as to make the deviation from the null hypothesis smaller. Thus a better approximation than (8.4.3) for the normal deviate is

$$u \simeq \frac{(p_1 - 0 \cdot 5/A) - (p_2 + 0 \cdot 5/B)}{\sqrt{[p(1-p)(1/A+1/B)]}}, \qquad (8.4.4)$$

where $p_1 > p_2$. Using the results in Table 8.3 again, gives $u = 2 \cdot 054$. Again using Table 1 of the Biometrika tables it is found that 4·0 per cent of the total area of the standard normal distribution lies outside $u = \pm 2 \cdot 054$ as shown in Fig. 8.4.1 (cf. Fig. 6.1.1). In other words, in repeated experiments it would be expected, *if* the null hypothesis were true, that in 2·0 per unit of experiments u would be less than $-2 \cdot 054$,

† This table actually gives the area below $u = +2 \cdot 468$, i.e. $1 - 0 \cdot 007 = 0 \cdot 993$. See §4.3 for details.

and in 2·0 per cent u would be greater than $+2·054$. This is a two-tail test (see § 6.1).

The result of the test. The probability of observing a difference (positive or negative) in success rate between the sample from population 1 (X-treated) and that from population 2 (Y-treated) as large as, or larger than, the observed sample difference, *if* there were no real difference between the treatments (populations), would be approximately 0.04, a 1 in 25 chance.

The corrected result, $P = 0·04$, is quite a good approximation to the exact probability, $P = 0·037$, found at the end of § 8.2. It is low

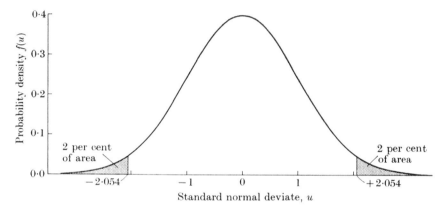

FIG. 8.4.1. Normal approximation to the binomial. Difference between two binomial proportions is converted to an approximate normal deviate, u, and referred to the standard Gaussian curve shown in the figure.

enough to make one suspect (without very great confidence) a real difference between the treatments.

Nonparametric nature of test. Although the normal distribution is used, the test just described is still essentially a nonparametric test. This is because the fundamental assumption is that the proportion of successes is binomially distributed and this can be assured by proper sampling methods. The normal distribution is used only as a mathematical approximation to the binomial distribution.

8.5. The chi-squared (χ^2) test. Classification measurements with two or more independent samples

The probability distribution followed by the sum of squares of n independent standard normal variates (i.e. $\Sigma(x_i-\mu_i)^2/\sigma_i^2$) *where x_i is normally distributed* with a mean of μ_i and a standard deviation of σ_i, see § 4.3), is called the chi-squared distribution with f degree of freedom, denoted $\chi^2_{(f)}$. As suggested by the definition, the scatter seen in estimates of the population variance, calculated from repeated samples from a *normally distributed* population, follows the χ^2 distribution. In fact $\chi^2 = fs^2/\sigma^2$ where s^2 is an estimate of σ^2 issued on f degrees of freedom. The consequent use of χ^2 for testing hypotheses about the true variance of such a population is described, for example, by Brownlee (1965, p. 282).

In the special case of $f = 1$ d.f., one has $\chi^2_{(1)} = u^2$, the square of a single standard normal variate. Tables of the distribution of chi squared with one degree of freedom can therefore be used, by squaring the values of u found in § 8.4, as an approximate test for the 2×2 table. In practice $\chi^2_{(1)}$ is not usually calculated by the method given for the calculation of u, but by another method which, although it does not look related at first sight, gives exactly the same answer, as will be seen. The conventional method of calculation to be described has the advantage that it can easily be extended to larger tables of classification measurements than 2×2. An example is given below.

The form in which χ^2 is most commonly encountered is that appropriate for testing (approximately) goodness of fit, and tables of classification measurements (contingency tables). If x_o is an observed frequency and x_e is the expected value of the frequency on some hypothesis, then it can be shown that the quantity

$$\sum \frac{(x_o - x_e)^2}{x_e}, \qquad (8.5.1)$$

which measure the discrepancy between observation and hypothesis, is distributed *approximately* like χ^2. This approach will be used to test the 2×2 table (Table 8.2.3) that has already been analysed in §§ 8.2 and 8.4.

The expected values, x_e, of the frequencies, given the null hypothesis that the proportion of successes is the same in both populations, are calculated as follows. The best estimate of this *proportion* of successes is, as found in § 8.4, $p = (a+b)/(A+B) = 9/27 = 0.3333$. Therefore, if the null hypothesis were true, the best estimate of the *number*† of successes in the sample from population 1 (e.g. number of patients improved on drug X) would be $0.3333 \times 15 = 5$, and similarly the expected number for population 2 (e.g. drug Y) would be 0.3333×12

---

† This need not be a whole number (see Table 8.5(b) for example). It is a predicted long-run *average* frequency. The individual frequencies must, of course, be integers.

$= 4$. The original table of observations, and the table of values expected on the null hypothesis are thus:

	Observed frequencies (x_o)			*Expected frequencies* (x_e)		
	success	failure	total	success	failure	total
Population 1	8	7	15	5	10	15
Population 2	1	11	12	4	8	12
Total	9	18	27	9	18	27

The summation in (8.5.1) is over all of the cells of the table. The differences $(x_o - x_e)$ are $8-5 = 3$, $1-4 = -3$, $7-10 = -3$, and $11-8 = 3$. Thus, from (8.5.1),

$$\chi^2 = \frac{3^2}{5} + \frac{(-3)^2}{4} + \frac{(-3)^2}{10} + \frac{3^2}{8} = 6 \cdot 075.$$

This is a value of χ^2 with one degree of freedom, because only one expected value need be calculated, the rest following by difference from the marginal totals. It is, as expected, exactly the square of the value of u found in § 8.4, $2 \cdot 4648^2 = 6 \cdot 075$.

Correction for continuity

As in § 8.4, this approximate test for association in a 2×2 table should not be applied without using the correction for continuity. Simply reduce the absolute values of the deviations by $0 \cdot 5$ giving

$$\chi^2_{(1)} = \frac{2 \cdot 5^2}{5} + \frac{(-2 \cdot 5)^2}{4} + \frac{(-2 \cdot 5)^2}{10} + \frac{2 \cdot 5}{8} = 4 \cdot 219.$$

Again it is seen that this is exactly the square of the (corrected) value of u found in § 8·4, $u^2 = 2 \cdot 054^2 = 4 \cdot 219 = \chi^2$. This can be referred directly to a table (e.g. Fisher and Yates, 1963, Table IV; or Pearson and Hartley, 1966, Table 8) of the chi-squared distribution which, for one degree of freedom, has the appearance shown in Fig. 8.5.1. It is found that 4·0 per cent of the area under the curve lies above $\chi^2 = 4 \cdot 219$ (because of the way the tables are constructed the most accurate value that can be found from them is that the area is a little less than 5·0 per cent, i.e. $0 \cdot 05 > P > 0 \cdot 025$). This is exactly the same as found in § 8.4 as it should be, since *the χ^2 test for the 2×2 table is just another way of writing the test using the normal approximation to the binomial.* The result of the test states that *if* the null hypothesis were true, a value of $\chi^2_{(1)}$ as large as 4·219 or larger would be found in 4·0 per cent of repeated experiments in the long run. This casts a certain amount of suspicion on the null hypothesis as explained in § 8.4.

It should be noticed that the probability found using χ^2 is that appropriate for a two-tail test of significance (as shown in § 8.4,

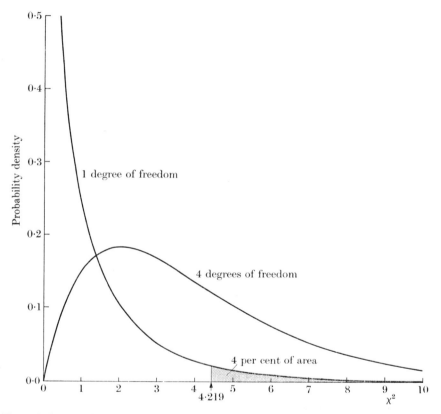

FIG. 8.5.1.   The distribution of chi-squared. The observed value, 4·219, for chi-squared with one degree of freedom (see text) would be exceeded in only 4 per cent of repeated experiments in the long run if the null hypothesis were true. The distribution for 4 degrees of freedom is also shown. (See Chapter 4 for explanation of probability density.)

Fig. 8.4.1, cf. § 6.1) in spite of the fact that only one tail of the χ^2 distribution is considered in Fig. 8.5.1. This is because χ^2 involves the *squares* of deviations, so deviations from the expected values in *either* direction increase χ^2 in the same direction.

Use of chi-squared for testing association in tables of classification measurements larger than 2×2

If the results of treatments X and Y had been classified in more than two ways, for example success, no change, or failure, the experiment shown in Table 8.2.3(b) might have turned out as in Table 8.5.1(a).

TABLE 8.5.1

	success	no change	failure		
Treatment X	8	3	4	15	
Treatment Y	1	5	6	12	(a) observed
	9	8	10	27	

	success	no change	failure		
Treatment X	5	4·4̇	5·5̇	15	
Treatment Y	4	3·5̇	4·4̇	12	(b) expected on null hypothesis
	9	8	10	27	

A proper randomization analysis could be done similar to that in § 8.2, but no tables exist to shorten the calculations for tables larger than 2×2. Often two or more columns or rows can be pooled (giving, for example, Table 8.2.3(b) again) to give 2×2 tables, which may answer the relevant questions. For example, is the proportion of success and of [no change or failure] the same for X and Y? This question is answered by the test of Table 8.2.3(b).

Table 8.5.1(a) itself can be tested using the χ^2 approximation, which is quite sufficiently accurate if all the expected frequencies are at least 5. (They are not in this case; the test is not really safe with such small numbers.) On the null hypothesis that the proportion of successes is the same from treatments X and Y this *proportion* would be estimated as $9/27 = 0·333\dot{3}$. So the *number* of successes expected on the null hypothesis, when 15 individuals are treated with X, is $0·333\dot{3} \times 15 = 5$. Proceeding similarly for 'no change' and 'failure' gives Table 8.5.1(b). Thus, from (8.5.1),

$$\chi^2_{(2)} = \frac{(8-5)^2}{5} + \frac{(3-4·\dot{4})^2}{4·\dot{4}} + \ldots + \frac{(6-4·\dot{4})^2}{4·\dot{4}} = 6·086.$$

Note that *no correction for continuity is used for tables larger than*
2×2. χ^2 has two degrees of freedom since only two cells can be filled
in Table 8.5.1(b), the rest then follow from the marginal totals. Consulting a table of the χ^2 distribution (e.g. Fisher and Yates 1963, Table
IV) shows that a value of χ^2 (with 2 d.f.) equal to or larger than 6·086
would occur in slightly less than 5 per cent of trials in the long run, *if*
the null hypothesis were true; i.e. for a two-tail test $0 \cdot 025 < P < 0 \cdot 05$.
This is small enough to cast some suspicion on the null hypothesis.

Independence of classifications (e.g. of treatment type and success
rate) is tested in larger tables in an exactly analogous way, χ^2 being the
sum of rk terms, and having $(r-1)(k-1)$ degrees of freedom, for a
table with r rows and k columns.

8.6. One sample of observations. Testing goodness of fit with chi-squared

In examples in §§ 8.2–8.5 *two* (r in general) samples (treatments)
were considered. Each sample was classified into two or more (k in
general) ways. The chi-squared approximation is the most convenient
method of testing a *single* sample of classification measurements, to
see whether or not it is reasonable to suppose that the number of
subjects (or objects, or responses) that are observed to fall in each of the
k classes is consistent with some hypothetical allocation into classes
that one is interested in.

For example, suppose that it were wished to investigate the (null)
hypothesis that a die was unbiased. If it were tossed say 600 times the
expected frequencies, on the null hypothesis, of observing 1, 2, 3, 4, 5, and 6
(the $k = 6$ classes) would all be 100, so x_e is taken as 100 each class in
calculating the value of eqn. (8.5.1). The observed frequencies are the
x_0 values. The value of eqn. (8.5.1) would have, approximately, the
chi-squared distribution with $k-1 = 5$ degrees of freedom *if* the null
hypothesis were true, so the probability of finding a discrepancy between observation and expectation at least as large as that observed
could be found from tables of the chi-squared distribution as above.
(See also numerical example below.)

As another example suppose that it were wished to investigate the
(null) hypothesis that all students in a particular college are equally
likely to have smoked, whatever their subject. Again the null hypothesis
specifies the number of subjects expected to fall into each of the k
classes (physics, medicine, law, etc.). If there are 500 smokers altogether
the observed numbers in physics, medicine, law, etc. are the x_0 values

in eqn. (8.5.1). The expected numbers, on the null hypothesis, are found much as above. The example is more complicated than in the case of tossing a die, because different numbers of students will be studying each subject and this must obviously be allowed for. The total number of smokers divided by the total number of students in the college gives the proportion of smokers that would be expected in each class is the null hypothesis were true, so multiplying this proportion by the number of people in each class (the number of physics students in the college, etc.) gives the expected frequencies, x_e, for each class. The value calculated from (8.5.1) can be referred to tables of the chi-squared distribution with $k-1$ degrees of freedom, as before.

A numerical example. Goodness of fit of the Poisson distribution

The chi-squared approximation, it has been stated, can be used to test whether the frequency of observations in each class differ by an unreasonable amount from the frequencies that would be expected if the observations followed some theoretical distribution such as the binomial, Poisson, or Gaussian distributions. In the examples just mentioned, the theoretical distribution was the rectangular ('equally likely') distribution, and only the total number of observations (e.g. number of smokers) was needed to find the expected frequencies. In § 3.6 the question was raised of whether or not it was reasonable to believe that the distribution of red blood cells in the haemocytometer was a Poisson distribution, and this introduces a complication. The determination of the expected frequencies, described in § 3.6, needed not only the total number of observations (80, see Table 3.6.1), but also the observed mean $\bar{r} = 6 \cdot 625$ which was used as an estimate of m in calculating the frequencies expected if the hypothesis of a Poisson distribution were true. The fact that an arbitrary parameter estimated *from the observations* ($\bar{r} = 6 \cdot 625$) was used in finding the expected frequencies gives them a better chance of fitting the observations than if they were calculated without using any information from the observations themselves, and it can be shown that this means that the number of degrees of freedom must be reduced by one, so in this sort of test chi-squared has $k-2$ degrees of freedom rather than $k-1$.

Categories are pooled as shown in Table 3.6.1 to make all calculated frequencies at least 5, because this is a condition (mentioned above) for χ^2 to be a good approximation. Taking the observed frequency as x_o and the calculated frequency (that expected on the hypothesis that cells are Poisson distributed as x_e gives, using (8.5.1) with the results in Table 3.6.1,

$$\chi^2 = \frac{(4-8)^2}{8} + \frac{(5-9)^2}{9} + \ldots + \frac{(3-5)^2}{5} + \frac{(0-5)^2}{5} = 14 \cdot 7.$$

The number of degrees of freedom, in this case, is the number of classes ($k = 9$ after pooling) minus two, as mentioned above. There are therefore 7 degrees of freedom. Looking up tables of the χ^2 distribution shows that $P[\chi^2_{(7)} \geqslant 14 \cdot 7] \simeq 0 \cdot 05$. This means that *if* the true distribution of red cells were a Poisson distribution, *then* an apparent deviation from the calculated distribution (measured by χ^2) as large as, or larger than, that observed in the present experiment would be expected to arise by random sampling error in only about 5 per cent of experiments. This is not low enough (see § 6.1) for one to feel sure that the premise that the distribution is Poissonian must be wrong, though it is low enough to suggest that further experiments *might* lead to that conclusion.

8.7. Related samples of classification measurements. Cross-over trials

Consider Table 8.7.1(a), which is based on an example discussed by Mainland (1963, p. 236). It looks just like the 2×2 tables previously presented. In fact it is not, because it was not based on two *independent* samples of 12. There were actually only 12 patients and not 24. Each patient was given both X and Y (in a random order). This is described as a cross-over trial because those (randomly chosen) patients who were given X first (period 1) were subsequently (period 2) given Y, and vice versa. Table 8.7.1(a) is an incorrect presentation of the results because it disguises this fact. Table 8.7.1(b) is a correct way of giving the results, and it contains more information since 8.7.1(a) can be constructed from it, whereas 8.7.1(b) cannot be constructed from 8.7.1(a). Table 8.7.1(b) cannot be tested in the way described for independent samples either. The 5 patients who reacted in the same way to both X and Y contribute no information about the difference between the drugs, only the 7 who reacted differently to X and Y do so. Furthermore, the possibility that the result depends on whether X or Y was given first can be taken into account. The correct method of analysis is described clearly by Mainland (1963 p. 236). The full results, which have been condensed into 8.7.1(b), were as given in Table 8.7.2. Note that of the 12 patients half (6 selected at random from the

12) have been assigned to the XY sequence, the other half to the YX sequence. These results can be arranged in a 2×2 table, 8.7.3(a) consisting of two independent samples. A randomization (exact) test or χ^2 approximation applied to this table will test the null hypothesis

TABLE 8.7.1

	improved (I)	not improved (N)		
Drug X	12	0	12	
Drug Y	5	7	12	
	17	7	24	8.7.1(a)

		Drug Y			
		I	N		
Drug X	I	5	7	12	
	N	0	0	0	
		5	7	12	8.7.1(b)

TABLE 8.7.2

Patients showing improvements	X in period (1)	X in period (2)	Totals
In both periods	3	2	5
In period (1) not in (2)	3	0	3
In period (2) not in (1)	0	4	4
In neither period	0	0	0
	6	6	12

that the proportion improving in the first period is the same whether X or Y was given in the first period, i.e. that the drugs are equi-effective. The test has been described in detail in § 8.2 (the table is the same as Table 8.2.1(a)), where it was found that P (one tail) $= 0 \cdot 029$ but that the sample is too small for a satisfactory two-tail test, which is what is needed. In real life a larger sample would have to be used.

The subjects showing the same response in both periods give no information about the difference between drugs but they do give information about whether the order of administration matters. Table 8.7.3(b) can be used to test (with the exact test or chi squared) the null hypothesis that the proportion of patients giving the same result in both periods does not depend on whether X or Y was given first. Clearly there is no evidence against this null hypothesis. If, for

TABLE 8.7.3

	Improved in (1) not (2)	Improved in (2) not (1)		
X in period (1)	3	0	3	
X in period (2)	0	4	4	
	3	4	7	8.7.3(a)

	Outcome in periods (1) and (2) same	different		
X in period (1)	3	3	6	
X in period (2)	2	4	6	
	5	7	12	8.7.3(b)

example, drug X had a very long-lasting effect, it would have been found that those patients given X first tended to give the same result in both periods because they would still be under its influence during period 2.

If the possibility that the order of administration affects the results is ignored then the use of the sign test (see § 10.2) shows that the probability of observing 7 patients out of 7 improving on drug X (on the null hypothesis that if the drugs are equi-effective there is a 50 per cent chance, $\mathscr{P} = 1/2$, of a patient improving) is simply $P = (1/2)^7 = 1/128$. For a two-tail test (including the possibility of 7 out of 7 *not* improving) $P = 2/128 = 0 \cdot 016$, a far more optimistic result than found above.

9. Numerical and rank measurements. Two independent samples

'Heiße Magister, heiße Doktor gar,
Und ziehe schon an die zehen Jahr
Herauf, herab und quer und krumm
Meine Schüler an der Nase herum—
Und sehe, daß wir nichts wissen können!
Das will mir schier das Herze verbrennen.'†

† 'They call me master, even doctor, and for some ten years now I've led my students by the nose, up and down, and around and in circles—and all I see is that we cannot know! It nearly breaks by heart.'

GOETHE
(*Faust*, Part 1, line 360)

9.1. Relationship between various methods

IN § 9.2 the randomization test (see §§ 6.3 and 8.2) is applied to numerical observations. In § 9.3 the Wilcoxon, or Mann–Whitney, test is described. This is a randomization test applied to the ranks rather than the original observations. This has the advantage that tables can be constructed to simplify calculations. These randomization methods have the advantage of not assuming a normal distribution; also they can cope with the rejection of particular allocations of treatments to individuals that the experimenter finds unacceptable, as described in § 8.3. They also emphasize the absolute necessity for random selection of samples in the experiment if any analysis is to be done. For large samples Student's *t* test, described in § 9.4, can be used, though how large is large enough is always in doubt (see § 6.2). At least four observations are needed in each sample, however large the differences, unless the observations are known to be normally distributed, as discussed in § 6.2.

9.2. Randomization test applied to numerical measurements

The principle involved is just the same as in the case of classification measurements and § 8.2 should be read before this section, as the arguments will not all be repeated.

Suppose that 4 patients are treated with drug A and 3 with drug B as in § 8.2, but, instead of each being classified as improved or not improved, a numerical measurement is made on each. For example, the reduction of blood glucose concentration (mg/100 ml) following treatment might be measured. Suppose the results were as in Table 9.2.1.

The numbering of the patients is arbitrary but notice that if a positive response is counted as an 'improved' and negative as 'not improved', Table 9.2.1 is the same as Table 8.2.1(b) so, if the *size* of the improvement is ignored, the results can be analysed exactly as in § 8.2.

However, with such a small sample it is easy to do the randomization test on the measurements themselves. The argument is as in

TABLE 9.2.1

Responses (glucose concentration, mg/100 ml) to two drugs. The ranks of the responses are given for use in § 9.3

Drug A			Drug B			
Patient number	Response (mg/100ml)	Rank	Patient number	Response (mg/100ml)	Rank	Total
1	10	5	4	5	4	
2	15	6	6	−3	2	
3	20	7	7	−5	1	
5	−2	3				
Total	43	21		−3	7	40

§ 8.2. See p. 117 for details. If the drugs were really equi-effective (the null hypothesis) each patient would have shown the same response whichever drug had been given, so the apparent difference between drugs would depend solely on which patients happened to be selected for the A group and which for the B group, i.e. on how the random numbers happened to come up in the selection of 4 out of the 7 for drug A. Again, as in § 8.2, the seven measurements could be written on cards from which 4 are selected at random (just as in the real experiment) and called A, the other 3 being B. The difference between the mean for A and the mean for B is noted and the process repeated many times. There is actually no need to calculate the difference between means each time. It is sufficient to look at the total response for drug B (taking the smaller group for convenience) because once this is known the total for A follows (the total of all 7 being always 40), and so the difference between means also follows. If the experimentally observed total response for B (-3 in the example), or a more extreme (i.e. smaller in this example) total, arises very rarely in the repeated randomizations it will be preferred to suppose that the difference between samples is caused by a real difference between drugs and the null hypothesis will be rejected, just as in § 8.2.

TABLE 9.2.2

Enumeration of all 35 possible ways of selecting a group of 3 patients from 7 to be given drug B. The response for each patient is given in Table 9.2.1. The total ranks for drug B are given for use in § 9.3

Patients given drug B			Total response (mg/100ml)	Total rank	Patients given drug B			Total response (mg/100ml)	Total rank
5	6	7	-10	6	1	2	5	23	14
					1	2	6	22	13
					1	2	7	20	12
1	5	6	5	10	1	3	5	28	15
1	5	7	3	9	1	3	6	27	14
1	6	7	2	8	1	3	7	25	13
2	5	6	10	11	1	4	5	13	12
2	5	7	8	10	1	4	6	12	11
2	6	7	7	9	1	4	7	10	10
3	5	6	15	12	2	3	5	33	16
3	5	7	13	11	2	3	6	32	15
3	6	7	12	10	2	3	7	30	14
4	5	6	0	9	2	4	5	18	13
4	5	7	-2	8	2	4	6	17	12
4	6	7	-3	7	2	4	7	15	11
					3	4	5	23	14
					3	4	6	22	13
					3	4	7	20	12
					1	2	3	45	18
					1	2	4	30	15
					1	3	4	35	16
					2	3	4	40	17

With such small samples the result of such a physical randomization can be predicted by enumerating all $7!/(3!4!) = 35$ possible ways (see eqn. (3.4.2)) of dividing 7 patients into samples of 3 and 4. This prediction depends on each of the possible ways being equiprobable, i.e. *the one used for the actual experiment must have been picked at random if the analysis is to valid.* The enumeration is done in Table 9.2.2. This table is exactly analogous to Table 8.2.2 but instead of counting the number improved, the total response is calculated. For example, if patients, 1, 5, and 6 had been allocated to drug B the total response would have been $10+(-2)+(-3) = 5$ mg/100 ml. The results from Table 9.2.2 are collected in Table 9.2.3, which shows the randomization distribution (on the null hypothesis) of the total response to drug B. This is exactly analogous to Fig. 8.2.1. The observed total (-3) and smaller totals (the only smaller one is -10) are seen to occur 2/35 $(= 0\cdot057)$ times, if the null hypothesis is true, and this is therefore the one-tail P. For a two-tail test (see § 6.1) an equal area can be cut off in the other tail (total for $B \geqslant 40$), so the result of the two-tail test is $P = 4/35 = 0\cdot114$. This is not small enough to cast much suspicion on the truth of the null hypothesis, but it is somewhat different from the $P = 0\cdot372$ (one tail) found in the analysis of Table 8.2.1(b), to which, as mentioned above, Table 9.2.1 reduces if the *sizes* of the improvements are ignored. In § 8.2 a one-tail $P = 0.372$ was found and a two-tail test was not possible. The reason for the difference is that in the results in Table 9.2.1 the 'improvements' on drug A are much greater in size than the (negative) 'non-improvements' on drug B. The two-tail test can be done since in § 8.2 all 35 randomizations yielded only 4 different possible results (Table 8.2.1) for the trial, but with numerical measurements the 35 randomizations have yielded 27 possible results, listed in Table 9.2.3, so it it is possible to cut off equal areas in each tail (cf. §6.1). Notice that if patient 3 had been in the B group and patient 4 in the A group (this leaves Tables 9.2.2 and 9.2.3 unchanged) the observed total for group B would have been $20+(-3)+(-5) = 12$ and it is seen from Table 9.2.3 that a total $\leqslant 12$ occurs in a proportion $13/55 = 0\cdot372$ of cases. This one-tail P (when a large improvement, patient 4, is seen with drug B) is as large as that found in § 8.2.

With larger samples there are too many permutations to enumerate easily. For two samples of 10 there are (by (3.4.2)) $20!/(10!10!)$ $= 184756$ ways of selecting 10 samples from 20 individuals. However it is not difficult for a computer to test a large sample of these possible allocations by simulating the physical randomization (random assortment of cards) mentioned at the beginning of this section, and of § 8.2. Programs for doing this do not seem to be widely available at the moment but will doubtless become more common. This method has the advantage that it can allow for the rejection of a random arrangement that the experimenter finds unacceptable (e.g. all men in one sample) as explained in § 8.3. The results in Table 9.2.4 are observations

TABLE 9.2.3

Randomization distribution of total response (mg/100ml) *of a group of 3 patients given drug B* (*according to the null hypothesis that A and B are equi-effective*). *Constructed from Table* 9.2.2

Total for drug B (mg/100ml)	Frequency
−10	1
−3	1
−2	1
0	1
2	1
3	1
5	1
7	1
8	1
10	2
12	2
13	2
15	2
17	1
18	1
20	2
22	2
23	2
25	1
27	1
28	1
30	2
32	1
33	1
35	1
40	1
45	1
Total	35

made by Cushny and Peebles (1905) on the sleep-inducing properties of (−)-hyoscyamine (drug A) and (−)-hyoscine (drug B). They were used in 1908 by W. S. Gosset ('Student') as an example to illustrate the use of his *t* test, in the paper in which the test was introduced.†

If two randomly selected groups of ten patients had been used, a

---

† In this paper the names of the drugs were mistakenly given as (−) hyoscyamine and (+)-hyoscyamine. When someone pointed this out Student commented in a letter to R. A. Fisher, dated 7 January 1935, 'That blighter is of course perfectly right and of course it doesn't really matter two straws . . .'

randomization test of the sort just described could be done as follows. (In the original experiment the samples were not in fact independent but related. The appropriate methods of analysis will be discussed in Chapter 10.) A random sample of 12000 from the 184756 possible permutations was inspected on a computer and the resulting randomization distribution of the total response to drug A is plotted in Fig. 9.2.1

TABLE 9.2.4

Response in hours extra sleep (compared with controls) induced by $(-)$*-hyoscyamine* (A) *and* $(-)$*-hyoscine* (B).

From Cushny and Peebles (1905)

Drug A	Drug B
$+0 \cdot 7$	$+1 \cdot 9$
$-1 \cdot 6$	$+0 \cdot 8$
$-0 \cdot 2$	$+1 \cdot 1$
$-1 \cdot 2$	$+0 \cdot 1$
$-0 \cdot 1$	$-0 \cdot 1$
$+3 \cdot 4$	$+4 \cdot 4$
$+3 \cdot 7$	$+5 \cdot 5$
$+0 \cdot 8$	$+1 \cdot 6$
$0 \cdot 0$	$+4 \cdot 6$
$+2 \cdot 0$	$+3 \cdot 4$
$\Sigma y_A = 7 \cdot 5$	$\Sigma y_B = 23 \cdot 3$
$n_A = 10$	$n_B = 10$
$\bar{y}_A = 0 \cdot 75$	$\bar{y}_B = 2 \cdot 33$

(cf. the distribution in Table 9.2.3 found for a very small experiment). Of the 12000 permutations 488 gave a total response to drug A of less than 7·5, the observed total (Table 9.2.4), so the result of a one-tail randomization test is $P = 488/12000 = 0 \cdot 04067$. With samples of this size there are so many possible totals that the distribution in Fig. 9.2.1 is almost continuous, so it will be possible to cut off a virtually equal area in the opposite (upper) tail of the distribution. Therefore the result of two-tail test can be taken as $P = 2 \times 0 \cdot 04067 = 0 \cdot 0813$. This is not low enough for the null hypothesis of equi-effectiveness of the drugs to be rejected with safety because the observed results would not be unlikely if the null hypothesis were true. The distribution in Fig. 9.2.1, unlike that in Table 9.2.3, looks quite like a normal (Gaussian) distribution, and it will be found that the t test gives a similar result to that just found.

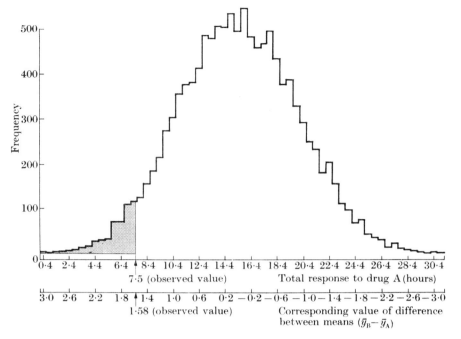

FIG. 9.2.1. Randomization distribution of the total response to drug A for Cushny and Peebles' results, when A and B are equi-effective (null hypothesis true). The values of the difference between means corresponding for each total for A is also shown on the abscissa (the total of all responses is 30·8 for every allocation so, for example, if the total for A were 10·4, the total for B must be 20·4 so the difference between means is 1·0). Constructed from a random sample of 12000 from the 184756 ways of allocating 10 patients out of 20 to drug A.

9.3. Two sample randomization test on ranks. The Wilcoxon (or Mann–Whitney) test

The difficulty with the method described in § 9.2 is that it is not possible to prepare tables for all possible sets of observations. However, if the observations are ranked in ascending order and each observation replaced by its rank before performing the randomization test it is possible to prepare tables, because now every experiment with N observations will involve the same numbers, 1, 2, . . ., N.

In addition to the fact that it is not necessary to assume a particular form of distribution of the observations, another advantage is that the method can be used for results that are themselves ranks, or results that are not quantitative numerical measurements but can be ranked in order of magnitude (e.g. subjective pain scores). Even with numerical measurements the loss of information involved in converting them to ranks is surprisingly small.

Assumptions. The null hypothesis is that both samples of observations come from the same population. If this is rejected, then, if it is wished to infer that the samples come from populations with different medians, or means, it must be *assumed* that the populations are the same in all *other* respects, for example that they have the same variance.

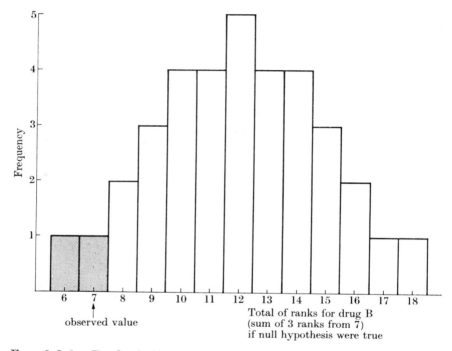

Fɪɢ. 9.3.1. Randomization distribution of the total of ranks for drug B (sum of 3 ranks from 7) if null hypothesis is true. From Table 9.2.2. The mean is 12 and the standard deviation is 2·828 (from eqns. (9.3.2) and (9.3.3)). This is the relevant distribution for *any* Wilcoxon two-sample test with samples of size 3 and 4.

The results in Table 9.2.1 will be analysed by this method. They have been ranked in ascending order from 1 to 7, the ranks being shown in the table. In Table 9.2.2 all 35 (equiprobable) ways of selecting 3 patients from the 7 to be given drug B are enumerated. And for each way the total rank is given; for example, if patients 1, 5, and 6 had had drug B then, on the null hypothesis that the response does not depend on the treatment, the total rank for drug B would be $5+3+2 = 10$.

The frequency of each rank total in Table 9.2.2 is plotted in Fig. 9.3.1, which shows the randomization distribution of the total rank for drug B (given the null hypothesis). This is exactly analogous to the distributions of total response shown in Table 9.2.3 and Fig. 9.2.1, but the distributions of total response depend on the particular numerical values of the observations, whereas the distribution of the rank sum (given the null hypothesis) shown in Fig. 9.3.1 is the same for *any* experiment with samples of 3 and 4 observations. The values of the rank sum cutting off 2·5 per cent of the area in each tail can therefore be tabulated (Table A3, see below).

The observed total rank for drug B was 7, and from Fig. 9.3.1, or Table 9.2.2, it can be seen that there are two ways of getting a total rank of 7 or less, so the result of a one-tail test is $P = 2/35 = 0.057$. An equal probability, 2/35, can be taken in the other tail (total rank of 17 or more) so the result of a two-tail test is $P = 4/25 = 0.114$. This is the probability that a random selection of 3 patients from the 7 would result in the potency of drug B (relative to A) appearing to be as small as (total rank = 7), or smaller than (total rank < 7), was actually observed, or an equally extreme result in the opposite direction, *if* A and B were actually equi-effective. Since such an extreme apparent difference between drugs would occur in 11·4 per cent of experiments in the long run, this experiment might easily have been one of the 11·4 per cent, so there is no reason to suppose the drugs really differ (see § 6.1). In this case, but not in general, the result is exactly the same as found in § 9.2.

A check can be applied to the rank sums, based on the fact that the mean of the first N integers, 1, 2, 3, . . . , N, is $(N+1)/2$ so therefore

$$\text{sum of the first } N \text{ integers} = N(N+1)/2. \qquad (9.3.1)$$

In this case $7(7+1)/2 = 28$, and this agrees with the sum of all ranks (Table 9.2.1), which is $21+7 = 28$.

The distribution of rank totals in Fig. 9.3.1 is symmetrical, and this will be so as long as there are no ties. The result of a two-tail test will therefore be exactly twice that for a one-tail test (see § 6.1).

The use of tables for the Wilcoxon test

The results of Cushny and Peebles in Table 9.2.4, which were analysed in § 9.2, are ranked in ascending order in Table 9.3.1. Where ties occur each member is given the average rank as shown. This method of dealing with ties is only an approximation if Table A3 is used because the table refers to the randomization distribution of integers, 1, 2, 3, 4, 5, 6, . . . 20, not the actual figures used, i.e. 1, 2, 3, $4\frac{1}{2}$, $4\frac{1}{2}$, 6, etc. Such evidence as there is suggests a moderate number of ties does not cause serious error.

The rank sum for drug A is $1+2+3+4\frac{1}{2}+6+8+9\frac{1}{2}+14+15\frac{1}{2}+17$ $= 80\frac{1}{2}$, and for drug B it is $129\frac{1}{2}$. The sum of these, $80\frac{1}{2}+129\frac{1}{2} = 210$,

TABLE 9.3.1

The observations from Table 9.2.4 ranked in ascending order

Drug	Observation (hours)	Rank	
A	−1·6	1	
A	−1·2	2	
A	−0·2	3	
B	−0·1	$4\frac{1}{2}$	$= \dfrac{4+5}{2}$
A	−0·1	$4\frac{1}{2}$	
A	0·0	6	
B	0·1	7	
A	0·7	8	
B	0·8	$9\frac{1}{2}$	$= \dfrac{9+10}{2}$
A	0·8	$9\frac{1}{2}$	
B	1·1	11	
B	1·6	12	
B	1·9	13	
A	2·0	14	
B	3·4	$15\frac{1}{2}$	$= \dfrac{15+16}{2}$
A	3·4	$15\frac{1}{2}$	
A	3·7	17	
B	4·4	18	
B	4·6	19	
B	5·5	20	
Total		210	

checks with (9.3.1), which gives $20(20+1)/2 = 210$. A randomization distribution could be found, just as above and in § 9.2, for the sum of 10 ranks picked at random from 20. The proportion of possible allocations of patients to drug A giving a total rank of $80\frac{1}{2}$ or less is the one-tail P, as above. The two-tail P may be taken as twice this value though, as mentioned this may not be exact when there are ties.

P (two tail) can be found (approximately) from Table A3 in which n_1 and n_2 are the sample sizes ($n_1 \leqslant n_2$). For each pair of sample sizes two figures are given. If the rank sum for sample 1 (that with n_1 observations) is equal to or less than the smaller tabulated figure, or if it is equal to or greater than the larger tabulated figure, then P (two tail) is not greater than the figure at the head of the column. In this case $n_1 = 10$, $n_2 = 10$ and the pair of tabulated figures is 82, 128† for $P = 0\cdot1$, and 78, 132 for $P = 0\cdot05$. The observed rank sum of $80\frac{1}{2}$ is less than 82 but greater than 78, so P is between $0\cdot1$ and $0\cdot05$. This means that if the null hypothesis of equi-effectiveness were true then the probability of observing a rank sum of $80\frac{1}{2}$ or less would be under $0\cdot05$, and the probability of observing a rank sum equally extreme in the other direction would also be under $0\cdot05$, so the total two-tail P (see § 6.1) is under $0\cdot1$. This result is similar to that found in § 9.2 using the slightly more powerful randomization test on the observations themselves. It is not small enough to provide evidence for a difference between the drugs.

How to deal with samples that are too large for Table A3

Table A3 only deals with samples containing up to 20 observations. For larger samples the randomization distribution of ranks (shown for a small sample in Fig. 9.3.1) is well approximated by a normal distribution. *If* the null hypothesis is true, the distribution of the rank sum, R_1 say, for the sample with n_1 observations can be shown (see, for example, Brownlee, 1965, p. 252) to have mean

$$\mu_1 = n_1(N+1)/2, \tag{9.3.2}$$

where $N = n_1 + n_2$ is the total number of observations. For example, in the first example discussed in this section, $n_1 = 3$, $N = 7$ so $\mu_1 = 3(7+1)/2 = 12$, as is obvious by inspection of Fig. 9.3.1. The standard deviation of R_1 is (loc. cit.)

$$\sigma = \sqrt{[n_1 n_2 (N+1)/12]}. \tag{9.3.3}$$

For the distribution in Fig. 9.3.1 the standard deviation is therefore $\sqrt{[3 \times 4(7+1)/12]} = 2\cdot828$. Using these values, an approximate standard normal deviate (see § 4.3) can be calculated from (4.3.1) as

$$u = \frac{R_1 - \mu_1}{\sigma}. \tag{9.3.4}$$

† These are the rank sums that cut off 5 per cent of the area in each tail (10 per cent, $P = 0\cdot1$, altogether), in the analogue of Fig. 9.3.1 for samples of size 10 and 10.

and the rarity of the result judged from tables of the standard normal distribution.

For example, the results in Table 9.3.1 gave $n_1 = 10$, $N = 20$, $R_1 = 80 \cdot 5$. Thus, from (9.3.2)–(9.3.4),

$$u = \frac{80 \cdot 5 - 10(20 + 1)/2}{\sqrt{[10 \times 10(20 + 1)/12]}} = -1 \cdot 85.$$

This value is found from tables (see § 4.3) to cut off an area $P = 0 \cdot 032$ in the lower tail of the standard normal distribution. The result of two-tail test (see § 6.1) is therefore this area, plus the equal area above $u = +1 \cdot 85$, i.e. $P = 2 \times 0 \cdot 032 = 0 \cdot 064$, in good agreement, even for samples of 10, with the exact result from Table A3. The two-tail result can be found directly by referring the value $u = 1 \cdot 85$ to a table of Student's t with infinite degrees of freedom (when t becomes the same as u, see § 4.4).

9.4. Student's t test for independent samples. A parametric test

This test, based on Student's t distribution (§ 4.4), assumes that the observations are normally distributed. Since this is rarely known it is safer to use the randomization test (§ 9.2) or, more conveniently, the Wilcoxon two-sample test (§ 9.3) (see §§ 4.2, 4.6 and 6.2). It will now be shown that when the results in Table 9.2.4, are analysed using t test the result is similar to that obtained using the more assumption-free method of §§ 7.2 and 7.3. But it cannot be assumed that the agreement between methods will always be so good with two samples of 10. It depends on the particular figures observed. If the observations were very non-normal the t test might be quite misleading with samples of 10. The assumptions of the test are explained in more detail in § 11.2 and there is much to be said for always writing the test as an analysis of variance as described at the end of § 11.4. There was no evidence that the assumptions were true in this example.

To perform the t test it is necessary to assume that the observations are independent, i.e. that the size of one is not affected by the size of others (this assumption is necessary for all the tests described), and that the observations are normally distributed, and that the standard deviation is the same for both groups (drugs). The scatter is estimated for each drug separately and the results pooled. The quantity of interest is the difference between mean responses $(\bar{y}_B - \bar{y}_A)$, so the object is to estimate the standard deviation of the difference, $s[\bar{y}_A-\bar{y}_B]$,† so that it can be predicted (see example in § 2.7) how much scatter would be seen in $(\bar{y}_A-\bar{y}_B)$ if it were determined many times (this prediction is likely to be optimistic, see § 7.2).

(1) For drug A the sum of squared deviations, using (2.6.5), is

$$\Sigma(y-\bar{y})^2 = \Sigma y^2 - \frac{(\Sigma y)^2}{n}$$

$$= 34{\cdot}43 - \frac{(7{\cdot}5)^2}{10} = 28{\cdot}805$$

with $n_A-1 = 10-1 = 9$ degrees of freedom (see § 2.6).

(2) For drug B the sum of squared deviations is similarly

$$\Sigma(y-\bar{y})^2 = 90{\cdot}37 - \frac{(23{\cdot}3)^2}{10} = 36{\cdot}081$$

with $n_B-1 = 10-1$ degrees of freedom.

(3) The pooled estimate of the variance of y (the response to either drug) is

$$s^2[y] = \frac{\text{total sum of squares}}{\text{total degrees of freedom}} = \frac{28{\cdot}805+36{\cdot}081}{9+9} = 3{\cdot}605$$

with $9+9 = 18$ degrees of freedom. As it is necessary to assume that the scatter of responses is the same for both groups, a singled pooled estimate of this scatter is made.

(4) Using (2.7.8), the variance of the mean of 10 observations on drug A is estimated to be

$$s^2[\bar{y}_A] = \frac{s^2[y]}{10} = 0{\cdot}3605,$$

and similarly the variance of the mean of 10 observations on drug B is estimated as

$$s^2[\bar{y}_B] = \frac{s^2[y]}{10} = 0{\cdot}3605.$$

† Note that this means the estimated standard deviation, s, of the random variable $(\bar{y}_A-\bar{y}_B)$. It is the functional notation described in § 2.1. It does not mean s *times* $(\bar{y}_A-\bar{y}_B)$.

(5) Using (2.7.3) the variance of the difference between two such means (assuming them to be independent, see also § 10.7) is

$$s^2[\bar{y}_A - \bar{y}_B] = s^2[\bar{y}_A] + s^2[\bar{y}_B] = 0\cdot3605 + 0\cdot3605 = 0\cdot7210.$$

The standard deviation of the difference between means is therefore $\sqrt{(0\cdot7210)} = 0\cdot8491$ hours $= s[\bar{y}_A - \bar{y}_B]$, with 18 degrees of freedom.

(6) The definition of t, given in (4.4.1), is $(x - \mu)/s(x)$ where x is normally distributed and $s(x)$ is its estimated standard deviation. In this case the normally distributed variable of interest is the difference between mean responses, $(\bar{y}_A - \bar{y}_B)$. It is required to test the null hypothesis that the drugs are equi-effective, i.e. that the population value of the difference between means is zero, $\mu = 0$, and therefore $\mu = 0$ is used in the expression for t because, as usual, it is required to find out *what would happen if the null hypothesis were true*. Inserting these quantities gives, on the null hypothesis,

$$t = \frac{(\bar{y}_A - \bar{y}_B) - 0}{s[\bar{y}_A - \bar{y}_B]} = \frac{2\cdot33 - 0\cdot75}{0\cdot8491} = 1\cdot861.$$

References to a table of the distribution of t (see § 4.4 p. 77) for 18 degrees of freedom, shows that 5 per cent of the area lies outside the $t = \pm2\cdot101$ and 10 per cent lies outside $t = \pm1\cdot734$ (cf. §§ 4.4 and 6.1). Therefore, for a two-tail test, P is between $0\cdot05$ and $0\cdot1$. This would be the probability of observing a value of t differing from zero by as much as, or more than, $1\cdot861$, if the null hypothesis were true, and if the assumptions of normality, etc. were correct. It is not small enough to make one reject the null hypothesis that the drugs are really equi-effective ($\mu = 0$). See also § 6.1.

In general, to compare two independent samples (A and B) of normally distributed mutually independent observations one calculates, condensing the above argument into a single formula,

$$t = \frac{|(\bar{y}_A - \bar{y}_B) - \mu|}{\sqrt{\left[\dfrac{\Sigma(y_A - \bar{y}_A)^2 + \Sigma(y_B - \bar{y}_B)^2}{n_A + n_B - 2}\left(\dfrac{1}{n_A} + \dfrac{1}{n_B}\right)\right]}}, \qquad (9.4.1)$$

where n_A and n_B are the numbers of observations in each sample (not necessarily equal); μ is the hypothetical value of the difference to be tested (most often zero, but see example in § 12.5 in which it is not), and the vertical bars round the numerator indicate that its sign is ignored, i.e. t is taken as positive. This quantity is referred to tables of the t distribution (with $n_A + n_B - 2$ degrees of freedom) in order to find P.

Use of confidence limits leads to the same conclusion as the t test

The variable of interest is the difference between means $(\bar{y}_A - \bar{y}_B)$ and its observed value in the example was $2 \cdot 33 - 0 \cdot 75 = 1 \cdot 58$ hours. The standard deviation of this quantity was found to be $0 \cdot 8491$ hours. The expression found in § 7.4 for the confidence limits for the population mean of any normally distributed available x, viz. $x \pm ts(x)$, will be used. For 90 per cent (or $P = 0 \cdot 9$) confidence intervals the $P = 0 \cdot 1$ value of t (with 18 d.f.) is found from tables. It is, as mentioned above, $1 \cdot 734$. Gaussian confidence limits for the population mean value of $(\bar{y}_A - \bar{y}_B)$ are therefore $1 \cdot 58 \pm (1 \cdot 734 \times 0 \cdot 8491)$, i.e. from $0 \cdot 11$ to $3 \cdot 05$ hours. Because these do not include the hypothetical value of zero (implying that the drugs are equi-effective) the observations are not compatible with this hypothesis, if $P = 0 \cdot 9$ is a sufficient level of certainty. For a greater degree of certainty 95 per cent confidence limits would be be found. The value of t for $P = 0 \cdot 05$ and 18 d.f. was found above to be $2 \cdot 101$ so the Gaussian confidence limits are $1 \cdot 58 \pm (2 \cdot 101 \times 0 \cdot 8491)$, i.e. from $-0 \cdot 2$ to $+3 \cdot 36$ hours. At this level of confidence the results are compatible with a population difference between means of $\mu = 0$, because the limits include zero. These results imply that confidence limits can be thought of in terms of a significance test. For any given probability level (α) the variable will be found 'significantly' different (at the $P = \alpha$ level) from any hypothetical value (zero in this case) of the variable that falls outside the $100(1 - \alpha)$ per cent confidence limits.

10. Numerical and rank measurements. Two related samples

10.1. Relationship between various methods

THE observations on the soporific effect of two drugs in Table 9.2.4 were analysed in §§ 9.2–9.4 as though they had been made on two independent samples of 10 patients. In fact both of the drugs were tested on each patient,† so there were only 10 patients altogether. The unit on which each pair of observations is made is called, in general, a *block* (= patient in this case). It is assumed throughout that observations are independent of each other. This may not be true when the pair of observations are both made on the same subject as in Table 10.1.1, rather than on two different subjects who have been matched in some way. The responses may depend on whether A or B is given first, for example, because of a residual effect of the first treatment, or because of the passage of time. It must be assumed that this does not happen. See § 8.6 for a discussion of this point. Appropriate analyses for related samples (see § 6.4) are discussed in this chapter. Chapters 6–9 should be read first.

Because results of comparisons on a single patient are likely to be more consistent than observations on two separate patients, it seems sensible to restrict attention to the difference in response between drugs A and B. These differences, denoted, d are shown in Table 10.1.1. The total for each pair is also tabulated for use in §§ 11.2 and 11.6.

These results will be analysed by four methods. The sign test (§ 10.2) is quick and nonparametric: and, alone in this chapter, it does not need quantitative numerical measurements; scores or ranks will do. The randomization test on the observed differences (§ 10.3) is best for quantitative numerical measurements. It suffers from the fact that, like the analogous test for independent samples (§ 9.2), it is impossible to construct tables for all possible observations; so, except in extreme cases (like this one), the procedure, though very simple, will be lengthy unless done on a computer. In § 10.4 this problem is overcome, as in

---

† Whether A or B is given first should be decided randomly for each patient. See §§ 6.4 and 2.3.

§ 9.3, by doing the randomization test on ranks instead of on the original observations—the Wilcoxon signed-ranks test. This is the best method for routine use (see § 6.2). In § 10.6 the test based on the

TABLE 10.1.1

The results from Table 9.2.4 presented in a way showing how the experiment was really done

Patient (block)	y_A	y_B	Difference $d = (y_B - y_A)$	Total $(y_B + y_A)$
1	+0·7	+1·9	+1·2	2·6
2	−1·6	+0·8	+2·4	−0·8
3	−0·2	+1·1	+1·3	0·9
4	−1·2	+0·1	+1·3	−1·1
5	−0·1	−0·1	0	−0·2
6	+3·4	+4·4	+1·0	7·8
7	+3·7	+5·5	+1·8	9·2
8	+0·8	+1·6	+0·8	2·4
9	0	+4·6	+4·6	4·6
10	+2·0	+3·4	+1·4	5·4
Totals	7·5	22·3	15·8	30·8
			mean 1·58	

assumption of a normal distribution, Student's paired t test, is described (see §§ 6.2 and 9.4). Unless the distribution of the observations is known to be normal, at least six pairs of observations are needed, as discussed in § 6.2 (see also § 10.5).

10.2. The sign test

This test is based on the proposition that the difference between the two readings of a pair is equally likely to be positive or negative if the two treatments are equi-effective (the null hypothesis). This means (if zero differences are ignored) that there is a 50 per cent chance (i.e. $\mathscr{P} = 0·5$) of a positive difference and a 50 per cent chance of a negative difference. In other words, the null hypothesis is that the population (true) median difference is zero. The argument is closely related to that in §§ 7.3 and 7.7 (see below). It is sufficient to be able to rank the members of each pair. Numerical measurements are not necessary.

Example (1). In Table 10.1.1 there are 9 positive differences out of 9 (the zero difference is ignored though a better procedure is probably to allocate to it a sign that is one the safe side, see footnote on p. 155).

If the probability of a positive difference is $1/2$ (null hypothesis) *then* the probability of observing 9 positive differences in 9 'trials of the event' (just like 9 heads out of 9 tosses of an unbiased coin) is given by the binomial distribution (3.4.3) as $(1/2)^9 = 1/512 \simeq 0.002$. For a two-tail test of significance (see § 6.1) equally extreme deviations in the opposite direction (i.e. 9 negative signs out of 9) must be taken into account and for this $P \simeq 0.002$ also, so the result of a two-tail sign test is $P \simeq 0.004$. This is substantially lower than the values obtained in Chapter 9 (when it was not taken into account that the samples were related) and suggests rejection of the null hypothesis because results deviating it by as much as was actually observed would be rare if it were true.

Example (2). If there had been one negative difference (however small) and 9 positive ones, then the one tail P (see § 6.1) would be the probability of observing 9 *or more* positive signs out of 10. This would be the situation if it were decided to count the zero difference in Table 10.1.1 as negative, to be on the safe side. From the binomial distribution, (3.4.3), the probability of observing 9 positive differences out of 10 is

$$P(9) = \frac{10!}{9!\,1!}(0.5)^9\,(0.5)^1$$

$$= \frac{10!}{9!\,1!}\left(\frac{1}{2}\right)^{10} = 0.00976,$$

and the probability of 10 positive differences out of 10 (the only more extreme result) is $P(10) = (1/2)^{10} = 0.000976$. Therefore the probability of observing 9 **or** 10 positive signs out of 10 is $0.00976 + 0.000976 = 0.0107$. The two-tail P (see § 6.1) includes equally extreme results in the other direction (1 or fewer positive signs out of 10, i.e. 9 or more negative signs) for which $P = 0.0107$ also, so the two-tail $P = 0.0107 + 0.0107 = 0.0214$.

This means, in words, that if the null hypothesis (that $\mathscr{P} = 0.5$, implying equi-effectiveness of the treatments) were true, then, in the long run, 2·14 per cent of repeated experiments would give results differing in either direction from the results expected on the null hypothesis (i.e. 5 negative signs out of 10) by as much as, or more than, was actually observed in the experiment. This is a sufficiently rare event to cast some doubt on the premise of equi-effectiveness (see § 6.1).

The general result

Generalizing the argument shows that if r_{obs} differences out of n are observed to be negative (or positive, if there are fewer positive signs than negative), then the result of a two-tail test of the hypothesis that the population median difference is zero is

$$P = 2 \sum_{r=0}^{r=r_{obs}} \frac{n!}{r!(n-r)!} \left(\frac{1}{2}\right)^n. \tag{10.2.1}$$

How to find the results without calculation

There are several ways of using tables to get the result.

Method (1). One way is to find confidence limits for the median difference (d value) from Table A1, as described in § 7.3. If the confidence limits do not include zero (or, more generally, any other hypothetical value that it is wished to test for consistency with the observations), then, as explained in § 9.4, the observations are not consistent with the null hypothesis. For example, the results (d values) in Table 10.1.1 consist of $n = 9$ non-zero differences.† The method of § 7.3 shows, using Table A1, that 99·60 per cent confidence limits for the population median are provided by the largest and smallest of the nine observations, i.e. $+0·8$ to $+4·6$. These limits do not include zero, so the results are not consistent with the null hypothesis and the result for a two-tail test is that P is not greater than $1-0·996 = 0·004$ (in this case $P = 0·004$ as shown above).

Putting the matter more precisely, the exact value of P for the confidence limits that just fail to include zero (e.g. such that the next smallest observation below the lower limit would be negative) will be the same as the exact value of P for a two-tail test (see method (2) below). By way of example suppose that patient 5 (Table 10.1.1) had given a difference of $-0·01$ (rather than zero) in Example (2) above. Table A1 shows that the 99·8 per cent confidence limits for the population median difference, based on a sample of $n = 10$ differences, are provided by the largest and smallest observations, $-0·01$ to $+4·6$ These limits include zero. The 97·86 per cent confidence limits, from

---

† This situation shows the difficulties that can be introduced by ties. There is no reason to exclude the zero difference when finding confidence limits for the median, but the results will only agree exactly with the sign test (from which the zero was omitted) if this is done. The best answer is probably to be on the safe side. This usually means counting the zero difference as though it had the sign least conducive to rejection of the null hypothesis. In the example discussed this means pretending that patient 5 actually gave a negative difference. Example (2) shows that $P = 0·0214$ in this case.

Table A1, are the next-to-smallest and next-to-largest observations, i.e. $+0\cdot8$ and $+2\cdot4$, which just fail to include zero. This agrees with the exact two-tail result, $P = 0\cdot0214$ ($= 1-0\cdot9786$), found by direct calculation above.

Method (2). The same result is obtained if Table A1 is entered with $r = r_{obs}+1$. This is obvious if (10.2.1) is compared with (7.3.3). Considerations of a few examples shows that if limits are taken as the $(r_{obs}+1)$ the observation from end of the ranked observations, the limits will just fail to include zero. For example, in Table 10.1.1, as just discussed, $r = r_{obs}+1 = 1$, gives $P = 1-0\cdot996 = 0\cdot004$. Likewise in the second example above, $r_{obs} = 1$ negative sign out of $n = 10$. Entering Table A1 with $n = 10$ and $r = r_{obs}+1 = 2$ gives the result of the two-tail significance test as $P = 1-0\cdot9786 = 0\cdot0214$, exactly as found from first principles above.

Method (3). As might be expected, the same result can be obtained by finding confidence limits, $\mathscr{P}_H$ and $\mathscr{P}_L$, for the population proportion ($\mathscr{P}$) of positive (or negative) differences and seeing whether these limits include $\mathscr{P} = 0\cdot5$ or not. The method has been described in § 7.7 and the the result can be obtained, as explained there, from Table A2. It will be left to the reader to improve his moral fibre by showing (by comparing (7.7.1), (7.7.2), and (7.3.2)) that if the upper confidence limit for the population median difference, found above, just fails to include zero, then it will be found that the upper confidence limit for $\mathscr{P}$, $\mathscr{P}_H$, is equal to or less than $0\cdot5$. Similarly, if the lower confidence limit for the population median just fails to include zero then it will be found that $\mathscr{P}_L \geqslant 0\cdot5$.

For example, in Table 10.1.1, $r_{obs} = 0$ out $n = 9$ differences were negative, so $100r/n = 0$ per cent negative differences were observed. Entering Table A2 with $r = 0$ and $n = 9$ shows that 99 per cent confidence limits for the population proportion of negative differences are $\mathscr{P}_L = 0$ and $\mathscr{P}_H = 0\cdot445$. These limits do not include $0\cdot5$ (as expected) and this implies that for a two-tail significance test $P < 0\cdot01$ (i.e. $1-0\cdot99$), as found above.

In the second example above ($r_{obs} = 1$ negative difference out of $n = 10$), consulting Table A2 with $r = 1$, $n = 10$, gives 95 per cent confidence limits for the population proportion ($\mathscr{P}$) of negative differences as $0\cdot0025$ and $0\cdot445$ which do not include $0\cdot5$. This is as expected from the fact that the $97\cdot86$ per cent (which is as near to 95 per cent as it is possible to get, see § 7.3) confidence limits for the population median difference, $+0\cdot8$ to $+2\cdot4$ found above, just fail to include zero. The 99 per cent confidence limits for $\mathscr{P}$ are 0·0005 to 0·5443, which *do* include the null hypothetical value, $\mathscr{P} = 0\cdot5$, as expected. These results imply that the result of a two-tail sign test is $0\cdot01 < P < 0\cdot05$. The exact result is 0·0214, found above.

10.3. The randomization test for paired observations

The principle involved is that described in §§ 6.3, 8.2, 8.3, and 9.2. As in § 9.2, it is not possible to prepare tables to facilitate the test when it is done on the actual observations. However, in extreme cases, like the present example, or when the samples are very small, as in § 8.2, the test is easy to do (see § 10.1).

As before, attention is restricted to the subjects actually tested. The members of a pair of observations may be tests at two different times on the same subject (as in this example), or tests on the members of a matched pair of subjects (see § 6.4). It is supposed that if the null hypothesis (that the treatments are equi-effective) were true then the observations on each member of the pair would have been the same even if the other treatment (e.g. drug) had been given (see p. 117 for details). In designing the experiment it was (or should have been) decided strictly at random (see § 2.3) which member of the pair received treatment A and which B, or, in the present case, whether A or B was given first. If A had been given instead of B, and B instead of A, the only effect on the difference in responses (d in Table 10.1.1), if the null hypothesis were true, would be that its sign would be changed. According to the null hypothesis then, the sign of the difference between A and B that was observed must have been decided by the particular way the random numbers came up during the random allocation of A and B to members of a pair. In repeated random allocations it would be equally probable that each difference would have a positive or a negative sign. For example, for patient 1 in Table 10.1.1, the randomization decided whether $+0\cdot7$ or $+1\cdot9$ was labelled A and hence, according to the null hypothesis, whether the difference was $+1\cdot2$ or $-1\cdot2$. It can therefore be found out whether (if the null hypothesis were true) it would be probable that a random allocation of drugs to these patients would give rise to a mean difference as large as (or larger than) that observed (1·58 hours), by inspecting the mean differences produced by all possible allocations, (i.e. all possible combinations of positive and negative signs attached to the differences). If this is sufficiently improbable the null hypothesis will be rejected in the usual way (Chapter

6). In fact it can be shown that the same result can be obtained by inspecting the sum of only the positive (or of only the negative) d values resulting from random allocation of signs to the differences, so it is not necessary to find the mean each time† (similar situations arose in §§ 8.2 and 9.2).

Assumptions. Putting the matter a bit more rigorously, it can be seen that the hypothesis that an observation (d value) is equally likely to be positive or negative, whatever its magnitude, implies that the distribution of d values is symmetrical (see § 4.5), with a mean of zero. The null hypothesis is therefore that the distribution of d values is symmetrical about zero, and this will be true *either* if the y_A and y_B values have identical distributions (not necessarily symmetrical), *or* the distributions of y_A and y_B values both have symmetrical distributions (not necessarily identical) with the same mean. This makes it clear that if the null hypothesis is rejected, then, if it is wished to infer from this that the distributions of y_A and y_B have different population means, it must be *assumed* either that their distributions both have the same shape (i.e. are identical apart from the mean), or that they are both symmetrical.

Note that when the analysis is done by enumerating possible allocations it is assumed that each is equi-probable, i.e. that *an allocation was picked at random for the experiment, the design of which is therefore inextricably linked with its analysis* (see § 2.3)

If there are n differences (10 in Table 10.1.1) then there are 2^n possible ways of allocating signs to them (because one difference can be + or −, two can be ++, +−, −+, or −−, and each time another is added the number of possibilities doubles). All of these combinations could be enumerated as in Table 8.2.2 and Fig. 8.2.1, and Tables 9.2.2 and 9.3. This is done, using ranks, in § 10.4. In the present example, however, only the most extreme ones are needed.

Example (1). In the results in Table 10.1.1 there are 9 positive differences out of 9 (the zero difference, even if included, would have no effect because the total is the same whatever sign is attached to it). The number of ways in which signs can be allocated is $2^9 = 512$. The observed allocation is the most extreme (no other can give a mean of 1·58 or larger) so the chance that it will come up is 1/512. For a two-tail test (taking into account the other most extreme possibility, all signs

---

† As before this is because the *total* of all differences is the same (15·8 in the example) for all randomizations, so specifying the sum of negative differences also specifies the mean difference.

negative, see § 6.1), the P value is therefore $2/512 \simeq 0.004$. In this most extreme case (though no other) the result is the same as given by the sign test (§ 10.2). Consider, for example, what would have happened if patient 5 had given a negative difference instead of zero. The result of the randomization test will, unlike that of the sign test, depend on how large the negative difference is.

Example (2). Suppose that patient 5 had given $d = -0.9$, the other patients being as in Table 10.1.1. There are now $2^{10} = 1024$ possible ways of allocating signs to the $n = 10$ differences. How many of these give a total for the negative differences (see above) equal to less than 0·9? Apart from the observed allocation, only two. That in which patient 8 is negative but 5 is positive giving a sum of negative differences of 0·8, and that in which all differences are positive giving a sum of negative differences of zero. The probability of observing, on the null hypothesis, a sum of negative differences as extreme as, or more extreme than 0·9 is thus $3/1024$. For a two-tail test (see § 6.1), therefore, $P = 6/1024 = 0.0059$† (see next example for the detailed interpretation).

Example (3). If, however, patient 5 had had $d = 2.0$, the mean difference, $\bar{d}$, would have been $13.8/10 = 1.38$. In this case a sum of negative differences equal to or less than 2 could arise in ten different ways, as well as that observed, so P (one tail) $= 11/1024$ and P (two tail) $= 22/1024 = 0.0225$.† The 11 possible ways are (a) all differences positive (sum $= 0$), (b) one difference negative (patient 8, 6, 1, 3, 4, 10, 7, or 5) giving a sum of 0·8, 1·0, 1·2, 1·3, 1·3, 1·4, 1·8, or 2·0, depending on which patient has the negative difference, (c) two differences negative, patients 6 and 8 giving a sum of negative differences of $1.0 + 0.8 = 1.8$, or patients 1 and 8 giving a sum of $1.2 + 0.8 = 2.0$.

This result means that *if* the null hypothesis were true *then* the probability would be only 0.0225 that the random numbers would come up, during the allocation of the treatments, in such a way as to give a sum of negative differences of 2·0 or less (i.e. a mean difference between B and A of 1·38 or more), or results equally extreme in the other direction (A giving larger responses than B). This probability is small enough to make one suspect the null hypothesis (see § 6.1).

---

† In general it is possible, though uncommon in this sort of test, that an exactly equal area could not be cut off in the opposite tail so twice the one-tail P may be a maximum value for the two-tail P (see § 6.1).

10.4. The Wilcoxon signed-ranks test for two related samples

This test works on much the same principle as the randomization test in § 10.3 except that ranks are used, and this allows tables to be constructed, making the test very easy to do. The relation between the methods of §§ 9.2 and 9.3 for independent samples was very similar. However, the signed-ranks test, unlike the sign test (§ 10.2) or the rank test for two independent samples (§ 9.3), will not work with observations that themselves have the nature of ranks rather than quantitative numerical measurements. The measurements must be such that the values of the differences between the members of each pair can properly be ranked. This would certainly not be possible if the observations were ranks. If the observations were arbitrary scores (e.g. for intensity of pain, or from a psychological test) they would be suitable for this test if it could be said, for example, that a pair difference of $80-70 = 10$ corresponded, in some meaningful way, to a smaller effect than a pair difference of $25-10 = 15$. Seigel (1956*a*, *b*) discusses the sorts of measurement that will do, but if you are in doubt use the sign test, and keep Wilcoxon for quantitative numerical measurements. Sections 9.2, 9.3, 10.3 and Chapter 6, should be read before this section. The precise nature of the assumptions and null hypothesis have been discussed already in § 10.3.

The method of ranking is to arrange all the differences in ascending order *regardless of sign*, rank them 1 to n and then attach the sign of the difference to the rank. Zero differences are omitted altogether. Differences equal in absolute value are allotted mean ranks as shown in examples (2) and (3) below (and in § 9.2). To use Table A4 find T, which is either the sum of the positive ranks or the sum of the negative ranks, whichever sum is smaller. Consulting Table A4 with the appropriate n and T gives the two-tail P at the head of the column. Examples are given below. Of course for simple cases the analysis can be done directly on the ranks as in § 10.3.

How Table A4 *is constructed*

Suppose that $n = 4$ pairs of observations were made, the differences (*d*) being $+0\cdot1$, $-1\cdot1$, $-0\cdot7$, $+0\cdot4$. Ranking, regardless of sign, gives

d	$+0\cdot1$	$+0\cdot4$	$-0\cdot7$	$-1\cdot1$
rank	1	2	-3	-4

The observed sum of positive ranks is $1+2 = 3$, and the observed sum of negative ranks is $3+4 = 7$. The sum of all four ranks, from (9.3.1), is $n(n+1)/2 = 4(4+1)/2 = 10$, which checks $(3+7 = 10)$. Thus $T = 3$, the smaller of the rank sums. Table A4 indicates that it is not possible to find evidence against the null hypothesis with a sample as small as 4 differences. This is because there are only $2^n = 2^4 = 16$ different ways in which the results could have turned out (i.e. ways of allocating signs to the differences, see § 10.3), when the null hypothesis is true.

TABLE 10.4.1

The 16 possible ways in which a trial on four pairs of subjects could turn out if treatments A and B were equi-effective, so the sign of each difference is decided by whether the randomization process allocates A or B to the member of the pair giving the larger response. For example, on the second line the smallest difference is negative and all the rest are positive, giving sum of negative ranks = 1

Rank	1	2	3	4	Sum of pos. ranks.	Sum of neg. ranks	T
	+	+	+	+	10	0	0
	−	+	+	+	9	1	1
	+	−	+	+	8	2	2
	+	+	−	+	7	3	3
	+	+	+	−	6	4	4
	−	−	+	+	7	3	3
	−	+	−	+	6	4	4
	−	+	+	−	5	5	5
	+	−	−	+	5	5	5
	+	−	+	−	4	6	4
	+	+	−	−	3	7	3
	+	−	−	−	1	9	1
	−	+	−	−	2	8	2
	−	−	+	−	3	7	3
	−	−	−	+	4	6	4
	−	−	−	−	0	10	0

Therefore, even the most extreme result, all differences positive, would appear, in the long run, in 1/16 of repeated random allocations of treatments to members of the pairs. Similarly 4 negative differences out of 4 would be seen in 1/16 of experiments. The result of a two tail test cannot, therefore, be less than $P = 2/16 = 0\cdot125$ with a sample of four differences, however large the differences (see, however, §§ 6.1 and

10.5 for further comments). With a small sample like this, it is easy to illustrate the principle of the method. More realistic examples are given below.

The $2^4 = 16$ possible ways of allocating signs to the four differences (i.e. the possible ways in which A and B could have been allocated to members of a pair, see §10.3 for a full discussion of this process), are listed systematically in Table 10.4.1, together with the sums of positive and negative ranks, and value of T, corresponding to each allocation. In Table 10.4.2, the frequencies of these quantities are listed from the

TABLE 10.4.2

The relative frequencies of observing various values of i.e. the distributions of, the rank sum, and T, with $n = 4$ pairs of observations when the null hypothesis is true. Constructed from Table 10.4.1

Rank sum	Frequency for pos. ranks	Frequency for neg. ranks	Frequency for T
0	1	1	2
1	1	1	2
2	1	1	2
3	2	2	4
4	2	2	4
5	2	2	2
6	2	2	
7	2	2	
8	1	1	
9	1	1	
10	1	1	
Total	16	16	16

results in Table 10.4.1, and in Fig. 10.4.1 the distribution of the sum of positive ranks is plotted (that for negative ranks is identical). (These are the paired sample analogues of the rank distribution worked out for two independent samples in Table 9.2.2 and Fig. 9.3.1.)

Now the observed sum of positive ranks was 3, and the probability of observing a sum of 3 or less is seen from Table 10.4.2 or Fig. 10.4.1, to be 5/16. The probability of an equally large deviation from the null hypothesis in the other direction (sum of positive ranks $\geqslant 7$) is also 5/16. (The distribution is symmetrical, like that in Fig. 9.3.1, unless there are ties, so the result of a two-tail test is twice that for a one-tail test. See §6.1.) The result of a two-tail significance test is therefore

$P = 10/16 = 0\cdot625$, so there is no evidence against the null hypothesis, because results deviating from it by as much as, or more than, the observed amount would be common if it were true. In other words, if the null hypothesis were true it would be rejected (wrongly) in 62·5 per cent of repeated experiments in the long run, if it were rejected whenever the sum of positive ranks was 3 or less, or when it was equally extreme in the other direction (7 or greater). A value of T equal to or less than the observed value (3) is seen, from Table 10.4.2, to occur in $4+2+2+2 = 10$ of the 16 possible random allocations. The probability (on the null hypothesis) of observing $T \leqslant 3$ is therefore $P = 10/16$

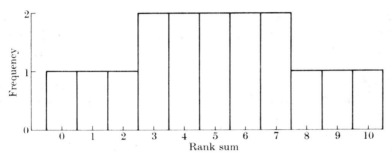

F IG. 10.4.1. Distribution of the sum of positive ranks when the null hypothesis is true for the Wilcoxon signed ranks test with four pairs of observations. The distribution is identical for sum of negative ranks. Plotted from Table 10.4.2.

$= 0\cdot625$ which is another way of putting the same result. As in § 9.3, the calculations in Tables 10.4.1 and 10.4.2 would be the same for *any* experiment with $n = 4$ pairs of observations. The values of T cutting off suitably small tail areas (1 per cent, 5 per cent, etc.) can therefore be tabulated for various sample sizes. (The smallest possible value, $T = 0$, cuts of an area of $P = 2/16 = 0\cdot125$ for the small sample in Table 10.4.2, as mentioned above.) This is what is given in Table A4.

Example (1). In Table 10.1.1, there are 9 positive differences out of 9 so all ranks are positive and $T = $ sum of negative ranks $= 0$. Consulting Table A3 with $n = 9$, $T = 0$, shows $P < 0\cdot01$ (because T is less than 2, the tabulated value for $P = 0\cdot01$). In fact doing the test directly it is seen that there is only one way (the observed one) of getting a sum of negative ranks as extreme as zero, out of $2^9 = 512$ ways of allocating signs (see § 10.3). So P (one tail) $= 1/512$, and P(two tail) $= 2/512$ $= 0\cdot004$ (exactly as in §§ 10.2 and 10.3 for this extreme case, but not in general). This is quite strong evidence against the null hypothesis.

This is, as usual, because *if* the null hypothesis were true, deviations from it (in either direction) as large as, or larger than, those observed in this experiment would occur only rarely ($P = 0.004$) because the random numbers happened to come up so that all the subjects giving big responses were given the same treatment.

Example (2). Suppose however, as in § 10.3, Example (2), that patient 5 had given $d = -0.9$ instead of zero. When the observations are ranked regardless of sign the result is as follows:

d	0·8	−0·9	1·0	1·2	1·3	1·3	1·4	1·8	2·4	4·6
rank	1	−2	3	4	$5\frac{1}{2}$	$5\frac{1}{2}$	7	8	9	10

Thus, $n = 10$, sum of negative ranks = 2, and sum of positive ranks = 53. Thus, $T = 2$, the smaller rank sum. The total of all 10 ranks should be, from (9.3.1), $n(n+1)/2 = 10(10+1)/2 = 55$, and in fact $53+2 = 55$. Consulting Table A4 with $n = 10$ and $T = 2$, again shows $P < 0.01$ (two tail). An exact analysis is easily done in this case. A sum of negative ranks of 2 or less could arise with only 2 combinations of signs, in addition to the observed one (rank 1 negative giving $T = 1$ or all ranks positive giving $T = 0$), and there are $2^n = 2^{10} = 1024$ possible ways of allocating signs (see § 10.3). Thus P (one tail) = $3/1024$ and P (two tail) = $6/1024 = 0.0059$. Again quite strong evidence against the null hypothesis.

Example (3). If patient 5 had had $d = -2.0$ (as discussed in § 10.3) the ranking process would be as follows:

d	0·8	1·0	1·2	1·3	1·3	1·4	1·8	−2·0	2·4	4·6
rank	1	2	3	$4\frac{1}{2}$	$4\frac{1}{2}$	6	7	−8	9	10

Consulting Table A4 with $n = 10$ and $T = 8$ gives $P = 0.05$. Enumeration of all possible ways of achieving a sum of negative ranks of 8 or less shows there to be 25 ways† so the exact two-tail P is $50/1024 = 0.049$.

Example (4). Consider the following 12 differences observed in a paired experiment, shown after ranking in ascending order, disregarding the sign.

d	0·1	−0·7	0·8	1·1	−1·2	1·5	−2·1	−2·3	−2·4	−2·6	−2·7	−3·1
rank	1	−2	3	4	−5	6	−7	−8	−9	−10	−11	−12

---

† This is found by constructing a sum of 8 or less from the integers from 1 to $n(= 10)$, as used in calculating Table A4. More properly it should be done with the figures 1, 2, 3, $4\frac{1}{2}$a, $4\frac{1}{2}$b, 6, 7, 8, 9, 10 and with these there are only 24 ways of getting a sum of 8 or less.

In this case most of the observed differences are negative. Is the population mean difference different from zero? The sum of the negative ranks is $2+5+7+8+9+10+11+12 = 64$, and the sum of the positive ranks is $1+3+4+6 = 14$, so $T = 14$, the smaller rank sum. An arithmetical check is provided by (9.3.1) which gives $n(n+1)/2 = 12(12+1)/2 = 78$, and, correctly, $64+14 = 78$. Table A4 shows that when $n = 12$, a value of $T = 14$ corresponds to $P = 0\cdot05$. Only marginal evidence against the null hypothesis (see § 6.1).

How to deal with samples too large for Table A4

Table A4 deals only with samples up up to $n = 25$ pairs of observations. For larger samples, as in § 9.3, it is a very good approximation to assume that the distribution of the rank sum, shown for a small sample in Fig. 10.4.1, is Gaussian (normal) with (given the null hypothesis) a mean of

$$\mu = n(n+1)/4 \qquad (10.4.1)$$

and standard deviation

$$\sigma = \sqrt{\left\{\frac{n(n+1)(2n+1)}{24}\right\}} \qquad (10.4.2)$$

(the derivations of μ and σ are given, for example, by Brownlee (1965, p. 258)). For example, for the distribution in Fig. 10.4.1, $n = 4$ so the mean is $\mu = 4(4+1)/4 = 5$, as is obvious from the figure, and $\sigma = \sqrt{\{4(4+1)(8+1)/24\}} = 2\cdot74$.

The results in Example (4) can be used to illustrate the normal approximation. In this example, $n = 12$, so $\mu = 12(12+1)/4 = 39$, and $\sigma = \sqrt{\{12(12+1)(24+1)/24\}} = 12\cdot75$. An approximate standard normal deviate (see § 4.3) can therefore be calculated from (4.3.1) as

$$u = \frac{|T-\mu|}{\sigma} = \frac{|14-39|}{12\cdot75} = \frac{25}{12\cdot75} = 1\cdot96. \qquad (10.4.3)$$

The vertical bars mean, as usual, that the numerator is taken as positive. The same value would be obtained if the sum of negative ranks, 64, were used in (10.4.3), because $64-39 = 25$. This value can now be referred to of the standard normal distribution (see § 4.3), or tables of t (with infinite degrees of freedom), (see § 4.4). A value of $u = 1\cdot96$ cuts off an area $P = 0\cdot05$ in the tails of the standard normal distribution, as explained in § 4.3. In other words of value of u above $+1\cdot96$, or less than $-1\cdot96$, would occur in 5 per cent of repeated experiments. In

this case the normal approximation gives the same value as the exact
result, $P = 0.05$, found above from Table A4.

10.5. A data selection problem arising in small samples

Consider the paired observations of responses to treatments A and B shown
in Table 10.5.1.

TABLE 10.5.1

Block	Treatment		Difference
	A	B	
1	1·7	0·5	+1·2
2	1·2	0·5	+0·7
3	1·8	0·9	+0·9
4	1·0	0·7	+0·3

The experiment was designed exactly like that in Table 10.1.1. All the differences
are positive so the three nonparametric tests described in §§ 10.2–10.4 all give
$P = 2/2^4 = 1/8$ for a two-tail test. In general, for n differences all with the
same sign, the result would be $2/2^n$.

It has been stated that the design of an experiment dictates the form its analysis must take. Selection of particular features after the results have been seen
(data-snooping) can make significance tests very misleading. Methods of dealing
with the sort of data-snooping† problem that arise when comparing more than
two treatments are discussed in § 11.9. Nevertheless, it seems unreasonable to
ignore the fact that in these results, the observations are completely separated,
i.e. the smallest response to A is bigger than the largest response to B, a feature
of the results that has not been taken into account by the paired tests. (In
general, the statistician is not saying that experimenters should not look too
closely at the results of their experiments, but that proper allowance should be
made for selection of particular features.) This feature means that *if* the results
could be analysed by the two nonparametric methods designed for independent
samples (described in §§ 9.2 and 9.3), both methods would give the probability
of complete separation of the results, if the treatments were actually equi-
effective, as $P = 2n!n!/(2n)! = 2(4!4!)/8! = 1/35$ (two-tail)—a far more 'significant' result! The naive interpretation of this is that it would have been better
not to do a paired experiment. This is quite wrong. It has been shown by Stone
(1969) that the probability (given the null hypothesis of equi-effectiveness of
A and B) of complete separation of the two groups as observed, would be 1/35
even if there were no differences between the blocks, and even less than 1/35 if
there were such differences. This is not the same as the $P = 1/8$ found using the
paired tests because it is the probability of a different event. *If* the null hypothesis were true, *then*, in the long run, 1 in 8 of repeated experiments would be
expected to show 4 differences all with the same sign out of 4, but only 1 in
35, or fewer, would have no overlap between groups as in this case.

It remains to be decided what should be done faced with observations such as

† This is statisticians jargon. 'Data selection' might be better.

those in Table 10.5.1. The snag is, of course, that *any* single specified arrangement of the results is improbable. If the treatments were equi-effective (and there were no differences between blocks) *any* of the $4!4!/8! = 70$ possible arrangements of the eight figures into two groups of 4 would have the same probability, $P = 1/70$, of occurring. It is only because the particular arrangement, with no overlap between A and B, corresponds to a preconceived idea, that it is thought unusual, and what constitutes 'correspondence with a preconceived idea' may be arguable. The problem is an old one:

'. . . . when Dr Beattie observed, as something remarkable which had happened to him, that he had chanced to see both No. 1 and No. 1000, of the hackney-coaches, the first and the last; "Why, Sir, (said Johnson,) there is an equal chance for one's seeing those two numbers as any other two". He was clearly right; yet the seeing of the two extremes, each of which is in some degree more conspicuous than the rest, could not but strike one in a stronger manner than the sight of any other two numbers'

(Boswell's *Life of Johnson*)

The only safe *general* rule that can be offered at the moment is to analyse the experiment as a paired experiment if it was designed in that way. In other words take $P = 1/8$ in the present case; not much evidence against the null hypothesis. The problem is, however, a complicated one that is still not fully solved.†

10.6. The paired *t* test

As for the two sample t test (§ 9.4), it is necessary to assume that the distributions of responses to the two treatments are Gaussian (normal) in form (see §§ 4.2 and 4.6), but it is no longer necessary to assume that they have identical variances. The assumptions are explained in more detail in § 11.2 and it would be preferable always to write the calculations in the form of analysis of variance as described in § 11.6.

The method will be applied to the results in Table 10.1 which have already been analysed properly in §§ 10.2–10.4 (and which were analysed as though the two samples were independent in Chapter 9) although there is no evidence that the assumptions are fulfilled.

The analysis is carried out on the differences, $d = y_{\mathrm{B}} - y_{\mathrm{A}}$. The variance of the differences is estimated to be, using (2.6.2) and (2.6 5),

$$s^2[d] = \frac{\Sigma(d - \bar{d})^2}{n - 1} = \frac{38 \cdot 58 - (15 \cdot 8)^2/10}{10 - 1}$$

$$= 1 \cdot 513 \text{ with } (n - 1) = 9 \text{ degrees of freedom.}$$

† In *some* cases, such as this one, when the smallest P value (given, in this case by the Wilcoxon two-sample test) is smaller than the P value that the other tests under consideration can ever reach, however large the difference between the treatments, Stone (1969) has argued that it is proper to quote the smaller P, i.e. $P \leqslant 1/35$ for the results in Table 10.5.1.

Stone's method also introduces another factor of $1/2$, i.e. he takes $P \leqslant 1/70$, but this feature has not yet come into wide use.

And the *variance of the mean difference* is, by (2.7.8),

$$s^2[\bar{d}] = \frac{s^2[d]}{n} = \frac{1\cdot513}{10} = 0\cdot1513 \text{ with 9 degrees of freedom.}$$

This should be carefully distinguished from the *variance of the difference between means* found in § 9.4, which was larger (0·7209) and had more degrees of freedom (18). The disappearence of 9 degrees of freedom will be explained when the results are looked at from the point of view of the analysis of variance in § 11.6.

The standard deviation of the mean difference is estimated as

$$s[\bar{d}] = \sqrt{0\cdot1513} = 0\cdot3890$$

and the null hypothesis is that the population (true) mean difference, μ, is zero. Thus, from (4.4.1),

$$t = \frac{\bar{d}-\mu}{s[\bar{d}]} = \frac{\bar{d}}{s[\bar{d}]} = \frac{\bar{d}}{\sqrt{[\{\Sigma d^2 - (\Sigma d)^2/n\}/\{n(n-1)\}]}}. \qquad (10.6.1)$$

In this example

$$t = \frac{1\cdot58}{0\cdot3890} = 4\cdot062.$$

Referring to a table of the t distribution† with $n-1 = 9$ degrees of freedom shows that a value of t (of either sign) as large as, or larger than 4·062 would be seen in less than 0·5 per cent of trials if the null hypothesis that the population (true) mean difference $\mu = 0$ were true, and if the assumptions of normality, etc. were true, i.e. P (two tail) < 0·005. This strongly suggests that the null hypothesis is not in fact true and that there is a real difference between the means.

This result is rather different from that found in § 9.4 and the other sections in Chapter 9, when the same results were analysed as though the drugs had been tested on independent samples, and the reasons for this are discussed in §§ 10.7, 11.2, 11.4, and 11.6. It is in reasonable agreement with the other analyses in this chapter but it cannot be assumed that the t test will always give similar results to the more assumption-free methods.

As in § 9.4, the same conclusion could be reached by calculating confidence limits for the mean difference. The 99·5 per cent confidence

† For example, Fisher and Yates (1963), Table 3, or Pearson and Hartley (1966), Table 12. Only the latter has $P = 0\cdot005$ values. See § 4.4 for details.

limits for μ would be found not to include zero, the null hypothetical value, but the 99·8 per cent confidence limits do include zero.

10.7. When will related samples (pairing) be an advantage?

In Chapter 9 the results in Table 9.2.4 and § 10.1 were analysed by several different methods and in no case was good evidence found against the hypothesis that the two drugs were equi-effective. The methods all assumed that the measurements had been made on two independent samples of ten subjects each. In § 10.1 it was revealed that in fact the measurements were paired and the same results were reanalysed making proper allowance for this in §§ 10.2–10.6. It was then found that the evidence for a difference in effectiveness of the drugs was actually quite strong. Why is this? On commonsense grounds the difference between responses to A and B is likely to be more consistent if both responses are measured on the same subject (or on members of a matched pair), than if they are measured on two different patients. This can be made a bit less intuitive if the *correlation* between the two responses from each pair is considered. The correlation coefficient, r (which is discussed in § 12.9, q.v.), is a standardized measure of the extent to which one value (say y_A) tends to be large if the other (y_B) is large (in so far as the tendency is linear, see § 12.9). It is closely related to the covariance (see § 2.6), the sample estimate of the correlation coefficient being $r = \text{cov}[y_A, y_B]/(s[y_A].s[y_B])$.

Now in § 9.4 the variance of the difference between two means was found as $s^2[\bar{y}_A - \bar{y}_B] = s^2[\bar{y}_A] + s^2[\bar{y}_B]$ using (2.7.3), which assumes that the two means are not correlated (which will be so if the samples are independent). When the samples are not independent the full equation (2.7.2) must be used, viz.

$$s^2[d] = s^2[\bar{y}_B - \bar{y}_A] = s^2[\bar{y}_A] + s^2[\bar{y}_B] - 2\,\text{cov}[\bar{y}_A, \bar{y}_B].\dagger$$

Using the correlation coefficient (r) this can be written

$$s^2[\bar{y}_A - \bar{y}_B] = s^2[\bar{y}_A] + s^2[\bar{y}_B] - 2r.s[\bar{y}_A].s[\bar{y}_B]. \qquad (10.7.1)$$

(This expression should ideally contain the *population* correlation coefficient. If an experiment is carried out on properly selected *independent* samples, this will zero, so the method given in § 9.4, which ignores r, is correct even if the *sample* correlation coefficient is not exactly zero.)

These relationships show that if there is a positive correlation between

† There are equal numbers in each sample so $(\bar{y}_A - \bar{y}_B) = (y_A - y_B) = d$.

the two responses of a pair $(r > 0)$ the variability of the difference between means will be reduced (by subtraction of the last term in (10.7.1)), as intuitively expected. In the present example $r = +0 \cdot 8$, and $s^2[\bar{y}_A - \bar{y}_B]$ is reduced from $0 \cdot 7210$ when the correlation is ignored (§§ 9.4 and 11.4), to $0 \cdot 1513$ when it is not (§§ 10.6 and 11.6). The correct value is $0 \cdot 1513$, of course.

Although correlation between observations is a useful way of looking at the problem of designing experiments so as to get the greatest possible precision, this approach does not extend easily to more than two groups and it does not make clear the exact assumptions involved in the test. The only proper approach to the problem is to make clear the exact mathematical model that is being assumed to describe the observations, and this is discussed in § 11.2.

11. The analysis of variance. How to deal with two or more samples

'. . . when I come to "Evidently" I know that it means two hours hard work at least before I can see way.'

W. S. GOSSET ('Student')
in letter dated June 1922, to R. A. Fisher

11.1. Relationship between various methods

THE methods in this chapter are intended for the comparison of two or more (k in general) groups. The methods described in Chapters 9 and 10 are special cases (for $k = 2$) of those to be described. The rationale and assumptions of the two sample methods will be made clearer during discussion of their k sample generalizations. The principles discussed in Chapter 6, although some were put in two-sample language, all apply here.

All that any of the methods will tell you is whether the null hypothesis (that all k treatments produce the same effect) is plausible or not. If this hypothesis is not acceptable, the analysis does not say anything about which treatments differ from which. Methods for comparing all possible pairs of the k treatments are described in § 11.9 and, references are given to other *multiple comparison* methods.

When the samples are independent (as in Chapter 9) the experimental design is described as a *one-way classification* because each experimental measurement is classified only according to the treatment given. Analyses are described in §§ 11.4 and 11.5. When the samples are related, as in Chapter 10, each measurement is classified according to the treatment applied and also according to the particular *block* (patient in Chapter 10) it occurs in. Such *two-way classifications* are discussed in §§ 11.6 and 11.7.

As in previous chapters nonparametric methods based on the randomization principle (see §§ 6.2, 6.3, 8.2, 8.3, 9.2, 9.3, 10.2, 10.3, and 10.4) are available for the simplest experimental designs. As usual, these experiments can also be analysed by methods involving the assumption (among others, see § 11.2) that experimental errors follow a normal (Gaussian) distribution (see § 4.2), but the nonparametric methods of §§ 11.5 and 11.7 should be used in preference to the Gaussian ones usually (see § 6.2). Unfortunately, nonparametric methods are not available (or at least not, at the moment, practicable) for analysis of the more complex and ingenious experimental designs (see § 11.8 and later chapters) that have been developed in the context of normal distribution theory. For this reason alone most of the chapters following this one will be based on the assumption of a normal distribution (see §§ 4.2 and 11.2).

When comparing two groups, the difference between their means or medians was used to measure the discrepancy between the groups. With more than two it is not obvious what measure to use and because of this it will be useful to describe the normal theory tests (in which a suitable measure is developed) before the nonparametric methods. This does not mean that the former are to be preferred. Tests of normality are discussed in § 4.6.

11.2. Assumptions involved in the analysis of variance based on the Gaussian (normal) distribution. Mathematical models for real observations

It was mentioned in § 10.7 that in order to see clearly the underlying principles of the t test (§ 9.4) and paired t test (§ 10.6) it is necessary to postulate that the observations can adequately be described by a simple mathematical model. Unfortunately it is roughly true that the more complex and ingenious the experimental design, the more complex, and less plausible, is the model (see § 11.8).

In the case of the two-sample t test (§§ 9.4 and 11.4) and its k sample analogue it is assumed that (1) the observations are normally distributed (§ 4.6), (2) the normal distributions have population means μ_j ($j = 1,2,...,k$) which may differ from group to group (e.g. μ_1 for drug A and μ_2 for drug B in § 9.4), (3) the population standard deviation, σ_j is the same for every sample (group, treatment), i.e. $\sigma_1 = \sigma_2 = ... = \sigma_k = \sigma$, say (it was mentioned in § 9.4 that it had to be assumed that the variability of responses to drug A was the same as that of responses to drug B), and (4) the responses are independent of each other (independence of responses in different groups is part of the assumption that the experiment was done on independent samples, but in addition the responses within each group must not affect each other in any way, cf. discussion in § 13.1, p. 286).

The additive model

These assumptions can be summarized by saying that the ith observation in the jth group (e.g. jth drug in § 9.4) can be represented as $y_{ij} = \mu_j + e_{ij}$ ($i = 1,2,...,n_j$, where n_j is the number of observations in the jth group—reread § 2.1 if the meaning of the subscripts is not clear). In this expression the μ_j (constants) are the population mean responses for the jth group, and e_{ij}, a random variable, is the error of the individual observation, i.e. the difference between the individual observation and the population mean. It is assumed that the e_{ij} are independent of each other and normally distributed with a population mean of zero (so *in the long run*† the mean $y_{ij} = \mu_j$) and standard deviation σ. Usually the population mean for the jth group, μ_j, is written as $\mu + \tau_j$ where μ is a constant common to all groups and τ_j is a constant (the *treatment effect*) characteristic of the jth group (treatment). The model is therefore usually written

$$y_{ij} = \mu + \tau_j + e_{ij}. \tag{11.2.1}$$

The paired t test (§§ 10.6 and 11.6), and its k sample analogue (§ 11.6), need a more elaborate model. The model must allow for the possibility that, as well as there being systematic differences between samples (groups, treatments), there may also be systematic differences between the patients in § 9.4, i.e. between *blocks* in general (see § 11.6). The analyses in § 11.6 assume that the observation on the jth sample (group, treatment) in the ith block can be represented as

$$y_{ij} = \mu + \beta_i + \tau_j + e_{ij}, \tag{11.2.2}$$

where μ is a constant common to all observations, β_i is a constant characteristic of the ith block, τ_j is a constant, as above, characteristic of the jth sample (treatment), and e_{ij} is the error, a random variable, values of which are independent of each other and are normally distributed with a mean of zero (so the long run average value of y_{ij} is $\mu + \beta_i + \tau_j$), and standard deviation σ. This model is a good deal more restrictive than (11.2.1) and its implications are worth looking at. Notice that the components are supposed to be *additive*. In the case of the example in § 10.6, this means that the *differences* between the responses of a pair (block in general) to drug A and drug B are supposed to be the same (apart from random errors) on patients who are very sensitive to the drugs (large β_i) as on patients who tend to give smaller

---

† In the notation of Appendix 1, $E(e) = 0$ so $E(y) = E(\mu_j) + E(e) = \mu_j$ from (11.2.1). And $E(y) = E(\mu) + E(\beta_i) + E(\tau_j) + E(e) = \mu + \beta_i + \tau_j$ from (11.2.2).

responses (small β_i). Likewise the *difference* in response between any two patients who receive the same treatment, and are therefore in different pairs or blocks, is supposed to be the same whether they receive a treatment (e.g. drug) producing a large observation (large τ_j) or a treatment producing a small observation (small τ_j). These remarks apply to *differences* between responses. It will not do if drug A always gives a response larger than that to drug B by a constant *percentage*, for example.

Consider the first two pairs of observations in Table 10.1.1. In the notation just defined (see also § 2.1) they are $y_{11} = +0{\cdot}7$, $y_{12} = +1{\cdot}9$, $y_{21} = -1{\cdot}6$, and $y_{22} = +0{\cdot}8$. The first difference is assumed, from (11.2.2), to be $y_{12}-y_{11} = (\mu+\beta_1+\tau_2+e_{12})-(\mu+\beta_1+\tau_1+e_{11})$ $= (\tau_2-\tau_1)+(e_{12}-e_{11}) = +1{\cdot}2$. That is to say, apart from experimental error, it measures *only* the difference between the two treatments (drugs), viz. $\tau_2-\tau_1$ whatever the value of β_1. Similarly, the second difference is $y_{22}-y_{21} = (\tau_2-\tau_1)+(e_{22}-e_{21}) = 2{\cdot}4$, which is also an estimate of exactly the same quantity, $\tau_2-\tau_1$, whatever the value of β_2, i.e. whatever the sensitivity of the patient to the drugs.

Looking at the difference in response to drug A (treatment 1) between patients (blocks) 1 and 2 shows $y_{11}-y_{21} = (\mu+\beta_1+\tau_1+e_{11})-(\mu+\beta_2+\tau_1+e_{21}) = (\beta_1-\beta_2)+(e_{11}-e_{21}) = +0{\cdot}7-(-1{\cdot}6) = +2{\cdot}3$, and similarly for drug B $y_{12}-y_{22} = (\beta_1-\beta_2)+(e_{12}-e_{22}) = 1{\cdot}9-0{\cdot}8 = 1{\cdot}1$. Apart from experimental error, both estimate only the difference between the patients, which is assumed to be the same whether the treatment is effective or not.

The best estimate, from the experimental results, of $\tau_2-\tau_1$ will be the mean difference, $d = 1{\cdot}58$ hours. If the treatment effect is not the same in all blocks then *block × treatment interactions* are said to be present, and a more complex model is needed (see below, § 11.6 and, for example, Brownlee (1956, Chapters 10, 14, and 15)).

This additive model is completely arbitrary and quite restrictive. There is no reason at all why any real observations should be represented by it. It is used because it is mathematically convenient. In the case of paired observations the addivity of the treatment effect can easily be checked graphically because the pair differences should be a measure of $\tau_2-\tau_1$, as above, and should be unaffected by whether the pair (patient in Table 10.1.1) is giving a high or a low average response. Therefore a plot of the difference, $d = y_A-y_B$, against the total, y_A+y_B, or equivalently, the mean, for each pair should be apart from random experimental errors, a straight horizontal line. This

plot, for the results in Table 10.1.1, is shown in Fig. 11.2.1. No systematic deviation from a horizontal line is detectable with the available
results but there are not enough observations to provide a good test
of the additive model. For methods of checking additivity in more
complex experiments see, for example, Bliss (1967, p. 323–41).

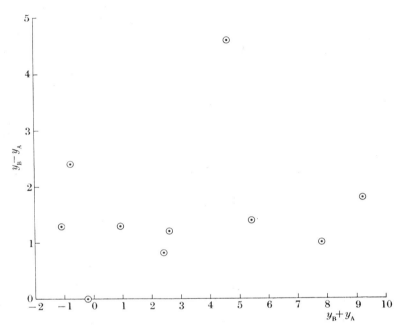

FIG. 11.2.1. Test of additive model for two way analysis of variance with
two samples (i.e. paired t test). Pair differences are plotted against pair sum
(or pair mean).

Homogeneity of error

In the models for the Gaussian analysis of variance all random errors
are pooled into the single quantity, represented by e in (11.2.1) and
(11.2.2), which is supposed to be normally distributed with a mean of
zero and a variance of σ^2. In other words, if observations could be
made repeatedly using a given treatment (e.g. drug) and block (e.g.
patient) the scatter of the results would be the same whatever the
size of the observation and whatever treatment was applied. This means
that the scatter of the observations must be the same for every group
(sample, treatment) for experiments with independent samples,
represented by (11.2.1).

13

To test whether the variance estimates calculated from each group can reasonably all be taken to be estimates of the same population value, a quick test is to calculate the ratio of the largest variance to the smallest one, s^2_{max}/s^2_{min}. This test assumes that the k samples are *independent* and that the variation within each follows a *normal* distribution. Under these conditions the distribution of s^2_{max}/s^2_{min} is known (when $k = 2$ it is the same as the variance ratio, see § 11.3). For the results in Table 11.4.1 it is seen that $k = 4$, $s^2_{max}/s^2_{min} = 158.14/34.29 = 4.61$, and each group variance is based on $7-1 = 6$ degrees of freedom. Reference to tables (e.g. Biometrika tables of Pearson and Hartley (1966 pp. 63–7 and Table 31)) shows that a value of s^2_{max}/s^2_{min} of 10.4 or larger would be expected to occur in 5 per cent of repeated experiments if the 4 independent samples of 7 observations were all from a single normally distributed population, and therefore all had the same population variance—the null hypothesis. Thus $P > 0.05$ and there are no grounds for believing that population variance is not the same for all groups, though the test is not very sensitive. The tables only deal with the case of k groups of equal size. If the sizes are not too unequal the average number of degrees of freedom can be used to get an approximate result.

Use of transformations

If the original observations do not satisfy the assumptions, some function of them may do so, although you will be lucky if you have enough observations to find out which function. Aspects of this problem are discussed in Bliss (1967 pp. 323–41), §§ 4.2, 4.5, 4.6 and 12.2.

For example, suppose the observations (1) were known to be log-normally distributed (see § 4.5) *and* (2) were represented by a multiplicative model (e.g. one treatment always giving say 50 *per cent* greater response than another, rather than a fixed increment in response) *and* (3) had standard deviations that were not constant, but which were proportional to the treatment mean (i.e. each treatment had the same coefficient of variation, eqn (2.6.4)). In this case the logarithms of the observations would be normally distributed with constant standard deviation, and would be represented by an additive model. The constancy of the standard deviation follows from eqn (2.7.14). Therefore the logarithm of each observation would be taken before doing any calculations.

If the standard deviation for each treatment group is plotted against the mean as in Fig. 11.2.2 the line should be roughly horizontal.

This can be tested rapidly using $s^2_{\max}/s^2_{\min}$ as above, given normality. If it is a straight line passing through the origin then the coefficient of variation is constant and the logs of the observations will have approximately constant standard deviation as just described. If the line is straight, but does not pass through the origin, as shown in Fig. 11.2.2, then $y_0 = a/b$ (where a is the intercept and b the slope of the

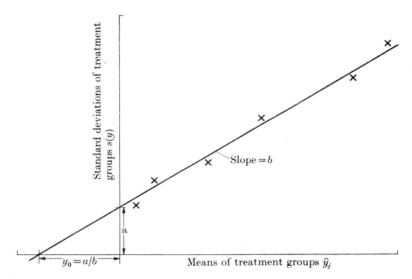

FIG. 11.2.2. Transformation of observations to a scale fulfilling the requirement of equal scatter in each treatment group. See text.

line, as shown) should be added to each observation before taking logarithms. It will now be found that $\log (y+y_0)$ has an approximately constant standard deviation, though this is no reason to suppose that this variable fulfils the other assumptions of normality and additivity.

It is quite possible that no transformation will simultaneously satisfy all the assumptions. Bartlett (1947) discusses the problem. A more advanced treatment is given by Box and Cox (1964). In the discussion of the latter paper, Nelder remarks 'Looking through the corpus of statistical writings one must be struck, I think, by how relatively little effort has been devoted to these problems [checking of assumptions]. The overwhelming preponderance of the literature consists of deductive exercises from *a priori* starting points . . . Frequently these prior assumptions are unjustifiably strong and amount to an assertion that the scale adopted will give the required additivity etc.' A good discussion is given by Bliss (1967 pp. 323–41).

What sort of model is appropriate for your experiment—fixed effects or random effects?

In the discussion above, it was stated that the values of β_i and τ_j were constants characteristic of the particular blocks (e.g. patients) and treatments (e.g. drugs) used in the experiment. This implies that when one speaks of what would happen in the long run if the experiment were repeated many times, one has in mind repetitions carried out with the same blocks and the same treatments as those used in the experiment (model 1). Obviously in the case of two drugs the repetitions would be with the same drugs, but it is not so obvious whether repetitions on the same patients are the appropriate thing to imagine. It would probably be preferred to imagine each repetition on a different set of patients, each set randomly selected from the same population, and in this case the β_i will not be constants but random variables (model 2). It is usually assumed that the β_i, as well as the e_{ij}, are normally distributed, and the variance of their distribution (ω^2, say) will then represent the variability of the population mean response for individual patients (blocks) about the population response for all patients (ω^2 is assumed to be the same for all treatments). Compare this with σ^2, which represents the variability of the individual responses of a patient about the true mean response for all responses on that patient (σ^2 is assumed to be the same for all patients and treatments).

The distinction between models based on fixed effects (model 1) and those based on random effects (model 2) will not affect the interpretation of the simple analyses described in this book *as long as the simple additive model (such as (11.2.2)) is assumed to be valid*, i.e. there are no interactions. But if interactions are present, and in more complex analyses of variance, it is essential for the interpretation of the result that the model be exactly specified because the interpretation will depend on knowing what the mean squares, which are all estimates of σ^2 when the null hypothesis is true, are estimates of when the null hypothesis is *not* true. In the language of Appendix 1, the first thing that must be known for the interpretation of the more complicated analyses of variance is the *expectations of the mean squares*. These are given, for various common analyses, by, for example, Snedecor and Cochran (1967, Chapters 10–12 and 16) or Brownlee (1965 Chapters 10, 14, and 15). See also § 11.4 (p. 186).

11.3. The distribution of the variance ratio *F*

This section describes the generalization of Student's t distribution (§ 4.4) that is necessary to extend the two-sample normal distribution-based tests (see §§ 9.4 and 10.6) to more than two (k in general) samples of observations with normally distributed errors (see §§ 11.4 and 11.6 and later chapters). The t tests of §§ 9.4 and 10.6 will be special cases ($k = 2$) of the more general methods.

In the case of the t test for two independent samples (§ 9.4), the discrepancy between the samples was measured by the difference between the sample means, and if this was large enough, compared with experimental errors, the null hypothesis that the two population means were equal, i.e. that both samples were samples from the same parent population, with variance σ^2, was rejected. If it is required to test the hypothesis that more than two samples are all from the same population, and therefore have the same mean and variance, σ^2, the first step is to find a measure of the discrepancy between the k sample means, to take the place of the simple difference that was used when $k = 2$. The usual measure of scatter in normal distribution theory is the standard deviation, and it turns out that the standard deviation of the k sample means is a suitable generalization of the difference between two means.

The sensibleness of this is made apparent when it is realized that the difference between two figures *is* their standard deviation, apart from a constant. Consider two observations, y_1 and y_2. What is their standard deviation? The variance is $\Sigma(y-\bar{y})^2/(n-1) = \Sigma(y-\bar{y})^2$ (because $n = 2$). By (2.6.5) this is $\Sigma(y-\bar{y})^2 = \Sigma y^2 - (\Sigma y)^2/n = y_1^2 + y_2^2 - (y_1+y_2)^2/2 = y_1^2 + y_2^2 - (y_1^2 + y_2^2 + 2y_1y_2)/2 = (y_1^2 + y_2^2 - 2y_1y_2)/2 = (y_1-y_2)^2/2$. The standard deviation of the two figures is the square root of this, viz.

$$s_1[y] = \frac{y_1 - y_2}{\sqrt{2}}, \qquad (11.3.1)$$

where the subscript 1 indicates that this is a standard deviation based on one ($= n-1$) degree of freedom. Now t is defined (see §§ 4.4, 9.4, and 10.6) as $(x-\mu)/s_f[x]$ where x is normally distributed, with mean μ, and sample standard deviation $s_f[x]$ based on, say, f degrees of freedom ($f = 18$ in § 9.4). If the variable of interest is $x = y_1 - y_2$, and one wishes to test the null hypothesis that the population values† of y_1 and y_2 are equal, i.e. $\mu = 0$, then one calculates $t = (y_1-y_2)/s[y_1-y_2]$. Now

---

† In the language of Appendix 1, $\mu = \mathrm{E}[y_1-y_2] = \mathrm{E}[y_1] - \mathrm{E}[y_2] = 0$.

y_1 and y_2 are assumed to be independent, and both are assumed to have the same variance, of which an estimate $s_f^2[y]$ based on f degrees of freedom is available (calculated, for example, from the variability within groups as in § 9.4 if y_1 and y_2 were group means). The estimated variance of $y_1 - y_2$ is therefore, by (2.7.3), $s^2[y_1 - y_2] = s^2[y_1] + s^2[y_2]$ $= 2s_f^2[y]$, so $s[y_1 - y_2] = \sqrt{2} s_f[y]$. Using these values gives

$$t_f = \frac{y_1 - y_2}{s[y_1 - y_2]} = \frac{y_1 - y_2}{\sqrt{2} s_f[y]}$$

$$= \frac{s_1[y]}{s_f[y]}, \text{ by } (11.3.1). \tag{11.3.2}$$

Thus, if the null hypothesis ($\mu = 0$) is true, t^2 is seen to be the ratio of two independent estimates of the same population variance σ^2 (see above), that in the numerator having one degree of freedom (in § 9.4 this was found from the difference *between* the sample means, $\bar{y}_{.1}$ and $\bar{y}_{.2}$), and that in the denominator having f degrees of freedom (in § 9.4 this was found from the differences *within* samples). Compare this approach with the discussion of predicted variances in § 2.7. In §§ 9.4 and 11.4 it is predicted from the observed scatter within samples what the scatter between samples could reasonably be expected, and the prediction compared with the observed scatter between samples.

Now the ratio of two independent estimates of the same population variance is called the *variance ratio* and is denoted F (after R. A. Fisher who discovered its distribution when the population is normally distributed). If the estimate in the numerator has f_1 degrees of freedom, and the estimate in the denominator has f_2 degrees of freedom then F is defined as

$$F(f_1, f_2) = \frac{s^2_{f_1}[y]}{s^2_{f_2}[y]}. \tag{11.3.3}$$

From (11.3.2) it is immediately seen that t^2 with f degrees of freedom is simply the special case† of the variance ratio with $f_1 = 1$ degree of freedom in the numerator, i.e. $t_f^2 \equiv s_1^2/s_f^2 \equiv F(1, f)$. Because the variance in the numerator can be found from (and used as a measure of the discrepency of) k sample means, F is the required generalization of Student's t. Numerical examples occur in §§ 11.4 and 11.6

† It is worth noting, in passing, that chi-squared distribution is another special case of the variance ratio. Since χ^2 with f degrees of freedom is the distribution of $f\, s^2/\sigma^2$ (see § 8.5) it follows that χ^2/f is simply $F(f, \infty)$, the population variance σ^2 being an estimate with ∞ d.f.

Imagine repeated samples drawn from a single population of normally distributed observations. From a sample of f_1+1 observations the sample variance, s_f^2, is calculated as an estimate of the population variance (with f_1 degrees of freedom). Another independent sample of f_2+1 observations is drawn from the same population and its sample variance, $s_{f_2}^2$, is also calculated. The ratio, F, of these two estimates of the population variance is calculated. If this process were repeated very many times the variability of the population of F values so

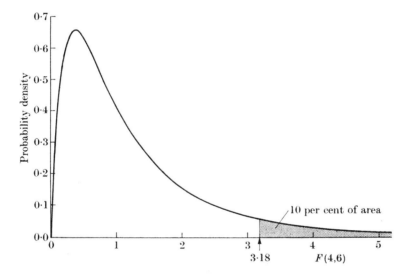

FIG. 11.3.1. Distribution of the variance ratio when there are 4 degrees of freedom for the estimate of s^2 in the numerator and 6 degrees of freedom for that in the denominator. In 10 per cent of repeated experiments in the long run, the ratio of two such estimates of the *same* variance (null hypothesis true) will be 3·18 or larger. The mode of the distribution is less than 1, and the mean is greater than 1.

produced would be described by the distribution of the variance ratio, tables of which are available.† The tables are more complicated than those for the distribution of t because both f_1 and f_2 must be specified as well as F and the corresponding P, so a three-dimensional table is needed. An example of the distribution of F for the case when $f_1 = 4$

† For example, (a) Fisher and Yates (1963), Table V. In these tables values of F are given for $P = 0.001$, 0.01, 0.05, 0.1, and 0.2 (the '0·1 per cent, etc. percentage points' of F). The degrees of freedom f_1 and f_2 are denoted n_1 and n_2, and the variance ratio F is, for largely historical reasons, called e^{2z} (the tables of z on the facing pages should be ignored). (b) Pearson and Hartley (1966), Table 18 give values of F for $P = 0.001$ 0.005, 0.01, 0.025, 0.05, 0.1, and 0.25. The degrees of freedom are denoted ν_1 and ν_2.

and $f_2 = 6$ is shown in Fig. 11.3.1. Reference to the tables shows that in 10 per cent of repeated experiments F will be 3.18 or larger, as illustrated. The distribution has a different shape for different numbers of degrees of freedom, but it is always positively skewed so mode and mean differ (see § 4.5). Since numerator and denominator are estimates of the same quantity values of F would be expected to be around one.

As in the case of χ^2 (see § 8.5), deviations from the null hypothesis in either direction tend to increase the value of F (because squaring makes all deviations positive), so the area in one tail of the F distribution, as in Fig. 11.3.1 is appropriate for a two-tail test (see § 6.1) of significance in the analysis of variance. This can be seen clearly in the case of the t test. In § 9.4 it was found that the probability that t_{18} will be either less than $-2 \cdot 101$ *or* greater than $+2.101$ was 0·05. Either of these possibilities implies that $t_{18}^2 \geqslant 2 \cdot 101^2$, i.e. $F(1,18) \geqslant 4 \cdot 41$. Reference to the tables of F with $f_1 = 1$ and $f_2 = 18$ shows that $F = 4 \cdot 41$ cuts off 5 per cent of the area in the upper tail of the distribution, the same result as the two-tail t test.

What to do if the variance ratio is less than one

When the null hypothesis is true F would be expected to be less than one quite often but the tables only deal with values of one or greater. In this case look up the reciprocal of the variance ratio which will now have f_2 degrees of freedom for the numerator and f_1 for the denominator. The resulting value of P is the probability of F equal to or *less than* the original observed value. For example, if $F(4,6) = 0 \cdot 25$, then look up $F(6,4) = 1/0 \cdot 25 = 4 \cdot 0$ with 6 d.f. for the numerator and 4 for the denominator. From the tables it is found that $P[F(6,4) \geqslant 4 \cdot 0] = 0 \cdot 1$, and therefore $P[F(4,6) \leqslant 0 \cdot 25] = 0 \cdot 1$. 10 per cent of the area lies below $F = 0 \cdot 25$ in Fig. 11.3.1. The probability required for the analysis of variance is $P[F(4,6 > 0 \cdot 25] = 1 - 0 \cdot 1 = 0 \cdot 9$. If a variance ratio is observed that is so *small* as to be very rare, it can only be assumed that either a rare event has happened, or else that the assumptions of the analysis are not fulfilled. Deviations from the null hypothesis can only result in a *large* variance ratio.

11.4. Gaussian analysis of variance for *k* independent samples (the one way analysis of variance). An illustration of the principle of the analysis of variance

The use of the variance ratio distribution to extend the method of § 9.4 to more than two samples will be illustrated using observations on the blood sugar level (mg/100ml) of 28 rabbits shown in Table 11.4.1. As usual, the rabbits are supposed to be randomly drawn from a population of rabbits, and divided into four groups in a strictly random way (see § 2.3). One of the four treatments (e.g. drug, type of diet, or environment) is assigned to each group. Is there any evidence that the treatments affect the blood sugar level? Or, in other words, do the

TABLE 11.4.1

Blood sugar level, (mg/100ml) — 100, in four groups of seven rabbits. See § 2.1 for explanation of notation. The figures in parentheses are the ranks and rank sums for use in § 11.5

	Treatment (j)				
	1	2	3	4	
	17 (10$\frac{1}{2}$)	37 (25)	35 (22$\frac{1}{2}$)	9 (5)	
	16 (9)	36 (24)	22 (15)	8 (4)	
	28 (18)	21 (13$\frac{1}{2}$)	35 (22$\frac{1}{2}$)	17 (10$\frac{1}{2}$)	
	4 (3)	13 (7)	38 (26)	18 (12)	
	21 (13$\frac{1}{2}$)	45 (28)	31 (19)	1 (2)	
	0 (1)	23 (16$\frac{1}{2}$)	34 (20$\frac{1}{2}$)	34 (20$\frac{1}{2}$)	
	23 (16$\frac{1}{2}$)	13 (7)	40 (27)	13 (7)	
Total					**Grand total**
$T._j = \sum\limits_{i=1}^{i=7} y_{ij}$ 109 (71·5)	188 (121)	235 (152·5)	100 (61)	$G = \sum\limits_{j=1}^{k} \sum\limits_{i=1}^{n_j} y_{ij}$ $= 632$
Mean					**Grand mean**
$\bar{y}._j = \dfrac{T._j}{n_j}$ 15·571	26·857	33·571	14·286	$\bar{y}.. = 632/28$ $= 22·5714$
Variance s_j^2 102·95	158·14	34·29	109·24	

four mean levels differ by more than reasonably could be expected if the null hypothesis that all 28 observations were randomly selected from the same population (so the population means are identical) were true?

The assumptions concerning normality, homogeneity of variances, and the model involved in the following analysis have been discussed in § 11.2 which should be read before this section. Although the largest group variance in Table 11.4.1 is 4·6 times the smallest, this ratio is not large enough to provide evidence against homogeneity of the variances, as shown in § 11.2. Tests of normality are discussed in § 4.6.

The nonparametric method described in § 11.5 should generally be preferred to the methods of this section (see § 6.2).

The following discussion applies to any results consisting of k independent groups, the number of observations in the jth group being n_j (the groups do not have to be of equal size). In this case $k = 4$ and all $n_j = 7$. The observation on the ith rabbit and jth treatment is denoted y_{ij}. The total and mean responses for the jth treatment are $T_{.j}$ and $y_{.j}$. See § 2.1 for explanation of the notation.

The arithmetic has been simplified by subtracting 100 from every observation. This does not alter the variances calculated in the analysis (see (2.7.7)), or the differences between means.

The problem is to find whether the means ($\bar{y}_{.j}$) differ by more than could be expected if the null hypothesis were true. There are four sample means and, as discussed in § 11.3, the extent to which they differ from each other (their scatter) can be measured by calculating the variance of the four figures.

The mean of the four figures is, in this case, the grand mean of all the observations (this is true only when the number of observations is the same in each group). The sum of squared deviations (SSD) is

$$\sum_{j=1}^{j=4} (\bar{y}_{.j} - \bar{y}_{..})^2 = (15 \cdot 571 - 22 \cdot 571)^2 + (26 \cdot 857 - 22 \cdot 571)^2 +$$
$$+ (33 \cdot 571 - 22 \cdot 571)^2 + (14 \cdot 286 - 22 \cdot 571)^2$$
$$= 257 \cdot 02.$$

The sum of squares is based on four figures, i.e. 3 degrees of freedom, so the variance of the group means, calculated directly (cf. § 2.7) from their observed scatter is

$$\frac{257}{4-1} = 85 \cdot 6733.$$

Is this figure larger than would be expected 'by chance', i.e. if the null hypothesis were true? It would be zero if all treatments had resulted in exactly the same blood sugar level, but of course this result would not be expected in an experiment, even if the null hypothesis were true. However, the result that *would* be expected can be predicted, because the null hypothesis is that all the observations come from a single Gaussian population. If the true mean and variance of this hypothetical population were μ and σ^2 then group means, which are means of 7 observations, would be predicted (from (2.7.8)) to have variance $\sigma^2/7$. If another estimate of σ^2, independent of differences between treatments, were obtainable then it would be possible to see whether this prediction was fulfilled. How this is done will be seen shortly.

With greater generality, suppose that k groups of n observations are to be compared. In the present example $k = 4$ and $n = 7$. If all the observations were from a single population with variance σ^2 then the variance of group means should be σ^2/n, so the variance of the means (85·673) in the present example), calculated directly from their observed scatter about the grand mean should be an estimate of σ^2/n, calculated from differences between groups. Multiplying through by n, it therefore follows that

$$\frac{n \sum_{j=1}^{j=k} (\bar{y}_{.j} - \bar{y}_{..})^2}{(k-1)}.$$

would be an estimate of σ^2, calculated from differences between groups, if the null hypothesis were true.

Thus in the present example $7 \times 85\cdot673 = 599\cdot71$ is an estimate of σ^2. It can be shown that if the number of observations were not equal, say n_j observations in the jth group, then this expression would become

$$\frac{\sum_{j=1}^{j=k} n_j(\bar{y}_{.j} - \bar{y}_{..})^2}{k-1}. \tag{11.4.1}$$

However, if the population means of the groups were *not* all the same, i.e. if the null hypothesis were not true, then the above expression would not be an estimate of σ^2. Its numerator would be inflated by the real differences between means so, on the average, the estimate calculated from differences *between groups* would be an estimate of something larger than σ^2. The *expectation* of the between-groups mean square will be greater than σ^2 if the null hypothesis is not true, see p. 186.

To test whether this has happened an independent estimate of σ^2, not dependent on the assumption that the true (population) group means are equal, is needed. This can be obtained from *differences within groups*. The estimate of σ^2 calculated from within the jth group— simply the estimated variance of the group—is, as usual, found by summing the squares of deviations (SSD) of the individual observations in a group from the *mean for that group*. Thus, for the jth group,

$$s^2[y] = \frac{\text{SSD within } j\text{th group}}{\text{degrees of freedom within } j\text{th group}} = \frac{\sum_{i=1}^{i=n_j}(y_{ij}-\bar{y}_{.j})^2}{n_j-1}.$$

Now, since all groups are assumed to have the same variance, the information about the variance from all groups can be pooled to give a single estimate. This is done by dividing the total sum of squares by the total degrees of freedom, as in § 9.4, so

$$s^2[y] = \frac{\sum_{j=1}^{j=k}(\text{SSD within the } j\text{th group})}{\sum_{j=1}^{j=k}(n_j-1)}$$

$$= \frac{\sum_{j=1}^{k}\sum_{i=1}^{n_j}(y_{ij}-\bar{y}_{.j})^2}{N-k} \quad (\text{using } (2.1.8) \text{ and } (2.1.7)), \quad (11.4.2)$$

where $N = \sum_{j=1}^{k} n_j$ is the total number of observations. This is the required estimate of σ^2 calculated from differences within groups. In the present example its value is $2427 \cdot 714/(28-4) = 101 \cdot 155$. An easy method of calculating the numerator is given below.

Furthermore, if all N observations were from the same population, σ^2 could be estimated from the sum of squared deviations (SSD) of all of the N ($= 28$ in this case) observations from their mean (the grand mean $\bar{y}_{..}$). Thus, using (2.6.5),

$$\text{Total SSD} = \sum_j \sum_i (y_{ij}-\bar{y}_{..})^2 = \Sigma \Sigma y_{ij}^2 - (\Sigma \Sigma y)^2/N \quad (11.4.3)$$

$$= (17^2+16^2+...+34^2+13^2) - (632)^2/28.$$

Tabulation of the results

It has been shown that (11.4.1) and (11.4.2) would both be estimates of σ^2 if the null hypothesis were true, so their ratio would follow the F distribution (see § 11.3). If there is a difference between groups then (11.4.1) will be enlarged† and the observed F would therefore be,

---

† The expectation (long-run mean, see Appendix 1) of MS_w in Table 11.4.2 is $E[MS_w]$ $= \sigma^2$, and in general $E[MS_b] = \sigma^2 + \sum_{j=1}^{k} n_j(\tau_j-\bar{\tau})^2/(k-1)$ for the fixed effect model (11.2.1) discussed in § 11.2, so if the null hypothesis, that all the τ_j are the same, is true, $E[MS_b] = \sigma^2$ also. For the random effects model (see p. 178) $E[MS_b] = \sigma^2 + n\omega^2$ when all groups are the same size (n). And the null hypothesis is that $\omega^2 = 0$, in which case again $E[MS_b] = \sigma^2$. (See Brownlee (1965, pp. 310 and 318).)

on the null hypothesis, improbably large. It is shown below that the sums of squares in the numerators of (11.4.1) and (11.4.2) add up to the total sum of squares (11.4.3). Furthermore, the number of degrees of freedom for between groups (11.4.1) comparisons, $k-1$, and that for

TABLE 11.4.2

General one-way analysis of variance. Notice that if the null hypothesis were true all the figures in the mean square column would be estimates of the same quantity (σ^2). Mean square is just another name for variance (see § 2.6). Putting in the figures in the example gives Table 11.4.3

Source of variation	d.f.	Sum of squares	Mean square (or variance)	Variance ratio	P
Between groups	$k-1$	$\sum\limits_{j}^{k} n_j(\bar{y}_{.j}-\bar{y}..)^2$	SSD/d.f. $= \mathrm{MS_b}$	$F = \mathrm{MS_b/MS_w}$	
Within groups	$N-k$	$\sum\limits_{j}^{k}\sum\limits_{i}^{n_j}(y_{ij}-\bar{y}_{.j})^2$	SSD/d.f. $= \mathrm{MS_w}$		
Total	$N-1$	$\sum\limits_{j}^{k}\sum\limits_{i}^{n_j}(y_{ij}-\bar{y}..)^2$			

TABLE 11.4.3

Analysis of the rabbit blood sugar level observations in Table 11.4.1

Source of variation	d.f.	Sum of squares	Mean square	Variance ratio	P
Between treatments	$4-1=3$	1799·143	$\dfrac{1799\cdot143}{3}=599\cdot71$	$\dfrac{599\cdot71}{101\cdot154}=5\cdot93$	$<0\cdot005$
Within treatments	$28-4=24$	2427·714	$\dfrac{2427\cdot714}{24}=101\cdot155$		
Total	$28-1=27$	4226·857			

within groups (11.4.2) comparisons, $\Sigma(n_j-1) = N-k$, add up to the total number of degrees of freedom.

These results can be written out as an *analysis of variance table*. All analysis of variance tables have the same column headings, but the sources of variation considered depend on the design of the experiment. For the one-way analysis the general result is as in Table 11.4.2.

If the null hypothesis were true 599·71 and 101·16 would both be estimates of the same variance (σ^2). Whether this is plausible or not is found by referring their ratio, 5·93, to tables of the distribution of F (see § 11.3) with $f_1 = 3$ and $f_2 = 24$ degrees of freedom. This shows (see Fig. 11.3.1) that a value of $F(3,24) = 5·93$ would be exceeded in only about 0·5 per cent of experiments in the long run (if the assumptions made are satisfied). Therefore, unless it is preferred to believe that a 1 in 200 chance has come off, the premise that both 599·71 and 101·16 are estimates of the same variances will be rejected, and this implies that all the observations cannot have come from the same population, i.e. that the treatments really differ in their effect on blood sugar level (see § 6.1).

Notice that this does not say anything about which treatments differ from which others—whether all treatments differ, or whether three are the same and one different for example. The answering of this question raises some problems, and a method of doing it is discussed in § 11.9. It is not correct to do t tests on all possible pairs of groups.

A practical method for calculating the sum of squares

A form of (11.4.1) that is more convenient for numerical computations can be derived along the same lines as (2.6.5). The SSD between groups, from (11.4.1), is

$$\sum_{j=1}^{j=k} n_j(\bar{y}_{.j} - \bar{y}_{..})^2 = \sum_{j=1}^{k} (n_j\bar{y}_{.j}^2 - 2n_j\bar{y}_{.j}\bar{y}_{..} + n_j\bar{y}_{..}^2)$$

$$= \sum_{j=1}^{k} n_j\bar{y}_{.j}^2 - 2\bar{y}_{..} \sum_{j=1}^{k} n_j\bar{y}_{.j} + \bar{y}_{..}^2 \sum_{j=1}^{k} n_j. \tag{11.4.4}$$

In this expression consider

(a) the first term: because $\bar{y}_{.j} = T_{.j}/n_j$ (see Table 11.4.1 and § 2.1) the first term can be written

$$\sum_{j=1}^{k} \left(\frac{T_{.j}^2}{n_j} \right);$$

(b) the second term: again substituting the definition of $\bar{y}_{.j}$ shows that the second term is $2\bar{y}_{..} \Sigma n_j\bar{y}_{.j} = 2\bar{y}_{..} \Sigma T_{.j} = 2\bar{y}_{..} G = 2G^2/N$, because $\sum_j T_{.j}$ = sum of group totals = grand total, G, and $\bar{y}_{..} = G/N$;

(c) the third term: because the sum of the group numbers $\sum_j n_j = N$, the total number of observations, and $\bar{y}_{..} = G/N$, the third term is $\bar{y}^2_{..} \sum_j n_j = N\bar{y}_{..}^2 = G^2/N$.

Substitution of these three results in (11.4.4) gives

$$\text{SSD between groups} = \sum_{j=1}^{k} \left(\frac{T_{.j}^2}{n_j}\right) - \frac{G^2}{N},$$

i.e., writing out the summation term by term,

$$\text{SSD between groups} = \frac{T_{.1}^2}{n_1} + \frac{T_{.2}^2}{n_2} + \ldots + \frac{T_{.k}^2}{n_k} - \frac{G^2}{N}, \qquad (11.4.5)$$

and this is the usual working formula. The formula for the total sum of squares, (2.6.5), can be regarded as the special case in which each total T contains $n = 1$ observation. In the present example

$$\text{SSD between groups} = \frac{109^2}{7} + \frac{188^2}{7} + \frac{235^2}{7} + \frac{100^2}{7} - \frac{(632)^2}{28}$$

$$= 1799 \cdot 143$$

as shown in Table 11.4.3.

The SSD within groups can now be found most easily by difference

$$\text{SSD within groups} = \text{total SSD} - \text{between-groups SSD} \qquad (11.4.6)$$

$$= 4226 \cdot 857 - 1799 \cdot 143 = 2427 \cdot 714,$$

as in Table 11.4.3.

A digression to show that sum of squares between and within groups must add up to the total sum of squares

Consider the total sum of squared deviations (SSD) of the observations about the grand mean (11.4.1), namely $\Sigma(y_{ij} - \bar{y}_{..})^2$. At first consider the SSD of the observations in the jth group about the grand mean, i.e.

$$\sum_{i=1}^{n_j} (y_{ij} - \bar{y}_{..})^2 = \sum_{i=1}^{n_j} [(y_{ij} - \bar{y}_{.j}) + (\bar{y}_{.j} - \bar{y}_{..})]^2$$

$$= \sum_{i=1}^{n_j} [(y_{ij} - \bar{y}_{.j})^2 + 2(y_{ij} - \bar{y}_{.j})(\bar{y}_{.j} - \bar{y}_{..}) + (\bar{y}_{.j} - \bar{y}_{..})^2]$$

$$= \sum_{i=1}^{n_j} (y_{ij} - \bar{y}_{.j})^2 + 2(\bar{y}_{.j} - \bar{y}_{..}) \sum_{i=1}^{n_j} (y_{ij} - \bar{y}_{.j}) + n_j(\bar{y}_{.j} - \bar{y}_{..})^2$$

$$= \sum_{i=1}^{n_j} (y_{ij} - \bar{y}_{.j})^2 + n_j(\bar{y}_{.j} - \bar{y}_{..})^2.$$

The last step in this derivation follows from (2.6.1), which shows that $\sum_i (y_{ij} - \bar{y}_{.j}) = 0$, so the central term disappears.

If the above result is summed over the k groups, the required result is obtained, thus

$$\sum_{j=1}^{k} \sum_{i=1}^{n_j} (y_{ij} - \bar{y}_{..})^2 = \sum_{j=1}^{k} \sum_{i=1}^{n_j} (y_{ij} - \bar{y}_{.j})^2 + \sum_{j=1}^{k} n_j (\bar{y}_{.j} - \bar{y}_{..})^2. \qquad (11.4.7)$$

Total SSD   = SSD within   +SSD between
                     groups   groups

Thus is a purely algebraic result and must hold for any set of numbers, but unless the observations can really be represented by the postulated model (see § 11.2) the components will not have the simple interpretation implied in the 'source of variation' column of Table 11.4.2.

The t test on the results of Cushny and Peebles written as an analysis of variance

The calculations in § 9.4 can usefully be written as an analysis of variance on the lines just described, with $k = 2$ independent groups. The results and necessary totals are given in Table 10.1.1. Again refer to § 2.1 if in doubt about the notation.

The first step is usually to calculate G^2/N as it appears several times in the calculations. This quantity is often called the *correction factor for the mean*, because, from (2.6.5), it corrects Σy^2 to $\Sigma (y - \bar{y})^2$.

From Table 10.1.1:

(a) correction factor $G^2/N = (30 \cdot 8)^2/20 = 47 \cdot 4320$; \qquad (11.4.8)

(b) total sum of squares, from (2.5.6) (cf. (11.4.3)),

$$\sum_{j=1}^{2} \sum_{i=1}^{10} (y_{ij} - \bar{y}_{..})^2 = \Sigma\Sigma y_{ij}^2 - G^2/N$$

$$= 0 \cdot 7^2 + 1 \cdot 6^2 + \ldots + 3 \cdot 4^2 - 47 \cdot 432$$

$$= 77 \cdot 3680; \qquad (11.4.9)$$

(c) sum of squares between columns (i.e. between drugs A and B), calculated from the working formula (11.4.5), is

$$\sum_{j=1}^{2} n_j (\bar{y}_{.j} - \bar{y}_{..})^2 = \frac{T_{.1}^2}{n_1} + \frac{T_{.2}^2}{n_2} - \frac{G^2}{N}$$

$$= \frac{7 \cdot 5^2}{10} + \frac{23 \cdot 3^2}{10} - 47 \cdot 432$$

$$= 12 \cdot 4820; \qquad (11.4.10)$$

and, as above, when divided by its number of degrees of freedom this would give an estimate of σ^2 if the null hypothesis (that all observations are from a single population with variance σ^2) were true;

(d)   the sum of squares within groups can now be found by difference, as in (11.4.6),

$$\sum_{j=1}^{2} \sum_{i=1}^{10} (y_{ij} - \bar{y}_{.j})^2 = 77 \cdot 3680 - 12 \cdot 4820 = 64 \cdot 8860.$$

These results can now be assembled in an analysis of variance table, Table 11.4.4, which is just like Tables 11.4.2 and 11.4.3.

TABLE 11.4.4

Source of variation	d.f.	Sum of squares	MS	F	P
Between drugs	1	12·4820	12·4820	3·463	0·1—0·05
Error (or within drugs)	18	64·8860	3·6047		
Total	19	77·3680			

Reference to tables of the distribution of F (see § 11.3) shows that, if the assumptions discussed in § 11.2 are true, a value of $F(1,18)$ equal to or greater than 3·46 would occur in between 5 and 10 per cent trials in the long run, if the null hypothesis, that the drugs are equi-effective, were true, and if the assumptions of normality, etc. were true. This is exactly the same result as found in § 9.4, and P is not small enough to reject the null hypothesis. Because there are only two groups ($k = 2$), F has one ($= k-1$) degree of freedom in the numerator, and is therefore (see § 11.3) a value of t^2. Thus $\sqrt{F} = \sqrt{(3 \cdot 463)} = 1 \cdot 861$ is a value of t with 18 d.f., and is in fact identical with the value of t found in § 9.4.

Furthermore, the error (within groups) variance of the observations, from Table 11.4.4, is estimated to be 3·605, exactly the same as the pooled estimate of $s^2(y)$ found in § 9.4. Table 11.4.4 is just another way of writing the calculations of § 9.4.

11.5. Nonparametric analysis of variance for independent samples by randomization. The Kruskal–Wallis method

As suggested in §§ 6.2 and 11.1, the methods in this section are to be preferred to that of § 11.4.

The randomization method

The randomization method of § 9.2 is easily extended to k samples and this is the preferred method, because it makes fewer assumptions than Gaussian methods. The disadvantage, as in § 9.2, is that tables cannot be prepared to facilitate the test, which will therefore be tedious (though simple) to calculate without an electronic computer. As in § 9.3, this can be overcome, with only a small loss in sensitivity, by replacing the original observations by their ranks, giving the Kruskal–Wallis method described below.

The principle of the randomization method is exactly as in §§ 9.2, 9.3, and 8.2 so the arguments will not all be repeated. If all four treatments were equi-effective then the observed differences between group means in Table 11.4.1, for example, must be due solely to the way the random numbers come up in the process of random allocation of a treatment to each rabbit. Whether such large (or larger) differences between treatments means are likely to have arisen in this way is again found by finding the differences between the treatment means that result from all possible ways of dividing the 28 observed figures into 4 groups of 7. On the assumption that, when doing the experiment, all ways were equiprobable, i.e. that the treatments were allocated strictly at random, the value of P is once again simply the proportion of possible allocations that give rise to discrepancies between treatment means as large as (or larger than) the observed differences. In § 9.2 the discrepancy between two means was measured by the difference between them. As explained in §§ 11.3 and 11.4, when there are more than two (say k) means, an appropriate thing to do is to measure their discrepancy by the variance of the k figures, i.e. by calculating, for each possible allocation the 'between treatments sum of squares' as described in § 11.4.

An approximation to the answer could be obtained by card shuffling, as in § 8.2. The 28 observations from Table 11.4.1 would be written on cards. The cards would then be shuffled, dealt into four groups of seven, and the 'between treatments sum of squares' calculated. This would be repeated until a reasonable estimate was obtained of the proportion (P) of shufflings giving a sum of squares equal to or larger than the value observed in the experiment. In fact, just as in § 9.2 it was found to be sufficient to calculate the total response for the smaller group, so, in this case, it is sufficient to calculate $\Sigma T_{.j}^{2}/n_j$ for each possible allocation, because once this is known the between treatments sum of squares, or between-treatments F ratio, follows from the fact that the total sum of squares is the same for every possible allocation.

By a slight extension of (3.4.3), the number of possible allocations of N objects into k groups of size $n_1, n_2, ...n_k$, $(\Sigma n_j = N)$ is $N!/(n_1!n_2!...n_k!)$.

In the case of Table 11.4.1 there are therefore $28!/(7!7!7!7!)$ $= 472518347558400$ possible allocations. This is rather too many to enumerate by hand (doing one every 5 minutes it would take about 20 thousand million normal working years), though it is easy to select a sufficiently large random sample of them with an electronic computer.

If a program for this is not available, the recommended procedure for analysis of k independent samples is, just as in § 9.3, to replace the observations by ranks allowing tables to be constructed. This is known as the Kruskal–Wallis method.

The k sample randomization test on ranks. The Kruskal–Wallis one-way analysis of variance

This is simply an extension to more than two (k, say) groups of the Wilcoxon two-sample test (see above, and § 9.3). As before, the null hypothesis is that all N observations are from the same population, and if this is rejected the conclusion will be that the populations differ. If it is wished to conclude that the population medians differ then it must be assumed that the underlying distributions (before ranking) for all groups are the same apart from the medians, though no particular

TABLE 11.5.1

	Treatment				
A		B		C	
Score	Rank	Score	Rank	Score	Rank
7	4	1	1	17	9
15	8	5	2	31	11
12	6	6	3	14	7
8	5			22	10
Totals $R_1 = 23$		$R_2 = 6$		$R_3 = 37$	

form of distribution is assumed except that it must be continuous (see § 4.1). This implies that the variance is assumed to be the same for all groups.

Again the method can be applied when the observations are themselves not numerical measurements but ranks, or arbitrary scores that must be reduced to ranks, as well as when they are numerical measurements.

All N observations are ranked in ascending order, ties being given the average rank as in § 9.3 (see Table 9.3.1). Table 11.5.1 shows the results of an experiment in which 11 patients were divided randomly into $k = 3$ groups, each being given a different analgesic drug (A, B, and C). In each group the figure recorded is the total subjective pain score recorded over a period of time by each patient. Such measurements should not be treated like numerical measurements but should be ranked. The ranks are shown in the table together with the rank sum, R_j, for each group ($j = 1,2,...k,$). The number of observations in each group is n_j and the total number is $N = \Sigma n_j$ as in § 11.4.

The measure of the extent to which the treatments differ, analogous to the rank sum of the smaller sample used in § 9.3, is the statistic H defined as

$$H = \frac{12}{N(N+1)}\sum_{j=1}^{k}\frac{R_j^2}{n_j} - 3(N+1) \qquad (11.5.1)$$

as long as there are not too many ties (see below). Notice that the term $\Sigma R_j^2/n_j$ makes H similar in character to a between-groups sum of squares (11.4.5). For the results in Table 11.5.1, $N = 11$, $n_1 = 4$, $n_2 = 3$, $n_3 = 4$, $R_1 = 23$, $R_2 = 6$, and $R_3 = 37$. Applying the check (9.3.1) gives the sum of all ranks as $N(N+1)/2 = 66$ and in fact $23 + 6 + 37 = 66$. Using these values with (11.5.1) gives

$$H = \frac{12}{11(11+1)}\left(\frac{23^2}{4} + \frac{6^2}{3} + \frac{37^2}{4}\right) - 3(11+1) = 8{\cdot}277.$$

Table A5 gives the exact distribution of H found, as in § 9.3, by the randomization method. It shows that for sample sizes 4, 4, and 3 (the order of these figures is irrelevant) a value of $H \geqslant 7{\cdot}1439$ would occur in 1 per cent of trials ($P = 0{\cdot}01$) in the long run *if* the null hypothesis were true, therefore $H = 8{\cdot}227$ must be even rarer, i.e. $P < 0{\cdot}01$. As in § 11.3 deviations from the null hypothesis in any direction increase the size of H. Again, as in all analyses of variance, this result does not give any information about differences between individual pairs of groups (see § 11.9).

Example with larger samples. Table A5 only deals with $k = 3$ groups with not more than 5 observations in any group. For larger experiments it is a sufficiently good approximation to assume that H is distributed like chi-squared with $k-1$ degrees of freedom. P can then be found from the chi-squared tables (see § 8.5). For example, the results in Table 11.4.1 have been converted to ranks, shown in parentheses. In this case $N = 28$, $n_1 = n_2 = n_3 = n_4 = 7$, $R_1 = 71{\cdot}5$, $R_2 = 121$, $R_3 = 152{\cdot}5$, and $R_4 = 61$. Applying the check $N(N+1)/2$

$= 28(28+1)/2 = 406$ and, correctly, $R_1+R_2+R_3+R_4 = 406$. Thus, from (11.5.1),

$$H = \frac{12}{28(28+1)}\left(\frac{71 \cdot 5^2}{7}+\frac{121^2}{7}+\frac{152 \cdot 5^2}{7}+\frac{61^2}{7}\right)-3(28+1) = 11 \cdot 66.$$

Consulting a table of the chi-squared distribution (Fisher and Yates (1963, Table IV) or Pearson and Hartley (1966, Table 8)) with $k-1 = 3$ degrees of freedom shows that $\chi^2_{(3)} = 11 \cdot 345$ would be exceeded in 1 per cent of experiments in the long run if the null hypothesis were true, so $P < 0 \cdot 01$ for the observed value of $11 \cdot 66$. This is somewhat larger than the value of $P < 0 \cdot 005$ found when the assumptions of the Gaussian analysis of variance were thought justified (Table 11.4.3), but it is still small enough to cast considerable doubt on the null hypothesis.

As in the Gaussian analysis, the finding that all k groups are unlikely to be the same says nothing about which groups differ from which others. A method for testing all pairs of groups to answer this question is described in § 11.9. It is not correct to do two-sample Wilcoxon tests on all possible pairs.

Correction for ties. Unless there is a very large number of ties the correction factor (described, for example, by Brownlee (1965, p. 256)) has a negligible effect. It always makes H larger and hence P smaller so there is no danger that neglecting the correction factor will lead to rejection of the null hypothesis when it would otherwise not have been rejected.

11.6. Randomized block designs. Gaussian analysis of variance for *k* related samples (the two-way analysis of variance)

In §§ 10.1 and 6.4 it was pointed out that if the experimental units (e.g. patients, periods of time) can be selected in some way to form groups that give more homogeneous and reproducible responses than units selected at random, then it will be advantageous if all the treatments (k in number, say) to be compared, are compared on the units of such a group. The group is known as a *block*. The units comprising the block are sometimes, because of the agricultural origins of the design, known as *plots*. It must clearly contain as many (k) experimental units as there are treatments, or at least a multiple of k. The k treatments must be allocated strictly randomly (see § 2.3) to the k units of each block. Because every treatment is tested in every block the blocks are described as *complete* (cf. § 11.8). This section deals with randomized complete block experiments when the observations are described by the single additive model with normally distributed error (11.2.2) described in § 11.2, which should be read before this section.

The analysis in § 10.6, Student's paired t test, was an example of a randomized block experiment with $k = 2$ treatments and 2 units (periods of time) in each block (patient). This test will now be reformulated as an analysis of variance.

The paired t test written as an analysis of variance

The observations of Cushney and Peebles in Table 10.1.1 were analysed by a paired t test in § 10.6. At the end of § 11.4 it was shown how a sum of squares for differences between drugs could be obtained, but no account was taken of the block arrangement actually used in the experiment. As before (see §§ 11.3 and 11.4) the calculations are such that *if* the null hypothesis (that all observations are from the same Gaussian population with variance σ^2) were true, *then* the quantities in the mean-square column of the analysis of variance would all be independent estimates of σ^2.†

Because of the symmetry of the design it is possible to obtain an estimate of σ^2, on the null hypothesis that there is no real difference between blocks (patients), by considering deviations of block means from the grand mean, i.e. by analogy with (11.4.1), from $\sum_i k(\bar{y}_{i.} - \bar{y}_{..})^2/$

$(n-1)$, where $k = $ number of treatments $= $ number of observations per block, and $n = $ number of blocks $= $ number of observations on each treatment. Unlike the one-way analysis, n must be the same for all treatments. $N = kn$ is the total number of observations. From (11.4.5), it can be seen that the numerator of this (sum of squares between blocks) can most simply be calculated from the block totals, as the sum of squares between treatments was found from treatment totals in (11.4.5). From the results in Table 10.1.1

$$\text{SSD between blocks} = \sum_{i=1}^{i=n} \frac{T_{i.}^2}{k} - \frac{G^2}{N} \tag{11.6.1}$$

$$= \frac{2 \cdot 6^2}{2} + \frac{0 \cdot 8^2}{2} + \dots + \frac{5 \cdot 4^2}{2} - 47 \cdot 4320$$

$$= 58 \cdot 0780.$$

---

† The expected value of the mean squares (see § 11.2, p. 173 and 11.4, p. 186) are derived by Brownlee (1965, Chapter 14). Often the mixed model in which treatments are fixed effects and blocks are random effects is appropriate (loc. cit., p. 498).

In this, the group (row, block) totals, $T_{i.}$, are squared and divided by the number of observations per total just as in (11.4.5). See §§ 2.1 and 11.4 if clarification of the notation is needed. Since there $n = 10$ groups (rows, blocks) this sum of squares has $n-1 = 9$ degrees of freedom.

The values of $G^2/N(47\cdot4320)$, and of the sum of squares between drugs (treatments, columns) (12·4820) and the total sum of squares (77·3680), are found exactly as in (11.4.8)–(11.4.10). The results are assembled in Table 11.6.1. The residual or error sum of squares is again

TABLE 11.6.1

The paired t test of § 10.5 written as an analysis of variance. The mean squares are found by dividing the sum of squares by their d.f. The F ratios are the ratio of each mean square to the error mean square

Source of variation	d.f.	Sum of squares	Mean square	F	P
Between treatments (drugs)	1	12·4820	12·4820	16·5009	0·001 − 0·005
Between blocks (patients)	9	58·0780	6·4531	8·531	0·001 − 0·005
Error	9	6·8080	0·7564		
Total	19	77·3680			

found by difference ($77\cdot3680 - 58\cdot0780 - 12\cdot4820 = 6\cdot8080$), and so is its number of degrees of freedom ($19 - 9 - 1 = 9$). The error mean square will be an estimate of the variance of the observations, σ^2, after the elimination of variability due to differences between treatments (drugs) and blocks (patients), i.e. the variance the observations would have because of sources of experimental variability if there were no such differences. This is only true if there are no interactions and the simple additive model (11.2.2) represents the observations. The other mean squares would also be an estimate of σ^2 if the null hypothesis were true, and therefore, when the null hypothesis is true the ratio of each mean square to the error mean square should be distributed like the variance ratio (see § 11.3). If the size of the F ratio is so large is to make its observance a very rare event, the null hypothesis will be abandoned in favour of the idea that the numerator of the ratio has been inflated by real differences between treatments (or blocks).

The variance ratio for testing differences between drugs is 16·5009 with one d.f. in the numerator and 9 in the denominator. Reference to tables of the F distribution (see § 11.3) shows that $F(1,9) = 13\cdot61$

would be exceeded in 0·5 per cent, and $F(1,9) = 22·86$ in 0·1 per cent of trials in the long run, if the null hypothesis were true. The observed F falls between these figures so $0·001 < P < 0·005$, just as in § 10.5.

As pointed out in § 11.3 (and exemplified in § 11.4), F with 1 d.f. for the numerator is just a value of t^2 so $\sqrt{[F(1,9)]} = \sqrt{(16·5009)}$ $= 4·062 = t(9)$, a value of t with 9 d.f.—exactly the same value, as was found in the paired t test (§ 10.6). Furthermore, the error variance from Table 11.6.1 is 0·7564 with 9 d.f. The error variance of the difference between two observations should therefore, by (2.7.3), be $0·7564 + 0·7564 = 1·513$, which is exactly the figure estimated directly in § 10.6.

If there were no real differences between blocks (patients) then 6·453 would be an estimate of the same σ^2 as the error mean square 0·756. Referring this ratio (8·531) to Tables (see § 11.3) of the distribution of the F ratio (with $f_1 = 9$ d.f. for the numerator and $f_2 = 9$ d.f. for the denominator) shows that the probability of an F ratio at least as large as 8·531 would be between 0·001 and 0·005, if the null hypothesis were true.

This analysis, and that in § 11.4, show clearly why t had 18 d.f. in the unpaired t test (§ 9.4) but only 9 d.f. in the paired t test (§ 10.6). In the latter, 9 d.f. were used up by comparisons between patients (blocks).

There is quite strong evidence (given the assumptions in § 11.2) that there are real differences between the treatments (drugs), as concluded in § 10.5. This is because an F ratio, i.e. difference between treatments relative to experimental error, as large as, or larger than, that observed (16·532) would be rare if there were no real (population) difference between the treatments (see §§ 6.1 and 11.3). Similarly, there is evidence of differences between blocks (patients).

An example of a randomized block experiment with four treatments

The following results are from an experiment designed to show whether the response (weal size) to intradermal injection of antibody followed by antigen depends on the method of preparation of the antibody. Four different preparations (A, B, C, and D) were tested. Each preparation was injected once into each of four guinea pigs. The preparation to be given to each of the four sites on each animal was decided strictly at random (see § 2.3). Guinea pigs are therefore blocks in the sense described above. The results are in Table 11.6.2. (This is actually an artificial example. The figures are taken from Table 13.11.1 to illustrate the analysis.)

TABLE 11.6.2

Weal diameters using four antibody preparations in guinea pigs

Guinea pig (block)	Antibody preparation (treatment)				Totals $(T_{i\cdot})$
	A	B	C	D	
1	41	61	62	43	207
2	48	68	62	48	266
3	53	70	66	53	242
4	56	72	70	52	250
Total $(T_{\cdot j})$	198	271	260	196	$G = 925$
Mean	49·5	67·7	65·0	49·0	

The calculations follow exactly the same pattern as above.

(1) *'Correction factor'* (see (11.4.8)). $\dfrac{G^2}{N} = \dfrac{(925)^2}{16} = 53\ 476\cdot5625.$

(2) *Between antibody preparations* (*treatments*). From (11.4.5),

$$\text{SSD} = \frac{198^2}{4} + \frac{271^2}{4} + \frac{260^2}{4} + \frac{196^2}{4} - 53\ 476\cdot5625 = 1188\cdot6875.$$

(3) *Between guinea pigs* (*blocks*). From (11.6.1),

$$\text{SSD} = \frac{207^2}{4} + \frac{226^2}{4} + \frac{242^2}{4} + \frac{250^2}{4} - 53\ 476\cdot5625 = 270\cdot6875.$$

(4) *Total sum of squared deviations.* From (2.6.5) (or (11.4.3)),

$$\text{SSD} = 41^2 + 48^2 + \ldots + 53^2 + 52^2 - 53\ 476\cdot5625 = 1492\cdot4375.$$

(5) *Error sum of squares found by difference.*

$$\text{SSD} = 1492\cdot4375 - (1188\cdot69 + 270\cdot687) = 33\cdot0625.$$

There are 3 d.f. for blocks and treatments (because there are 4 blocks and 4 treatments) and the total number of d.f. is $N-1 = 15$, so, by difference, there are $15-(3+3) = 9$ d.f. for error. These results are assembled in Table 11.6.3. Comparison of each mean square with the error mean square gives variance ratios (both with $f_1 = 3$ and $f_2 = 9$ d.f.) which, according to tables of the F distribution (see § 11.3), would be very rare if the null hypothesis were true. It is concluded that there is evidence for real differences between different antibody preparations, and between different animals, for the same reasons as in the previous example.

TABLE 11.6.3

Source of variation	d.f.	Sum of squares	Mean square	F	P
Between antibody preps. (treatments)	3	1188·6875	396·229	107·86	< 0·001
Between guinea pigs (blocks)	3	270·6875	90·229	24·56	< 0·001
Error	9	33·0625	3·674		
Total	15	1492·4375			

Multiple comparisons. As in §§ 11.4 and 11.5, if it is wished to go further and decide which antibody preparations differ from which others the method described in § 11.9 must be used. It is *not* correct to do paired t tests on all possible pairs of treatments.

11.7. Nonparametric analysis of variance for randomized blocks. The Friedman method

Just as in §§ 9.2, 10.3, and 11.5 the best method of analysis is to apply the randomization method to the original observations. The principles of the method have been discussed in §§ 6.3, 8.2, 9.2, 9.3, 10.3, 10.4, and 11.5. Reasons for preferring this sort of test are discussed in § 6.2. As before, the drawback of the method is that tables cannot be prepared to the calculation will be tedious without a computer, though they are very simple; and as before, this disadvantage can be overcome by using ranks in place of the original observations (the Friedman method).

The randomization method

The argument, simply an extension to more than two samples of that in § 10.3, is again that if the treatments were all equi-effective each subject would have given the same measurement whichever treatment had been administered (see p. 117 for details), so the observed difference between treatments would be merely a result of the way the random numbers came up when allocating the k treatments to the k units in each block (see § 11.6). There are $k!$ possible ways (permutations) of administering the treatments in each block, so if there are n

blocks there are $(k!)^n$ ways in which the randomization could come out (an extension to k treatments of the 2^n ways found in § 10.3). If the randomization was done properly these would all be equi-probable, and if the F ratio for 'between treatments' is calculated for all of these ways (cf. § 11.5), the proportion of cases in which F is equal to or larger than the observed value is the required P, as in § 10.3. As in § 11.5 it will give the same result if the sum of squared treatment totals, rather than F, is calculated for each arrangement.

As in previous cases an approximation to this result could be obtained by writing down the observations on cards. The cards for each block would be separately shuffled and dealt. The first card in each block would be labelled treatment A, the second treatment B, and so on. If this process were repeated many times, an estimate of the proportion (P) of cases giving 'between treatments' F ratios as large as, or larger than the observed ratio, could be found. If this proportion was small it would indicate that it was improbable that a random allocation would give rise to the observed result, if the observation was not dependent on the treatment given. In other words, an allocation that happened to put the same treatment group subjects that were, despite any treatment, going to give a large observation, would be unlikely to turn up.

The analysis of randomized blocks by ranks. The Friedman method

If the observations are replaced by ranks, tables can be constructed to make the randomization test very simple.

The null hypothesis is that the observations in each block are all from the same population, and if this is rejected it will be supposed that the observations in any given block are not all from the same popula-tion, because the treatments differ in their effects. If it is wished to conclude that the median effects of the treatments differ it must be assumed that the underlying distributions (before ranking) of the observations is the same for observations in any given block, though the form of the distribution need not be known, and it need not be the same for different blocks.

As in the case of the sign test for two treatments (§ 10.2) the observa-tions within each block (pair, in the case of the sign test) are ranked. If the observations in each block are themselves not proper measure-ments but ranks, or arbitrary scores which must be reduced to ranks, the Friedman method, like the sign test, is still applicable. In fact the Friedman method, in the special case of $k = 2$ treatments, becomes identical with the sign test. Compare this with the Wilcoxon signed-ranks test (§ 10.4) in which proper numerical measurements were necessary because differences had to be formed between members of a pair before the differences could be ranked.

Suppose, as in § 11.6, that k treatments are compared, in random order, in n blocks. The method is to rank the k observations in each block from 1 to k, if the observations are not already ranks. The rank totals, R_j (see §§ 2.1, 11.4, and 11.5 for notation), are then found for each treatment. If there were no difference between treatments these totals would be approximately the same for all treatments. The sum of the ranks (integers 1 to k) in each block should be $k(k+1)/2$, by (9.3.1), and because there are n blocks the sum of the rank sums should be

$$\sum_{j=1}^{k} R_j = nk(k+1)/2. \tag{11.7.1}$$

As a measure of the discrepancy between rank sums now simply calculate S, the sum of squared deviations of the rank sums for each treatment from their mean (cf. (11.4.1)). From (2.6.5), this is

$$S = \Sigma(R - \bar{R})^2 = \sum_{j=1}^{k} R_j^2 - \frac{(\Sigma R_j)^2}{k}. \tag{11.7.2}$$

The exact distribution of this quantity, calculated according to the randomization method—see sections referred to at start of this section, is given in Table A6, for various numbers of treatments and blocks. For experiments with more treatments or blocks than are dealt with in Table A6, it is a sufficiently good approximation to calculate

$$\chi^2_{rank} = \frac{12S}{nk(k+1)} \tag{11.7.3}$$

and find P from tables of the chi-squared distribution (e.g. Fisher and Yates (1963, Table IV) or Pearson and Hartley (1966, Table 8)) with $k-1$ degrees of freedom.

As an example, consider the results in Table 11.6.2, with $k = 4$ treatments and $n = 4$ blocks. If the observations in each block (row) are ranked in ascending order from 1 to 4, the results are as shown in Table 11.7.1. Ties are given average ranks as in Table 9.3.1. This is an approximation but it is not thought to affect the result seriously if the number of ties is not too large.

Applying the check (11.7.1), shows that ΣR_j should be $4.4.(4+1)/2 = 40$, as found in Table 11.7.1. Now calculate, from (11.7.2),

$$S = 6^2 + 15^2 + 13^2 + 6^2 - \frac{(40)^2}{4} = 66 \cdot 00.$$

Consulting Table A6 shows that when $k = 4$ and $n = 4$, $S = 64$ corresponds to $P = 0 \cdot 0069$. So the observed $S = 66$ corresponds to $P < 0 \cdot 0069$. This means that *if* the treatments were equi-effective (null hypothesis) then in less than $0 \cdot 69$ per cent of experiments in the

TABLE 11.7.1

The observations within each block in Table 11.6.2 *reduced to ranks*

Guinea pig (block)	(Antibody preparation (treatment)			
	A	B	C	D
1	1	3	4	2
2	$1\frac{1}{2}$	4	3	$1\frac{1}{2}$
3	$1\frac{1}{2}$	4	3	$1\frac{1}{2}$
4	2	4	3	1
Rank sum(R_j)	$R_1 = 6$	$R_2 = 15$	$R_3 = 13$	$R_4 = 6$ $\Sigma R_j = 40$

long run would a random allocation of treatments to the units of each block be chosen that gave differences between treatment rank sums (i.e. a value of S) as large as, or larger than, that observed ($S = 66$). The null hypothesis of equi-effectiveness is therefore rejected, though not with as much confidence as when the same results were analysed by the Gaussian analysis of variance in § 11.6. In Table 11.6.3, it was seen that if the assumptions made (see § 11.2) were correct, $P \ll 0 \cdot 001$, was much lower than found by the present method.

If the experiment had been outside the scope of Table A6 then (11.7.3) would have been used giving $\chi^2_{rank} = 12.66/4.4(4+1) = 9 \cdot 90$. Consulting tables of the chi-squared distribution (see above) with $k-1 = 3$ degrees of freedom shows that a value of $9 \cdot 837$ would be exceeded in 2 per cent of experiments in the long run so $P \simeq 0 \cdot 02$. Not a very good approximation, in such small samples, to the exact value of P (just less than $0 \cdot 0069$) found from Table A6.

If it were of interest to find out whether there was a difference between blocks, exactly the same method would be used (e.g. interchange the words block and treatment throughout this section).

Multiple comparisons. As in §§ 11.4–11.6, the conclusion that the treatments do not all have the same effect says nothing about which ones differ from which others. It would not be correct to perform sign tests on all possible pairs of treatment groups, in order to find out, for example, whether treatment B differs from treatment D. A method of answering this question is given in § 11.9.

11.8. The Latin square and more complex designs for experiments

There is a vast literature describing ingenious designs for experiments but the analysis of almost all of these depends on the assumption of a normal distribution of errors and on elaborations of the models described in § 11.2. If the experiments are large there is, in some cases, some evidence that the methods will not be very sensitive to the assumptions. As the assumptions are rarely checkable with the amount of data available it may be as well to treat these more complex designs with caution (see comments below about use of small Latin squares). Certainly if they are used the advice of a *critical* professional statistician should be sought about the exact nature of the assumptions being made, and the interpretation of the results in the light of the mathematical model (see § 11.2).

To emphasize the point it should be sufficient to quote Kendall and Stuart (1966, p. 139): 'The fact that the evidence for the validity of normal theory tests in randomized Latin squares is flimsy, together with the even greater paucity of such evidence for most other, more complicated, experiment designs, leads one to doubt the prevailing serene assumption that randomization theory will always approximate normal theory.'

The Latin square

The experiment summarized in Table 11.6.2, was actually arranged so that each of the four injection sites (e.g. anterior and posterior on each side), received every treatment once, according to the design shown in Table 11.8.1(a). The measurements, from Table 11.6.2, are given in Table 11.8.1(b).

In the randomized block design (§ 11.6) each treatment appeared once, in random order, in each block (row). In the design shown in Table 11.8.1, which is called a Latin square, there is the additional restriction that each treatment appears once in each column so that the column totals are comparable. The number of columns (injection sites) as well as the number of blocks (rows, guinea pigs) must be the same as (or a multiple of) the number of treatments. If a model like (11.2.2), but with another additive component characteristics of each column (injection site), is supposed to represent the real observations, then a sum of squares (see §§ 11.2, 11.3, 11.4, and 11.6) can be found from the observed scatter of the column totals (the corresponding mean square, would, as usual, estimate σ^2, if the null hypothesis were true), and used

TABLE 11.8.1

The Latin square design

Row	Column 1	2	3	4
1	A	B	C	D
2	C	A	D	B
3	D	C	B	A
4	B	D	A	C

(a)

Guinea pig	Injection site 1	2	3	4	Total
1	41	61	62	43	207
2	62	48	48	68	226
3	53	66	70	53	242
4	72	52	56	70	250
Total	228	227	236	234	925

(b)

in the Gaussian analysis of variance to eliminate errors due to systematic differences between columns (injection sites). The sum of squares is found from column totals, and number of observations per total (4 in this case), using (11.4.5) again.

SSD between injection sites (columns)

$$= \frac{228^2}{4}+\frac{227^2}{4}+\frac{236^2}{4}+\frac{234^2}{4}-53\ 476\cdot5625=14\cdot6875 \quad (11.8.1)$$

TABLE 11.8.2

Analysis of variance for the Latin square

Source of variation	d.f.	Sums of squares	Mean square	F	P
Between antibody preparation (treatments)	3	1188·6875	396·23	129·4	<0·001
Between guinea pigs (rows)	3	270·6875	90·23	29·5	<0·001
Between sites (columns)	3	14·6875	4·89	1·6	>0·2
Error	6	18·3750	3·06		
Total	15	1492·4375			

with 3 degrees of freedom (because there are 4 columns). The sums of squares for differences between treatments and between guinea pigs, and the total sum of squares, are exactly as in § 11.6. When these results are filled into Table 11.8.2, the error sum of squares and degrees of freedom can be found by difference, and the rest of the table completed as in § 11.6. Referring the variance ratio, $F(3,6) = 1\cdot6$, to tables (see § 11.3) shows that there is no evidence for a population difference between injection sites ($P > 0\cdot2$).

Choosing a Latin square at random

As usual it is essential that the treatments be applied randomly, given the restraints of the design. This means the design, Table 11.8.1 (a), actually used in the experiment had the same chance of being chosen as each of the 575 other possible 4×4 Latin squares. The selection of a square at random is not as straightforward as it might appear at first sight and is frequently not done correctly. Fisher and Yates (1963) give a catalogue of Latin squares (Table 25), and instructions for choosing a square at random (introduction, p. 24).

Are Latin squares reliable?

The answer is that *if* the assumptions of the mathematical model are true, then they are an excellent way of eliminating experimental errors to two sorts (e.g., guinea pigs and injection sites) from the comparisons between treatments which is of primary interest. However, as usual, it is very rare that there is any information about whether the model is correct or not. In the case of the *t* test the Gaussian approach could be justified because it has been shown to be a good approximation to the randomization method (§ 9.2) if the samples are large enough. However, there is much less information on the sensitivity of Latin squares (and more complex designs) to departures from the assumptions. In the case of the 4×4 Latin square the randomization method does *not*, in general, give results in agreement with the Gaussian analysis so one is totally reliant on the assumptions of the latter being sufficiently nearly true. It is thus doubtful whether Latin squares as small as 4×4 should be used in most circumstances, though the larger squares are thought to be safer (see Kempthorne 1952).

Incomplete block designs

In § 11.6, the randomized block method was described for eliminating errors due to differences between blocks, from comparisons between treatments. Sometimes it may not be possible to test every treatment on every block as when, for example, four treatments are to be compared, but each patient = block is only available for long enough to receive two. It is sometimes still possible to eliminate differences between blocks even when each block does not contain every treatment. Catalogues of designs are given by Fisher and Yates (1963, pp. 25, 91–3) and Cochran and Cox (1957).

A nonparametric analysis of incomplete block experiments has been given by Durbin (1951).

Examples of the use of balanced incomplete block designs for biological assays (see § 13.1) have been given by, for example, Bliss (1947) and Finney (1964). General formulas for the simplest analysis of biological assays based on balanced incomplete blocks are given by Colquhoun (1963).

11.9. Data snooping. The problem of multiple comparisons

In all forms of analysis of variance discussed, it has been seen that all that can be inferred is whether or not it is plausible that all of the k treatments (or blocks, etc.) are really identical. If there are more than two treatments the question of which ones differ from which others is *not* answered. The obvious answer is never to bother with the analysis of variance but to test all possible pairs of treatments by the two sample methods of Chapters 9 and 10. However, it must be remembered that it is expected that the null hypothesis will sometimes be rejected even when it is true (see § 6.1), so if a large number of tests are done some will give the wrong answer. In particular, if several treatments are tested and the results inspected for possible differences between means, and the likely looking pairs tested ('data selection', or as statisticians often call it 'data snooping'), the P value obtained will be quite wrong.

This is made obvious by considering an extreme example. Imagine that sets of, say, 100 samples are drawn repeatedly from the *same* population (i.e. null hypothesis true), and each time the sample out of the set of 100 with largest mean is tested, using a two-sample test, against the sample with the smallest mean. With 100 samples the largest mean is likely to be so different from the smallest that the null hypothesis (that they come from the same population) would be rejected (wrongly) almost every time the experiment was repeated, not only in 1 or 5 per cent (according to what value of P is chosen as low enough to reject the null hypothesis) of repeated experiments as it should be (see § 6.1). If the particular treatments to be compared are not chosen *before* the results are seen, allowance must be made for data snooping. There are various approaches.

One way is to compare all possible pairs of treatments. This is probably the most generally useful, and methods of doing it for both nonparametric and Gaussian analysis of variance are described below.

Another case arises when one of the treatments is a control and it is required to test the difference between each of the other treatment means and the control mean. In the Gaussian analysis of variance this is done by finding confidence intervals for the jth difference as *difference* $\pm ds\sqrt{(1/n_c + 1/n_j)}$ where n_c is the number of control observations, n_j the number of observations on the jth treatment, s is the square root of the error mean square from the analysis of variance, and d is a quantity (analogous to Student's t) tabulated by Dunnett (1964). Tables for doing the same sort of thing in the nonparametric analyses of variance are given by Wilcoxon and Wilcox (1964).

A third possibility is to ask whether the largest of the treatment means differs from the others. Nonparametric tables are given by McDonald and Thompson (1967).

The critical range method for testing all possible pairs in the Kruskal–Wallis nonparametric one way analysis of variance (§ 11.5)

Using this method, which is due to Wilcoxon, all possible pairs of treatments can be compared validly using Table A7, though the table only deals with equal sample sizes. The procedure is very simple. Just calculate the difference between the rank sums for any pair of groups that is of interest. If this difference is equal to (or larger than) the critical range given in Table A7, the P value is equal to (or less than) the value given in the table. For small samples exact probabilities are given in the table (they cannot be made exactly the same as the approximate P values at the head of the column because of the discontinuous nature of the problem, as in § 7.3 for example). For larger samples use the approximate P value at the head of the column.

The first example of the Kruskal–Wallis analysis given in § 11.5 cannot be used to illustrate the method because it has unequal groups. The second example in § 11.5, based on the (parenthesized) ranks in Table 11.4.1, will be used. In this example there were $k = 4$ treatments and $n = 7$ replicates, and evidence was found in § 11.5 that the treatments were not equi-effective. Consulting Table A7 shows that a difference between two rank sums (selected from four) of 79·1 or larger would occur in about 5 per cent of random allocations of the 28 subjects to 4 groups (i.e. in about 5 per cent of repeated experiments if the null hypothesis were true), that is to say $P \simeq 0.05$ for a difference of 79·1. Similarly $P \simeq 0.01$ for a difference of 95·8.

The simplest way of writing down the differences between all six possible pairs of rank sums is to construct a table of differences, with the rank sums from § 11.5 (or Table 11.4.1), as in Table 11.9.1. The treatments have been arranged in ascending order so the largest differences occur together in the bottom left-hand corner of the table.

TABLE 11.9.1

Treatment		4	1	2	3
	rank sum	61	71·5	121	152·5
4	61				
1	71·5	11·5			
2	121	60·0	49·5		
3	152·5	91·5*	81·0*	31·5	

The differences marked with an asterisk in Table 11.9.1 are larger than 79·1 but less than 95·8. So P is somewhere between 0·01 and 0·05 for these differences suggesting (see § 6.1) that there is a real difference between treatments 3 and 1, and between treatments 3 and 4. All other differences are less than 79·1 so there is little evidence ($P > 0.05$) for any other treatment differences.

The critical range method for testing all possible pairs in the Friedman nonparametric two way analysis of variance (§ 11.7)

This method, also due to Wilcoxon, allows valid comparison of any pair of treatments in the Friedman method (§ 11.7), using Table A8 in much the same way as just described for the one way analysis.

The results in Table 11.7.1 will be used to illustrate the method. There are $k = 4$ treatments and $n = 4$ blocks (replicates), so reference to Table A8 shows that a difference (between any two treatment rank sums selected from the four) as large as, or larger, than 11 would be expected in only 0·5 per cent of repeated experiments if the null hypothesis (see § 11.7) were true, i.e. if ranks were allocated randomly within blocks. Similarly a difference of 10 would correspond to $P = 0.026$.

A table of all possible pair differences between the rank sums from Table 11.7.1 can be constructed as above in Table 11.9.2.

All the six differences are less than 10, i.e. none reaches the $P = 0·026$ level of significance. Despite the evidence (in § 11.7) that the four treatments are not equi-effective, it is not, in this case, possible to detect with any certainty which treatments differ from which others. This is not so surprising looking at the ranks in Table 11.7.1, but looking at the original figures in Table 11.6.2 suggests strongly that

TABLE 11.9.2

Treatment	rank sum	A 6	D 6	C 13	B 15
A	6				
D	6	0			
C	13	7	7		
B	15	9	9	2	

treatments B and C give larger responses than A and D. In fact, *if* the assumptions of Gaussian methods (see § 11.2) are though justifiable, the Scheffé method could be used, and it is shown below that it gives just this result. The reason for the apparent lack of sensitivity of the rank method with the small samples is similar to that discussed in § 10.5 for the two-sample case.

Scheffé's method for multiple comparisons in the Gaussian analysis of variance

The Gaussian analogue of the critical range methods just described is the method of Tukey (see, for example, Mood and Graybill (1963, pp. 267–71)), but Scheffé's method is more general.

Suppose there are k treatments. Define, in general, a *contrast* (see examples below, and also § 13.9) between the k means as

$$L = \sum_{j=1}^{k} a_j \bar{y}_j \tag{11.9.1}$$

where $\Sigma a_j = 0$. The values of a_j are constants, some of which may be zero. When the $\bar{y}_j$ are the means of independent samples the estimated variance of this contrast follows from (2.7.10) and is

$$\text{var}(L) = \sum_{j=1}^{k} a_j^2 \frac{s^2}{n_j} = s^2 \sum \frac{a_j^2}{n_j} \tag{11.9.2}$$

where s^2 is the variance of y (the error mean square from the analysis of variance) and n_j is the number of observations in the jth treatment mean, $\bar{y}_j$. The method is to construct confidence limits (see Chapter 7) for the population (true) mean value of L as

$$L \pm S\sqrt{[\mathrm{var}(L)]} \tag{11.9.3}$$

where $S = \sqrt{[(k-1)F]}$, and F is the value of the variance ratio (see § 11.3) for the required probability. For the numerator F has $(k-1)$ degrees of freedom, and for the denominator the number of degrees of freedom associated with s^2. If the confidence limits include any hypothetical value of L the observations cannot be considered incompatible with this value, as explained in § 9.4.

Three numerical examples

Example 1. Suppose that it were decided to test whether the largest mean in Table 11.4.1 ($\bar{y}_3 = 33 \cdot 57$) really differs from the smallest ($\bar{y}_4 = 14 \cdot 29$), this pair being chosen after the results were known. If in (11.9.1) we take $a_1 = 0$, $a_2 = 0$, $a_3 = +1$ and $a_4 = -1$ then $L = \bar{y}_3 - \bar{y}_4 = 19 \cdot 28$, the difference between means. From Table 11.4.3 it is seen that $s^2 = 101 \cdot 15$ with 24 degrees of freedom. Thus, by (11.9.2), $\mathrm{var}(L) = 101 \cdot 15(0^2/7 + 0^2/7 + 1^2/7 + -1^2/7) = 28 \cdot 90$. There are $k = 4$ treatments so to find the 99 per cent confidence limits, the variance ratio for $P = 0 \cdot 01$ ($= 1 - 0 \cdot 99$) with 3 and 24 degrees of freedom is required. From the tables (see § 11.3) this is found to be $4 \cdot 72$. Thus $S = \sqrt{[(4-1).4 \cdot 72]} = 3 \cdot 763$. The $P = 0 \cdot 99$ confidence limits are $19 \cdot 28 \pm 3 \cdot 763\sqrt{(28 \cdot 90)} = 19 \cdot 28 \pm 20 \cdot 23$, i.e. $-0 \cdot 95$ to $+39 \cdot 51$. The limits include 0, so the difference between the two means cannot be considered to differ from 0 at the $P = 0 \cdot 99$ level of confidence. In other words a significance test (see § 6.1) for the difference between the largest and smallest means would give $P > 0 \cdot 01$ (compare § 9.4). And, because $S\sqrt{[\mathrm{var(L)}]}$ is the same for any pair of means, the same can be said of any pair of means differing by less than $20 \cdot 23$ (see Example (3) also).

Now try the 97·5 per cent limits. From the Biometrika tables (see § 11.3) the value of $F(3, 24)$ for $P = 0 \cdot 025$ is $3 \cdot 72$, so $S\sqrt{[\mathrm{var}(L)]} = \sqrt{(3 \times 3 \cdot 72 \times 28 \cdot 90)} = 17 \cdot 96$. This is *less* than the observed difference, $19 \cdot 28$, so the result of a significance test would be that P is between $0 \cdot 025$ and $0 \cdot 01$, suggesting, though not with great confidence, a real difference.

Example 2. As another example, suppose that it were wished to test the null hypothesis that mean of the two more effective treatments (2 and 3) is equal to the mean of the other two treatments in Table

11.4.1. To do this, take $a_1 = -1$, $a_2 = +1$, $a_3 = +1$ and $a_4 = -1$ so $L = (\bar{y}_2 + \bar{y}_3) - (\bar{y}_1 + \bar{y}_4)$. The true (population) value of this will be zero if the hypothesis to be tested is true. The sample value is $L = -15 \cdot 57 + 26 \cdot 86 + 33 \cdot 57 - 14 \cdot 29 = 30 \cdot 57$. From (11.9.2) $\text{var}(L) = 101 \cdot 15 \; (-1^2/7 + 1^2/7 + 1^2/7 + -1^2/7) = 57 \cdot 80$. $S = 3 \cdot 763$ exactly as above, so the 99 per cent ($P = 0 \cdot 99$) confidence limits for the population value of L are $30 \cdot 57 \pm 3 \cdot 763 \sqrt{(57 \cdot 80)} = 30 \cdot 57 \pm 28 \cdot 61$, i.e. $+1 \cdot 96$ to $+59 \cdot 18$. The limits do not include zero so the null hypothesis that the true (population) value of L is zero would be rejected if $P < 0 \cdot 01$ were considered sufficiently small (see § 6.1). The same could be said of any difference (between the sum of any two means and the sum of the other two), that exceeded $28 \cdot 61$.

Example 3. The method can be used, at least as an approximation, for randomized block experiments also. For the results in Table 11.6.2, $s^2 = 3 \cdot 674$ with 9 d.f. (from Table 11.6.3). To test $\bar{y}_2$ against $\bar{y}_1$ take $a_1 = -1$, $a_2 = +1$, $a_3 = 0$, $a_4 = 0$, as in Example (1). There are $n = 4$ replicates, so $\text{var}(L) = 3 \cdot 674 \, (-1^2/4 + 1^2/4 + 0^2/4 + 0^2/4) = 1 \cdot 837$, from (11.9.2). And this value will be the same for the difference between any two means. There are $k = 4$ treatments so values of $F(3,9)$ are required. From the Biometrika tables (see § 11.3) the $P = 0 \cdot 25$ value is $1 \cdot 63$ and the $P = 0 \cdot 001$ value is $13 \cdot 90$. Thus, $S = \sqrt{(3 \times 1 \cdot 63)} = 2 \cdot 211$ and $S \sqrt{[\text{var}(L)]} = 2 \cdot 211 \sqrt{(1 \cdot 837)} = 2 \cdot 996$ for $P = 0 \cdot 25$. And for $P = 0 \cdot 001$, $S = \sqrt{(3 \times 13 \cdot 90)} = 6 \cdot 457$, so $S \sqrt{[\text{var}(L)]} = 8 \cdot 749$. The differences between the six possible pairs of means from Table 11.6.2 are, tabulating as above, shown in Table 11.9.3.

TABLE 11.9.3

Treatment	Mean	D 49·0	A 49·5	C 65·0	B 67·7
D	49·0				
A	49·5	0·5			
C	65·0	16·0*	15·5*		
B	67·7	18·7*	18·2*	2·7	

The four differences marked by an asterisk in Table 11.9.3 are greater than $8 \cdot 749$ so the null hypothesis that the true values of these differences are zero can be 'rejected' at $P < 0 \cdot 001$ (see § 6.1). The other differences are less than $2 \cdot 996$ so there is no evidence ($P > 0 \cdot 25$) that these differences are due to anything but experimental error. It is concluded that treatments B and C both give larger responses than treatments A and D, but that no difference can be detected between A and D, or between B and C. Compare this result with rank analysis of the same observations (§§ 11.7 and 11.9, above). Remember that a normal (Gaussian) distribution has been assumed throughout these calculations despite the fact that no evidence was presented, in any of the examples, to suggest that this assumption was justified.

12. Fitting curves. The relationship between two variables

12.1. Nature of the problem

IN all the examples discussed so far, measurements of only one variable have been involved (e.g. blood sugar level or change in duration of sleep). However, experiments are often concerned with the relationship between two (or more) variables; for example, dose of drug and response, concentration and optical density, time and extent of chemical reaction, or school and university examination results. The last of these examples is rather different from the others and suggests that two sorts of situation occur.

(a) One variable can be measured accurately and its value chosen by the experimenter, for example the time when a measurement is taken, or the dose of a drug. This sort of variable is called an *independent variable* (notice that *independent* in this context has a different meaning from that encountered in §§ 2.4 and 2.7). The other variable, called the *dependent variable*, is subject to experimental error, and its value depends on the value chosen for the independent variable. For example, response is a dependent variable which is related to dose, the independent variable (as long as dose can be measured with negligible error).

(b) Quite often the value of neither variable can be chosen by the experimenter, or measured without error. For example ability before and after university (as measured by school and university exam results) are *both* measured inaccurately.

In both of these cases the first thing usually done is to plot the results and draw some sort of line through them.

Case (a) is described (for historical reasons, now irrelevant) as a *regression problem*. The line fitted to the points is called a *regression line*, the formula for calculating it being the *regression equation*. This sort of problem is dealt with in §§ 12.1–12.8.

The second type of problem, case (b), is a *correlation problem* and the graph of the results is often called a *scatter diagram* (see §§ 1.2 and 12.9).

The expression 'fitting a curve to the observed points' means the process of finding estimates of the parameters of the fitted equation that result in a calculated curve which fits the observations 'best' (in a sense to be specified below and in § 12.8). For example if a straight line is to be fitted the 'best' estimates of its arbitrary parameters (the slope and intercept) are wanted. The method of fitting the straight line is discussed in detail in §§ 12.2 to 12.6 because it is the simplest problem. But very often, especially if one has an idea of the physical mechanism underlying the observations, the observations will not be represented by a straight line and a more complex sort of curve must be fitted. This situation is discussed in §§ 12.7 and 12.8. Often some way of transforming the observations to a straight line is adopted, but this may have a considerable hazards as explained in §§ 12.2 and 12.8.

It is, however, usually (see § 13.14) not justified to fit anything but a straight line if the deviations from the fitted line are no greater than could reasonably be expected by chance (i.e. than could be expected if the *true* line were straight). In general it is usually reasonable to use the simplest relationship consistent with the observations. By simplest is meant the equation containing the smallest number of arbitrary parameters (e.g. slope), the values of which have to be estimated from the observations. This is an application of 'Occam's razor' (one version of which states 'It is vain to do with more what can be done with fewer': William of Occam, early fourteenth century). The reason for doing this is not that the simplest relationship is likely to be the true one, but rather because the simplest relationship is the easiest to refute should it be wrong. (The opposite would, of course, be true if the parameters were not arbitrary, and estimated from the observations, but were specified numerically by the theory.)

The role of statistical methods

Statistical methods are useful

(1) for finding the best estimates of the parameters (see §§ 12.2, 12.7, and 12.8) of the chosen regression equation, and confidence limits (Chapter 7 and §§ 12.4–12.6) for these estimates,

(2) to test whether the deviations of the observed points from the calculated points (the latter being obtained using the best estimates of the parameters) are greater than could reasonably be expected by chance, i.e. to test whether the type of curve chosen fits the observations adequately. It is important to remember (see § 6.1) that if observations do not deviate 'significantly' from, say, a straight line, this does not

mean that the true relationship can be inferred to be straight (see § 13.14 for an example of practical importance).

The *best fitting* curve (see § 12.8) is usually found using the *method of least squares*. This means that the curve is chosen that minimizes the 'badness of fit' as measured by the sum of the squares of the deviations of the observations (y) from the calculated values (Y) on the fitted curve. In other words, the values of the parameters in the regression equation must be adjusted so as to minimize this sum of squares. In the case of the straight line and some simple curves such as the parabola the best estimates of the parameters can be calculated directly (see §§ 12.2 and 12.7). The principle of least squares can be applied to any sort of curve but for non-linear problems (see § 12.8; but note that fitting *some* sorts of curve is a linear problem in the statistical sense, as explained in § 12.7) it may not have the optimum properties that it can be shown to have for linear problems (those of providing unbiased estimates of the parameters with minimum variance, see § 12.8, and Kendall and Stuart (1961, p. 75)). For linear problems these optimum properties are, surprisingly, not dependent on any assumption about the distribution of the observations, but the construction of confidence limits and all the analyses of variance depend on the assumption that the errors of the observations follow the Gaussian (normal) distribution, so all regression methods must (unfortunately) be classed as parametric methods (see § 6.2). Tests for normality are discussed in § 4.6.

12.2. The straight line. Estimates of the parameters

It is assumed throughout this discussion of *linear regression* that the *independent variable*, x (e.g. time, concentration of drug, see § 12.1) can be measured reproducibly and its value fixed by the experimenter. The experimental errors are assumed to be in the observations on the *dependent variable*, y (e.g. response, see § 12.1). Suppose that several (k, say) values of the independent variable , $x_1, x_2, \ldots, x_k$, are chosen and that for each observations are made on the dependent variable, $y_1, y_2, \ldots, y_N$ (there being N observations altogether; N may be bigger than k if several observations are made at each value of x, as in § 12.6).

In order to find the 'best' straight line by the method of least squares (see §§ 12.1, 12.7, and 12.8) it is necessary to find the line that will minimize the badness of fit as measured by the sum of squares of deviations of the *dependent* variable from the line, as shown in Fig. 12.2.1.

Thus the best fitting straight line is the one that minimizes the sum of squared deviations

$$S = \sum_{j=1}^{j=N} d_j^2 = \Sigma(y_j - Y_j)^2 \qquad (12.2.1)$$

where y_j is the observed, and Y_j the calculated, value of the dependent variable corresponding to x_j. The resulting line is called *the regression line of y on x*. If the deviations of points from the fitted line were not measured vertically as in (12.2.1), but, say, horizontally, the least squares line would be different from that found in the way just described

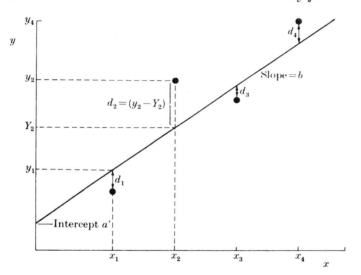

FIG. 12.2.1. The dependent variable y, plotted against the independent variable x. Definition of y_j, x_j, Y_j and d_j for discussion of curve fitting.

(it would be called the regression line of x on y), but this would not be the correct approach when the experimental errors are supposed to affect y only.

The general equation for a straight line can be written $Y = a' + bx$ where a' is the intercept (i.e. the value of Y when $x = 0$). It will be more convenient (for reasons explained in § 12.7) to write this in a slightly different form, viz.

$$Y = a + b(x - \bar{x}) \qquad (12.2.2)$$

where b is the slope, and a is the value of Y when $x = \bar{x}$ (so that $a - b\bar{x}$, which is a constant, is the same as a'). The left-hand side is written as capital Y to emphasize that the evaluation of the equation gives the calculated value of the dependent variable, which will in general, differ from the observations (y) at the same value of x, unless the observation happens to lie exactly on the calculated line.

The true (population) regression equation, assuming the line to be really straight, can be written, for any specified value of x,

$$\mu = \text{population value of } y = \alpha + \beta(x - \bar{x}) \qquad (12.2.3)$$

where α and β are the true parameters, of which the statistics a and b are estimates made from a sample of observations. Because the independent variable, x, is assumed to be measured with negligible error there is no distinction between the observed and true values of x.

The problem is now to find the least squares estimates of α and β, from the observations. This will be done algebraically for the moment. In § 12.7 the geometrical meaning of the algebra is explained. First substitute the calculated value of Y at the jth value of x, which, from (12.2.2), is $Y_j = a + b(x_j - \bar{x})$, into (12.2.1) giving

$$S = \sum_{j=1}^{j=N} [y_j - a - b(x_j - \bar{x})]^2. \qquad (12.2.4)$$

Squaring the term in brackets gives

$$S = \sum_{j=1}^{N} [y_j^2 + a^2 + b^2(x_j - \bar{x})^2 - 2ay_j - 2y_j b(x_j - \bar{x}) + 2ab(x_j - \bar{x})]$$

and therefore, using (2.1.5), (2.1.6), and (2.1.8),

$$S = \sum_{j=1}^{N} y_j^2 + Na^2 + b^2 \sum_{j=1}^{N} (x_j - \bar{x})^2 - 2a \sum_{j=1}^{N} y_j - 2b \sum_{j=1}^{N} y_j(x_j - \bar{x}) + 2ab \sum_{j=1}^{N} (x_j - \bar{x}).$$
$$(12.2.5)$$

The last term in this equation is zero because, by (2.6.1), $\Sigma(x - \bar{x}) = 0$. The object is to find, for the particular values of x used and the particular values of y observed, the values of a and b which make S as small as possible. For a particular set of results we are, for the moment regarding x and y values as fixed and a and b as variables. The usual procedure in calculus for finding a minimum is to differentiate† and equate the result to zero as illustrated (for a) in Fig. 12.2.2 (see Thompson (1965, p. 78 et seq.)) *A fuller explanation of this process is given in* § 12.7.

---

† Because there are two variables, a and b, partial differential coefficients (with curly ∂) are used. This makes the differentiation of (12.2.5) even simpler because it means that when differentiating with respect to a, b is treated as a constant (and vice versa). See § 12.7.

It is shown below (see (12.2.10)) how the least squares estimates can be derived without using calculus at all. Thus, to find the least squares value of a, differentiate (12.2.5) treating b as a constant

$$\frac{\partial S}{\partial a} = 2Na - 2\sum_{j=1}^{N} y_j = 0$$

therefore $Na = \Sigma y_j$

so
$$a = \frac{\Sigma y}{N} = \bar{y}. \tag{12.2.6}$$

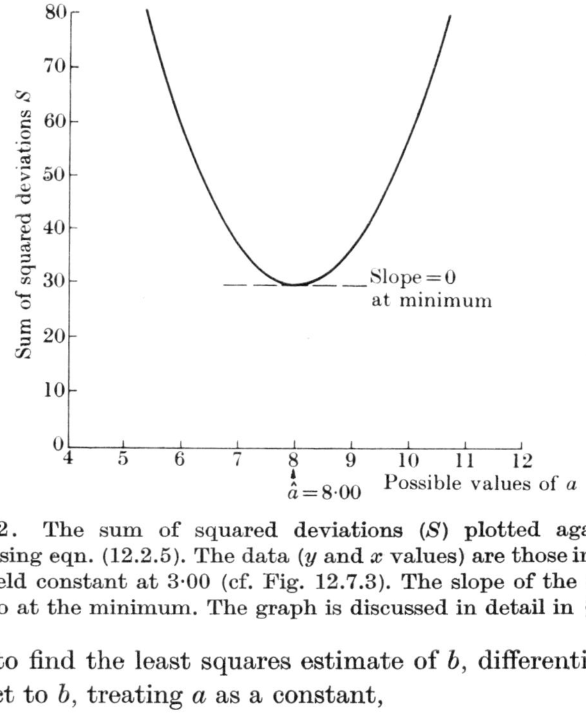

FIG. 12.2.2. The sum of squared deviations (S) plotted against various values of a using eqn. (12.2.5). The data (y and x values) are those in Table 12.7.1 and b was held constant at 3·00 (cf. Fig. 12.7.3). The slope of the curve, $\partial S/\partial a$, is zero at the minimum. The graph is discussed in detail in § 12.7.

Similarly, to find the least squares estimate of b, differentiate (12.2.5) with respect to b, treating a as a constant,

$$\frac{\partial S}{\partial a} = 2b\sum_{j=1}^{N}(x_j - \bar{x})^2 - 2\sum_{j=1}^{N} y_j(x_j - \bar{x}) = 0$$

therefore $2b\Sigma(x_j - \bar{x})^2 = 2\Sigma y_j(x_j - \bar{x})$

so
$$b = \frac{\Sigma y_j(x_j - \bar{x})}{\Sigma(x_j - \bar{x})^2} \tag{12.2.7}$$

or
$$b = \frac{\Sigma(y_j - \bar{x})(x_j - \bar{x})}{\Sigma(x_j - \bar{x})^2}. \tag{12.2.8}$$

Although it is not immediately obvious, the numerators of these two expressions for b are identical, as shown by (12.2.9) below.

Using (2.6.2) and (2.6.6) shows that the estimated slope, (12.2.8), can be written $b = \text{cov}(x, y)/\text{var}(x)$. It was shown in § 2.6 that $\Sigma(y - \bar{y})$ $(x - \bar{x})$, and hence the covariance, measure the extent to which y tends to increase when x is increased.

Proof that $\Sigma(y - \bar{y})(x - \bar{x})$ can be written $\Sigma y(x - \bar{x})$

$$\sum_{j=1}^{N} (y_j - \bar{y})(x_j - \bar{x}) = \Sigma[y_j x_j - y_j \bar{x} - \bar{y} x_j + \bar{y}\bar{x}]$$

$$= \Sigma[y_j(x_j - \bar{x}) - \bar{y}(x_j - \bar{x})]$$

$$= \Sigma y_j(x_j - \bar{x}) - \bar{y}\Sigma(x_j - \bar{x})$$

$$= \sum_{j=1}^{N} y_j(x_j - \bar{x}) \tag{12.2.9}$$

because the last term in the penultimate equation is, by (2.6.1), zero.

How to find the least squares estimates without using calculus

The argument is exactly analogous to that already used for the arithmetic mean in § 2.5 (p. 27). The sum of squares to be minimized, (12.2.4), can be written

$$S = \Sigma[y_j - a - b(x_j - \bar{x})]^2 = \Sigma[y_j - \bar{y} - \hat{b}(x_j - \bar{x})]^2 + N(a - \bar{y})^2 + (b - \hat{b}) \cdot \Sigma(x_j - \bar{x})^2, \tag{12.2.10}$$

where a and b denote possible estimates of α and β, and $\hat{b}$ denotes, as in § 12.7, the estimate of β given by (12.2.7). This expression is analogous to (2.5.6). It is easy to see that the values of a and b that minimize this are $a = \bar{y}$ and $b = \hat{b}$ (the same estimates as found above), because in this case the last two terms will be zero, their minimum possible value. It can quite easily be shown that (12.2.10) is an algebraic identity by inserting $\hat{b}$ from (12.2.7) and expanding the right side, in the same sort of way as shown in detail for (2.5.6).

Assumptions made in find the least squares fitting and analysis of straight lines

(1) The standard deviation of y was assumed to be a constant. That is to say, that the observations have the same scatter at all points along the curve so that equal weight can be attached to all observations (as has been done in the above derivations, cf. (2.5.1) and (2.5.2)). When this condition is fulfilled the observations are described as *homoscedastic*. Quite often this condition is not fulfilled as illustrated in Fig. 12.2.3. For instance, it is quite commonly found in practice that there is a tendency for the smaller observations to have less scatter, in a way that the *relative* scatter (e.g. the coefficient of variation,

(2.6.4)) is more nearly constant than the *absolute* scatter (e.g. the standard deviation). If this is the case the observations (which are said to show *heteroscedasticity*) should not be given equal weight, and this makes the calculations more complicated (cf. Chapter 14).

(2) The *population* (*true*) relation between y and x has been assumed to be a straight line. In § 12.6 it will be shown how it can be judged whether deviations from linearity can reasonably be ascribed to experimental error.

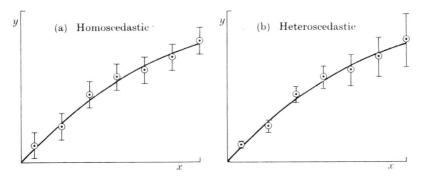

FIG. 12.2.3.   (a) A homoscedastic curve-fitting problem (idealized). (b) An example of heteroscedastic observations.

(3) The independent variable has been assumed to be measured with negligible error. For a discussion of what to do when it is not see, for example, Brownlee (1965, p. 391).

(4) The analyses to be described will all assume that the errors in the observations (y, the dependent variable) at each of the selected x values follow Gaussian (normal) distributions (see §§ 4.2, 4.6 and 12.1).

The use of transformations

This discussion is closely related to that in § 11.2, in which a method of choosing a transformation to equalize variances was described. Transformation (e.g. logarithm, square root, reciprocal) may be used to make results conform with the above assumptions. For example, if the observations are described by an exponential relationship, $Y = Y_0 e^{-kx}$, then taking natural logarithms gives $\log Y = \log Y_0 - kx$. The regression of $\log y$ on x should therefore be a straight line with intercept $= \log Y_0$ and slope $= -k$. An example is worked out in § 12.6. Notice, however, that if y were homoscedastic and normally distributed then $\log y$ would be neither, so it may not be possible to satisfy all the assumptions simultaneously (see §§ 11.2, 12.8 and Bartlett (1947)). Tests for normality are discussed in § 4.6.

It is important to distinguish between the effects of transformations of the dependent variable, y, on one hand, and on the independent variable, x, on the other. Transformations of x are often used to make a line straight (e.g. response, y, is often plotted against the log of the dose, x, in pharmacology). This merely alters the spacing at which points are plotted along the abscissa in Fig. 12.3, but cannot have any effect on the homoscedasticity or distribution of errors of the observations, y. Transformations of y, on the other hand, affect these as well as linearity.

12.3. Measurement of the error in linear regression

Consider the straight line fitted to the results in Fig. 12.3.1. As before, y stands for the observation at a particular value of x, and Y

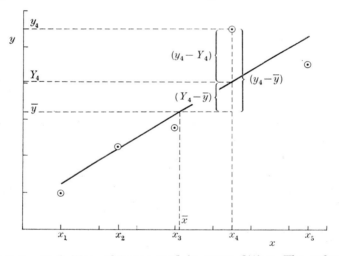

FIG. 12.3.1. Definition of terms used in curve fitting. The values of the dependent variable are plotted on the ordinate and the independent variable (x) on the abscissa (see §§ 12.1 and 12.2). The five observed values (○), y_1 to y_5, have been plotted against the corresponding x values, x_1 to x_5, and a straight line fitted to them. The nature of the terms $(y-Y)$, $(Y-\bar{y})$, and $(y-\bar{y})$ occurring in eqns. (12.3.2) and (12.3.3), is illustrated for the fourth x value.

for the predicted value of the dependent variable (i.e. that calculated from the estimated line) at a particular x. The equation for the estimated line, $Y = a+b(x-\bar{x})$, can be written, using (12.2.6),

$$Y = \bar{y}+b(x-\bar{x}), \tag{12.3.1}$$

from which it can be see that the line must go through the point $(\bar{y}, \bar{x})$ because $Y = \bar{y}$ when $x = \bar{x}$ (i.e. when $x - \bar{x} = 0$).

This section is concerned only with errors in y, because x has been assumed to be measured without error (§§ 12.1, 12.2). The total deviation of the observed point from the mean, in Fig. 12.3.1, can be divided into two parts: $(y - Y)$ = deviation of observed value from the line, and $(Y - \bar{y})$ = deviation of predicted value on the line from the mean of all observations. This can be written

$$(y - \bar{y}) \quad = \quad (y - Y) \quad + \quad (Y - \bar{y}) \qquad (12.3.2)$$

$$
\begin{array}{ccc}
\text{total deviation} & \text{deviation from} & \text{part of the total} \\
& \text{straight line} & \text{deviation accounted} \\
& & \text{for by the linear} \\
& & \text{relation between} \\
& & y \text{ and } x
\end{array}
$$

It is now possible to use the analysis of variance approach. The total sum of squared deviations (SSD) of each observation from the grand mean of all observations is $\Sigma(y_j - \bar{y})^2$, and this total SSD can be divided into two components (compare (12.3.2))

$$\Sigma(y_j - \bar{y})^2 = \Sigma(y_j - Y_j)^2 + \Sigma(Y_j - \bar{y})^2 \qquad (12.3.3)$$

in which the first term on the right-hand side measures the extent to which the observations deviate from the line and is called the *SSD for deviations from linearity*. It is this that is minimized in finding least squares estimates, see § 12.2. The second term on the right-hand side measures the amount of the total variability of y from $\bar{y}$ that is accounted for by the linear relation between y and x, and is called the *SSD due to linear regression*. That (12.3.3) is merely an algebraic identity following from (12.3.2) will now be proved.

Digression to prove (12.3.3), *and to obtain a working formula for the sum of squares due to linear regression*

(1) *To show that* (12.3.3) *follows from* (12.2.2). The summations are, as before, over all N observations of y (there may be more than one y at each x value). From (12.2.2),

$$
\begin{aligned}
\text{total SSD} &= \Sigma(y - \bar{y})^2 = \Sigma[(y - Y) + (Y - \bar{y})]^2 \\
&= \Sigma[(y - Y)^2 + 2(y - Y)(Y - \bar{y}) + (Y - \bar{y})^2] \\
&= \Sigma(y - Y)^2 + 2\Sigma(y - Y)(Y - \bar{y}) + \Sigma(Y - \bar{y})^2 \\
&= \Sigma(y - Y)^2 + \Sigma(Y - \bar{y})^2 \qquad \text{Q.E.D.}
\end{aligned}
$$

$$
\begin{array}{cc}
\text{deviations} & \text{due to linear} \\
\text{from} & \text{regression} \\
\text{linearity} &
\end{array}
$$

The central term in the penultimate equation is zero because

$$2\Sigma(y-Y)(Y-\bar{y})$$
$$= 2\Sigma[y-\bar{y}-b(x-\bar{x})][\bar{y}+b(x-\bar{x})-\bar{y}] \quad \text{(from (12.3.1))}$$
$$= 2\Sigma[yb(x-\bar{x})-\bar{y}b(x-\bar{x})-b^2(x-\bar{x})^2]$$
$$= 2b\Sigma y(x-\bar{x})-2b^2\Sigma(x-\bar{x})^2 \quad \text{(from (2.6.1))}$$
$$= 2b\Sigma y(x-\bar{x})-2b\Sigma y(x-\bar{x}) = 0. \quad \text{Q.E.D.}$$

$$\text{(from (12.2.7))}$$

(2) *A working formula for the sum of squares due to linear regression*

As usual, it is inconvenient to calculate the individual deviations $(Y-\bar{y})$, and a more convenient working formula is used. As before, the summations are over all N observations.

$$\Sigma(Y-\bar{y})^2 = \Sigma[\bar{y}+b(x-\bar{x})-\bar{y}]^2 \quad \text{(from (12.3.1))}$$
$$= \Sigma[b(x-\bar{x})]^2 = b^2\Sigma(x-\bar{x})^2 \quad \text{(from (2.1.5))}.$$

Substituting (12.2.8) for the slope, b, gives the alternative forms:

$$\text{SSD due to linear regression} = b^2\Sigma(x-\bar{x})^2 \text{ or } \frac{[\Sigma(y-\bar{y})(x-\bar{x})]^2}{\Sigma(x-\bar{x})^2}. \quad (12.3.4)$$

12.4. Confidence limits for a fitted line. The important distinction between the variance of *y* and the variance of *Y*

It was stated in § 12.2 that the method of fitting a straight line there described involves the assumption that the scatter of the observations does not depend on their size (see Fig. 12.2.3), i.e. that their population (true) variance, $\sigma^2[y]$, is a constant, independent of the value of x (and of y) The estimated value of σ^2 from a sample is $s^2[y]$ or var[y], the error mean square from the analysis of variance table (see §§ 11.4, 12.5, and 12.6). The width of the confidence interval for the population value of an observation is therefore the same ($\pm ts[y]$, see (7.4.2)) whatever the size of the observation.

In practice a straight line is usually fitted for one of the following reasons.

(a) To estimate the slope or intercept and their confidence limits (see §§ 7.2, 7.9, 12.5, and 12.6).

(b) To predict values of y for a given x. For example, it may be required to predict from the fitted line what response (Y) would be produced by a particular dose (x), or what optical density (Y) a solution of a particular concentration (x) will have. The error of such a prediction is discussed in this section. There are two forms of the problem, as in § 7.4.

(c) To predict the value of x required to produce a given (observed or hypothetical) value of y. For example, the prediction of the dose (x) needed for a given response, or of the concentration (x) of a solution of a particular optical density. This sort of problem is probably the most important in practice but its solution is rather more complicated than for (a) and (b). Its solution will be given in § 13.14.

In case (b) confidence limits are required for a value of Y calculated from the fitted line, rather than for an observed value, y (see § 12.2); and to find these limits an estimate of the variance of Y will now be found. For the meaning and interpretation of confidence limits see §§ 7.2 and 7.9.

Equation (12.2.2) for the fitted line is $Y = a+b(x-\bar{x})$ where $a = \bar{y}$ and $b = \Sigma y(x-\bar{x})/\Sigma(x-\bar{x})^2$ (from (12.2.6) and (12.2.7)). Because the independent variable (x) is assumed to be measured without error (§ 12.2), terms involving only x can be treated, for the purposes of assessing error, as constants. Because, as shown below, a and b, and hence Y, are linear functions of the observations, it follows that if the observations are normally distributed then a, b, and Y will be normally distributed. Imagine the experiment being repeated many times on repeated random samples from the same population, using the same x values. From each experiment a and b are estimated, and Y, for example the response for a particular dose, is calculated from the fitted line (12.2.2). The variation of the repeated a, b, and Y values should follow normal distributions with means α, β, and μ (see (12.2.3)), and variances $var[a]$, $var[b]$, and $var[Y]$ say. Compare the non-normal distribution of parameter estimates found in the non-linear problem discussed in § 12.8. If Y is normally distributed confidence limits can be found as in § 7.4 once var[Y] is known. To find var[Y] it is necessary to know the variances of a and b.

The variance of the estimated slope, b

The least squares estimate (b) of the true slope (β) given in eqn. (12.2.7) can be written out term by term in the form

$$b = \frac{\Sigma y_j(x_j-\bar{x})}{\Sigma(x_j-\bar{x})^2} = y_1\frac{(x_1-\bar{x})}{\Sigma(x_j-\bar{x})^2}+y_2\frac{(x_2-\bar{x})}{\Sigma(x_j-\bar{x})^2}+...+y_N\frac{(x_N-\bar{x})}{\Sigma(x_j-\bar{x})^2}. \quad (12.4.1)$$

Therefore b is a linear function of the observations, i.e. it can be written in the form $\Sigma c_j y_j = c_1 y_1+c_2 y_2+...+c_N y_N$ where the c_j are constants (for more comments on this use the word 'linear' see § 12.7). In this case the constants are $c_j = (x_j-\bar{x})/\Sigma(x_j-\bar{x})^2$. The variance of b now follows directly from (2.7.10) and is $\text{var}[y].\Sigma c_j^2$. Now $c_j^2 = (x_j - \bar{x})^2/[\Sigma(x_j - \bar{x})^2]^2$ so $\Sigma c_j^2 = \Sigma(x_j - \bar{x})^2/[\Sigma(x_j - \bar{x})^2]^2$ and therefore, cancelling,

$$\text{var}[b] = \frac{\text{var}[y]}{\Sigma(x_j - \bar{x})^2}. \qquad (12.4.2)$$

This gives a *prediction* of what the variance of repeated estimates of b should be, based on the scatter of the observations, $\text{var}[y]$, seen in the one experiment actually done (see §§ 2.7 and 7.2). Notice that the slope will be most accurately determined (var $[b]$ smallest) when the values of x are widely spaced making the values of $(x - \bar{x})$ large, as common sense suggests.

Confidence limits for the slope can be obtained just as in § 7.4 (because b is normally distributed when the observations are, see above) as

$$b \pm t\sqrt{(\text{var}[b])}, \qquad (12.4.3)$$

where t is Student's t for the required P and with the number of d.f. associated with $\text{var}[y]$. See §§ 12.5 and 12.6 for examples.

The variance of a

By (12.2.6) $a = \bar{y}$ so $\text{var}[a] = \text{var}[\bar{y}] = \text{var}[y]/N$, by (2.7.8).

Confidence limits for the true (population) straight line

The value of Y estimated from the line, $Y = a + b(x - \bar{x})$, is a linear function of the observations because, as above, both a and b are. It will therefore be normally distributed when the observations are. The population mean value of Y at any given value of x is μ (see (12.2.3)), so the error of a value of Y is $Y - \mu$ which has a population mean value† of $\mu - \mu = 0$ and variance $\text{var}[Y]$ (because μ is a constant). The variance of Y is

$$\text{var}[Y] = \text{var}[\bar{y} + b(x - \bar{x})] \qquad \text{(by (12.3.1))}$$

$$= \text{var}[\bar{y}] + \text{var}[b(x - \bar{x})] \qquad \text{(by (2.7.3))}$$

$$= \text{var}[\bar{y}] + (x - \bar{x})^2 \text{var}[b] \qquad \text{(by (2.7.5))}$$

$$= \frac{\text{var}[y]}{N} + (x - \bar{x})^2 \frac{\text{var}[y]}{\Sigma(x_j - \bar{x})^2} \qquad \text{(by (2.7.8) and (12.4.2))}$$

$$= \text{var}[y].\left(\frac{1}{N} + \frac{(x - \bar{x})^2}{\Sigma(x_j - \bar{x})^2}\right). \qquad (12.4.4)$$

---

† See Appendix 1 for a rigorous definition. $\text{E}[Y - \mu] = \text{E}[Y] - \text{E}[\mu] = \mu - \mu = 0$.

Notice that the use of (2.7.3) assumes that $\bar{y}$ and b are uncorrelated (i.e. in repeated experiments there will be no tendency for y to be large in experiments when b is). This has not been proved but a similar relationship is discussed in greater detail in §§ 13.8 and 13.10. See also § 12.7.

The confidence limits for μ (at a particular value of x), when it is estimated by calculating Y, are, by (7.4.2),

$$Y \pm t\sqrt{(\text{var}[Y])}. \tag{12.4.5}$$

Several points about (12.4.4) are worth noticing. First, although the term $\Sigma(x_j - \bar{x})^2$ is a constant (depending on the particular values of x chosen) for a given experiment, the term $(x - \bar{x})^2$ is *not*. The presence of this latter term shows that *the variance of Y, unlike that of y, is dependent on the value of x*. The variance of Y will be at a minimum when $x = \bar{x}$ because at this point the second term, which involves $x - \bar{x}$, disappears leaving $\text{var}[Y] = \text{var}[\bar{y}]$ as expected (because at this point $Y = \bar{y}$, see § 12.3). It can also be seen that the variance of Y (and hence the width of the confidence limits) increases as x deviates in *either* direction from $\bar{x}$, because the deviation $(x - \bar{x})$ is squared and therefore always positive.

The common sense of these results is discussed further when they are illustrated numerically in §§ 12.5 and 13.14 (and plotted in Figs. 12.5.1 and 13.14.1).

Confidence limits for new observations

Just as in § 7.4 the situation is changed if instead of wishing to find confidence limits for, say, the population (true) response, μ, produced by a particular concentration of drug, x, it is wished to find confidence limits within which the mean $(\bar{y}_m)$ of m observations of the response to concentration x would be expected to lie. The best estimate of y_m is the same as the best estimate of μ, viz. $Y = a + b(x - \bar{x})$; but, as in § 7.4, its error is different. The error of the prediction is $Y - \bar{y}_m$, which will be normally distributed with a population mean of $\mu - \mu = 0$. Because the m new observations are supposed to be independent observations from the same Gaussian population (2.7.3) can be used giving $\text{var}[Y - \bar{y}_m] = \text{var}[Y] + \text{var}[\bar{y}_m]$. Now $\text{var}[Y]$ is given by (12.4.4), and, by (2.7.8), $\text{var}[\bar{y}_m] = \text{var}[y]/m$, so, by exactly the same argument as used to find (7.4.3), the confidence limits for $\bar{y}_m$ will be

$$Y \pm t\sqrt{\left[\text{var}[y]\left(\frac{1}{N} + \frac{1}{m} + \frac{(x - \bar{x})^2}{\Sigma(x_j - \bar{x})^2}\right)\right]}. \tag{12.4.6}$$

As expected (and as in § 7.4) this reduces to (12.4.5) when m is very large so $\bar{y}_m$ becomes the same as μ. The prediction is that if repeated experiments are conducted and in each experiment the limits calculated, then in 95 per cent of experiments (or any other chosen proportion, depending on the value chosen for t) the mean of m new observations will fall within the limits. The limits, and $\bar{y}_m$, will of course vary from experiment to experiment—see § 7.9. This prediction is, as usual, likely to be optimistic (see § 7.2). The use of this method is illustrated on § 13.14 (and plotted in Fig. 13.14.1).

12.5. Fitting a straight line with one observation at each x value

The results in Table 12.5.1 show a single observation on the dependent variable, y, at each value of the independent variable, x. For example, y might be the plasma concentration of a drug at a precisely measured time x after administration. The common sense supposition that the times have not been chosen sensibly will be confirmed by the analysis. The assumptions necessary for the analysis have been discussed in §§ 11.2, 12.1, and 12.2, and the meaning of confidence limits has been discussed in §§ 7.2 and 7.9. These should be read before this section.

TABLE 12.5.1

	x	y
	160	59
	165	54
	169	64
	175	67
	180	85
	188	78
Totals	1037	407

There is a tendency for y to increase as x increases. Is this trend statistically significant? To put the question more precisely, does the estimated slope of this line, b, differ from zero to an extent that would be likely to occur by random experimental error if the *true* slope of the line, β, were in fact zero? In other words it is required to test the null hypothesis that $\beta = 0$.

Fitting the straight line $Y = a + b(x - \bar{x})$

The least squares estimate of α in (12.2.3) is, by (12.2.6),

$$a = \bar{y} = 407/6 = 67{\cdot}833.$$

The least squares estimate of the slope, β, is, by (12.2.8),

$$b = \Sigma(y-\bar{y})(x-\bar{x})/\Sigma(x-\bar{x})^2.$$

First calculate, by (2.6.5),

$$\Sigma(x-\bar{x})^2 = \Sigma x^2 - (\Sigma x)^2/N = 160^2 + 165^2 + \ldots + 188^2 - 1037^2/6 = 526\cdot833.$$

The sum of products is found using (2.6.7),

$$\begin{aligned}\Sigma(y-\bar{y})(x-\bar{x}) &= \Sigma yx - \Sigma y.\Sigma x/N \\ &= (59 \times 160) + \ldots + (78 \times 188) - (407 \times 1037)/6 \\ &= 511\cdot833.\end{aligned}$$

Thus $b = 511\cdot833/526\cdot833 = 0\cdot9715$. Also $\bar{x} = 1037/6 = 172\cdot833$.

Inserting these values in (12.2.2) gives the equation for the least squares straight line.

$$Y = 67\cdot833 + 0\cdot9715\ (x - 172\cdot833) \qquad (12.5.1)$$

This line is plotted in Fig. 12.5.1 together with the observed values.

Does the estimated slope, $b = 0\cdot9715$, differ from zero by more than could reasonably be expected if the population slope, β, were zero?

The analysis of variance

The analysis is performed with the observations (y values). The independent variable, x, only comes in incidentally. The principle of the method is described in § 12.3.

The total sum of squares, by (2.6.5), is

$$\Sigma(y-\bar{y})^2 = \Sigma y^2 - (\Sigma y)^2/N = 59^2 + \ldots + 78^2 - 407^2/6 = 682\cdot833.$$

The sum of squares due to linear regression, by (12.3.4), is

$$(511\cdot833)^2/526\cdot833 = 497\cdot260.$$

The sum of squares due to deviations from linearity is found, using (12.3.3), by difference, as $682\cdot833 - 497\cdot260 = 185\cdot573$.

There are 6 values of y so the total number of degrees of freedom is 5. The sum of squared deviations (SSD) due to linear regression has one d.f. because it corresponds to the calculation of one statistic (b) from the observations (this is made obvious by the identity of the analysis with a t test, shown at the end of this section). The analysis summarized by (12.3.3) is tabulated in Table 12.5.2, which is completed and interpreted in the same way as previous analyses of variance (e.g. Table 11.3, 11.4, and 11.5). The two figures in the mean square column would be independent estimates of the *same* quantity (σ^2) if all 6 observations were from a single population (with mean μ and variance σ^2). This way of stating the null hypothesis implies that the population mean of the observations is always μ (whatever the x value), i.e. it implies that $\beta = 0$, the way in which the null hypothesis was put above. The probability that the ratio of two independent estimates of the *same* variance will be 10·72 (as observed, Table 12.5.2), or larger, is 0·02 to 0·05 (see §§ 11.3 and 11.4), i.e. 10·72 would be exceeded in something

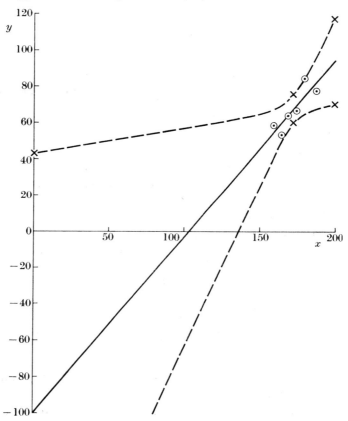

FIG. 12.5.1.   Observed points from Table 12.5.1.
————Least squares estimate of straight line (eqn. (12.5.1).
– – – 95 per cent confidence limits for Y, i.e. for the fitted line.
×Particular values of confidence limits calculated in the text.

between 2 and 5 per cent of experiments in the long run (the limitations of the tables of F, see § 11.3, prevent P being found to any greater accuracy).

In this analysis there is no good estimate of the experimental error. because only one observation was made at each value of x. This analysis should be compared with that in § 12.6, in which replication of the observations gives a proper estimate of σ^2. The best that can be done in this case is to *assume* that the line is straight, in which case the mean square for deviations from linearity, 46·393, will be an estimate of

TABLE 12.5.2

Source of variation	d.f.	SSD	MS	F	P
Linear regression	1	497·260	497·260	10·72	0·02–0·05
Deviations from linearity	$N-2=4$	185·573	46·393		
Total	$N-1=5$	682·833			

the error variance (see § 12.6). Following this procedure shows that a value of b differing from zero by as much as or more than that observed (0·9715) would be expected in between 2 and 5 per cent of repeated experiments *if β were zero*. This suggests, though not very conclusively, that y really does increase with x (see § 6.1).

Gaussian confidence limits for the population line

The error variance of the 6 observations (the part of their variance not accounted for by a linear relationship with x) is estimated, from Table 12.2, to be var[y] $= 46·393$ with 4 d.f. The value of Student's t for $P = 0·95$ and 4 d.f. is 2·776 (from tables, see § 4.4). The confidence limits for the population value of Y at various values of x can be found from (12.4.5). To evaluate var[Y] from (12.4.4) at each value of x, the values var[y] $= 46·393$, $N = 6$, $\bar{x} = 172·833$ and $\Sigma(x-\bar{x})^2 = 526·833$ are needed. These have already been calculated and are the same at all values of x. Enough values of x must be used in (12.4.4) to allow smooth curves to be drawn (the broken lines in Fig. 12.5.1). Three representative calculations are given.

(a) $x = 200$. At this point, (by 12.5.1), the estimated value of Y is

$Y = 67·833 + 0·9715 \ (200 - 172·833) = 94·23$
and, by (12.4.4),

$$\text{var}[Y] = 46·393\left(\frac{1}{6} + \frac{(200 - 172·833)^2}{526·833}\right) = 72·72.$$

The Gaussian confidence limits for the population value of Y at $x = 200$ are thus, by (12.4.5), $94\cdot23 \pm 2\cdot776\sqrt{(72\cdot72)}$, i.e. from $Y = 70\cdot56$ to $Y = 117\cdot90$; these are plotted in Fig. 12.5.1 at $x = 200$.

(b) $x = 172\cdot833 = \bar{x}$. At the point $(x-\bar{x}) = 0$ so, from (12.5.1), $Y = 67\cdot833 = \bar{y}$. From (12.4.4) $\mathrm{var}[Y] = \mathrm{var}(y)(1/N) = 46\cdot393/6 = 7\cdot73$, and the confidence limits for the population value of Y are, by (12.4.5), $67\cdot833 \pm 2\cdot776\sqrt{(7\cdot73)}$, i.e. from $60\cdot11$ to $75\cdot55$.

(c) $x = 0$. At this point, the intercept on the y axis, $Y = 67\cdot833 + 0\cdot9715\,(0-172\cdot833) = -100\cdot1$. This is, of course, a considerable extrapolation beyond the range of the experimental results. From (12.4.4),

$$\mathrm{var}[Y] = 46\cdot393\left(\frac{1}{6}+\frac{(0-172\cdot833)^2}{526\cdot833}\right) = 2638,$$

which is far larger than when x is nearer $\bar{x}$. The confidence limits are $-100\cdot1 \pm 2\cdot776\sqrt{(2638)}$, i.e. from -243 to $+42\cdot5$.

The confidence limits, are much wider at the ends than at the central portion of the curve which illustrates the grave uncertainty involved in extrapolation beyond the observations. Moreover it must be remembered that these confidence limits *assume* that the population (true) line is really straight in the region of extrapolation. There is, of course no reason (from the evidence of this experiment) to assume this. In fact with only one observation at each x value linearity could not be tested even *within* the range of the observations. The uncertainty in the extrapolated intercept, $Y = -100\cdot1$, at $x = 0$ is therefore really even greater than indicated by the very wide confidence limits which extend from -243 to $+42\cdot5$ (even apart from the further uncertainties discussed in § 7.2). The intercept does not differ 'significantly' from zero (or even from $+40$ or -240) 'at the $P = 0\cdot05$ level'.

Testing a hypothetical value with the t test

As in § 9.4 the confidence limits can be interpreted as a t test, and this will make it clear that the (rather undesirable) expression 'not significant at the $P = 0\cdot05$ level' means the result of the test is $P > 0\cdot05$. For example to test the hypothesis that the population value of the intercept is $\mu = +40$, calculate, from (4.4.1),

$$t = (Y-\mu)/\sqrt{\mathrm{var}[Y]} = (-100\cdot1-40)/\sqrt{(2638)} = -2\cdot728$$

with 4 degrees of freedom. Referring $t = 2\cdot728$ to a table (see § 4.4) of Student's t distribution shows $P > 0\cdot05$ (two tail; see § 6.1).

The curvature of the confidence limits for the population line is only common sense because there is uncertainty in the value of a, i.e. in the vertical position of the line, as well as in b, its slope. If lines with the steepest and shallowest reasonable slopes (confidence limits for β) are drawn for the various reasonable values of a the area of uncertainty will have the outline shown by the broken lines in Fig. 12.5.1. Another numerical example (with unequal numbers of observations at each point) is worked out in § 13.14.

Confidence limits for the slope. Identity of the analysis of variance with a t test

In § 12.4 it was mentioned that the slope will be normally distributed if the observations are, with variance given by (12.4.2). In this example $b = 0.9715$ and $\text{var}[b] = 46.393/526.833 = 0.08806$. The 95 per cent confidence limits, using $t = 2.776$ as above, are thus, by (12.4.3), $0.9715 \pm 2.776\sqrt{(0.08806)}$, i.e. from 0.15 to 1.80. These limits do not include zero, indicating that b 'differs signficantly from zero at the $P = 0.05$ level'.

As above, and as in § 9.4, this can be put as a t test. The Gaussian (normal) variable of interest is b, and the hypothesis is that its population value (β) is zero so, by (4.4.1),

$$t = \frac{b - \beta}{\sqrt{\text{var}[b]}} = \frac{0.9715 - 0}{0.08806} = 3.274$$

with 4 degrees of freedom (the number associated with var[y], from which var[b] was found). Referring to tables (see § 4.4) of Student's t distribution shows that the probability of a value of t as large as, or larger than, 3.274 occurring is between 0.02 and 0.05, as inferred from the confidence limits.

It was mentioned in § 11.3 (and illustrated in § 11.4) that the variance ratio, F, with 1 d.f. for the numerator and f for the denominator is simply a value of t^2 with f degrees of freedom. In this case t^2 with 4 d.f. $= 3.274^2 = 10.72 = F$ with 1 d.f. for the numerator and 4 for the denominator. This is *exactly* the value of F found in Table 12.5.2 (and $P = 0.02 - 0.05$ exactly as in Table 12.5.2). It is easy to show that this t test is, in general, algebraically identical with the analysis of variance, so the component in the analysis of variance labelled 'linear regression' is simply a test of the hypothesis that the population value of the straight line through the results is zero. This approach also, incidentally, makes it clear why this component in the analysis of variance should have one degree of freedom.

12.6. Fitting a straight line with several observations at each *x* value. The use of a linearizing transformation for an exponential curve and the error of the half-life

The figures in Table 12.6.1 are the results of an experiment on the destruction of adrenaline by liver tissue *in vitro*. Three replicate determinations ($n = 3$) of adrenaline concentration (the dependent variable, y) were made at each of $k = 5$ times (the independent variable, x). The figures are based on the experiments of Bain and Batty (1956).

TABLE 12.6.1

Values of adrenaline (epinephrine) concentration, y (μg/ml)

	6	18	Time, x (min) 30	42	54	Total
	30·0	8·9	4·1	1·8	0·8	
	28·6	8·0	4·6	2·6	0·6	
	28·5	10·8	4·7	2·2	1·0	
Totals	87·1	27·7	13·4	6·6	2·4	137·2

The decrease in adrenaline concentration with time plotted in Fig. 12.6.1 is apparently not linear. Because there is more than one observation at each point in this experiment it is possible to make an estimate of experimental error without assuming the true line to be straight (cf. § 12.5). *Therefore it is possible to judge whether or not it is reasonable to attribute the observed deviations from linearity to experimental error.* The assumptions of the analysis have been discussed in §§ 6.1, 7.2, 11.2, 12.1, and 12.2 which should be read first. *There are not enough results for any of the assumptions to be checked satisfactorily* (see §§ 4.6 and 11.2).

The basic analysis is exactly the same as the one way analysis of variance described in § 11.4, the 'treatments' in this case being the different x values (times). As in § 11.4, it is not necessary to have the same number of observations in each sample (at each time). If the three rows in Table 12.6.1 had corresponded to three blocks (e.g. if three different observers had been responsible for the observations in the first, second, and third rows) then the two-way analysis described in § 11.6 would have been appropriate, with a between-rows (between blocks, between observers) component in the analysis of variance. The additional factor, compared with the one-way analysis in § 11.4, is

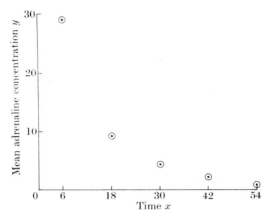

FIG. 12.6.1. Observed mean adrenaline concentration $(\bar{y}_{.j})$ plotted against time (x). Data of Bain and Batty (1956) from Table 12.6.1.

that part of the differences 'between treatments' (i.e. between the mean concentrations at the five different times) can be accounted for by a linear change of concentration with time (see § 12.3).

Calculating the analysis of variance of y

The first part is exactly as in § 11.4 (where more details will be found).

(1) *Correction factor* $G^2/N = (137 \cdot 2)^2/15 = 1254 \cdot 923$.

(2) *Total sum of squares*, from (2.6.5) (cf. (11.4.3) and refer to § 2.1 if you are confused by the notation).

$$\sum_{j=1}^{5} \sum_{i=1}^{3} (y_{ij} - \bar{y}_{..})^2 = \Sigma\Sigma y_{ij}^2 - G^2/N$$
$$= 30 \cdot 0^2 + 8 \cdot 9^2 + ... + 2 \cdot 2^2 + 1 \cdot 0^2 - 1254 \cdot 923$$
$$= 1612 \cdot 037.$$

(3) *Sum of squares (SSD) between columns* (i.e. between the concentrations at different times), by (11.4.5), is

$$\text{SSD between times} = \frac{87 \cdot 1^2}{3} + \frac{27 \cdot 7^2}{3} + ... + \frac{2 \cdot 4^2}{3} - 1254 \cdot 923 = 1605 \cdot 937.$$

This SSD can be split into two components, just as in § 12.5. In this case the calculations could be made easier by transforming the independent variable (x), as shown at the end of this section. But, for generality, the full calculation will be given first.

(a) *Sum of squares due to linear regression.* This is found from (12.3.4) as

$$\text{SSD} = [\overset{N}{\sum}(y-\bar{y})(x-\bar{x})]^2\Big/\overset{N}{\sum}(x-\bar{x})^2.$$

It is easy to make a mistake at this stage by supposing that there are only five x values, when in fact there are $N = 15$ values. This will be avoided if the $N = 15$ pairs of observations are written out in full, as in Table 12.5.1, rather than in the condensed form shown in Table 12.6.1. This is shown in Table 12.6.2.

TABLE 12.6.2

	x	y
	6	30·0
	6	28·6
	6	28·5
	18	8·9
	·	·
	·	·
	·	·
	54	0·8
	54	0·6
	54	1·0
Totals	450	137·2

Firstly find the sum of products using (2.6.7) and Table 12.6.2:

$$\overset{N}{\sum}(y-\bar{y})(x-\bar{x}) = \overset{N}{\sum}xy - \frac{(\Sigma x)(\Sigma y)}{N}$$

$$= (6\times 30\cdot 0)+(6\times 28\cdot 6)+\ldots$$

$$+(54\times 1\cdot 0) - \frac{(450)(137\cdot 2)}{15}$$

$$= -2286\cdot 000.$$

Notice that $\Sigma x = 3(6+18+30+42+54) = 450$ (compare Table 12.6.2), because each x occurs three times; and also that the calculation of the sum of products can be shortened by using the group totals from Table 12.6.1 giving

$$(6\times 87\cdot 1)+(18\times 27\cdot 7)+\ldots+(54\times 2\cdot 4) - \frac{(450)(137\cdot 2)}{15}$$

$$= -2286\cdot 000. \tag{12.6.1}$$

Secondly, find the sum of squares for x. From (2.6.5)

$$\sum_{}^{N}(x-\bar{x})^2 = \sum_{}^{N}x^2 - \frac{(\Sigma x)^2}{N} = 6^2+6^2+...+54^2+54^2-\frac{450^2}{15}$$

$$= 3(6^2+18^2+...+54^2)-\frac{450^2}{15}$$

$$= 4320{\cdot}000. \qquad (12.6.2)$$

From (12.3.4) the SSD due to linear regression now follows

$$\text{SSD} = \frac{(-2286{\cdot}00)^2}{4320{\cdot}000} = 1209{\cdot}68.$$

(b) *SSD for deviations from linearity.* As in § 12.4 this is most easily found by difference (cf. (12.3.3))

SSD due to deviations from linearity = SSD between x values— SSD due to linear regression (12.6.3)

$$= 1605{\cdot}937 - 1209{\cdot}68$$
$$= 396{\cdot}26.$$

(4) *SSD for error.* This is simply the within groups SSD of § 11.4. The experimental error is assessed from the scatter of replicate observations of each x value. It has $N-k = 15-5 = 10$ degrees of freedom

TABLE 12.6.3

Gaussian analysis of variance of y

Source	d.f.	SSD	MS	F	P
Linear regression	1	1209·68	1209·68	1983	≪0·001
Deviations from linearity	$k-2 = 3$	396·26	132·09	216·5	≪0·001
Between x values (times)	$k-1 = 4$	1605·937	401·48	658·2	≪0·001
Error (within x values (times))	$N-k = 10$	6·100	0·6100		
Total	$N-1 = 14$	1612·037			

and is most easily found by difference as in § 11.4, thus $1612{\cdot}037 -160{\cdot}937 = 6{\cdot}100$.

These results can now be assembled in an analysis of variance table (Table 12.6.3) the bottom part resembling Table 11.4.2, the top part resembling Table 12.5.2 (except that the number of different x values, k, is no longer the same as the total number of observations, N).

The table is completed as described in Chapter 11 and § 12.5. Each mean square would be an estimate of σ^2 if the null hypothesis that all 15 observations came from a single normal population (with variance σ^2) were true. The ratio of each mean square to the error mean square is referred to tables of the F ratio (see § 11.3), to see whether it is larger than could be expected by chance. Although a considerable part of the differences between the mean adrenaline concentrations at different times ('between times') is accounted for by a linear relationship between concentration (y) and time (x), the remainder ('deviations from linearity') is still much larger than could be reasonably expected if the *true* line were straight. $P \ll 0.001$, i.e. deviations from linearity as large as those observed, or larger, would occur in far fewer than 1 in 1000 repeated experiments if the true line were straight, and if the assumptions about normal distributions, etc (see § 12.2), made in the calculations are sufficiently nearly true.

There are now two possibilities. Either a curve can be fitted directly to the observations (see §§ 12.7 and 12.8), or a transformation can be sought that converts the graph to a straight line. The latter approach is now described.

A linearizing transformation. Does the catabolism of adrenaline follow an exponential time course?

If the rate of catabolism of adrenaline by liver tissue at any given moment were proportional to the concentration of adrenaline (y) present at that time (x) than the concentration of adrenaline (Table 12.6.1) would be expected to fall exponentially, i.e.

$$y = y_0 e^{-kx}, \tag{12.6.4}$$

where y_0 is the concentration present at time $x = 0$ and k is the rate constant.† The reciprocal of k, the time constant, is the time taken for the concentration to fall to $100/e \simeq 36.8$ per cent of its original value (when $x = 1/k$, it follows from (12.6.4) that $y = y_0/e$). Taking natural logarithms (logs to base e) of (12.6.4) gives

$$\log_e y = \log_e y_0 - kx. \tag{12.6.5}$$

† The symbol k has already been used for the number of treatments (times), but there should be no risk of confusion between its two meanings.

(Remember the log is the power to which the base must be raised to give the argument, so $\log_e e^{-kx} = -kx$.) Therefore there is a straight line relation between $\log y$ and x, with slope $-k$ and intercept $\log y_0$. The half-life of adrenaline is related to the rate constant in a simple way. Putting $y = y_0/2$ in (12.6.5) gives the half-life as

$$x_{0\cdot 5} = \frac{\log_e 2}{k} = \frac{0\cdot 69315}{k}. \tag{12.6.6}$$

The interpretation of the rate constant in molecular terms is discussed in § A2.3.

Common logarithms (to base 10) are more easily available than natural logarithms so it will be convenient to write (12.6.5) in terms of common logarithms. Dividing though by $\log_e 10 \simeq 2\cdot 3026$ gives, using (13.3.5),

$$\log_{10} y = \log_{10} y_0 - \frac{kx}{2\cdot 3026}, \tag{12.6.7}$$

a straight line with slope $= -k/2\cdot 3026$ and an intercept $\log_{10} y_0$.

In order to do the following analysis it is necessary to assume that the values of $\log y$ at each x value are normally distributed (i.e. that y is lognormal, § 4.5) and homoscedastic (see § 12.2). These assumptions,

TABLE 12.6.4

Values of $\log_{10} y$ found from Table 12.6.1

| | Time, x (min) | | | | | Total |
	6	18	30	42	54	
	1·4771	0·9494	0·6128	0·2553	−0·0969	
	1·4564	0·9031	0·6628	0·4150	−0·2218	
	1·4548	1·0334	0·6721	0·3424	0·0000	
Total	4·3883	2·8859	1·9477	1·0127	−0·3187	9·9159

of course, contradict those just made in doing the analysis of variance of y (Table 12.6.3), when y itself was supposed normal and homoscedastic (see the discussions of transformations in §§ 11.2 and 12.2; this problem would not arise if the transformation was made on the independent variable x). There is no way of telling how likely it is that this contradiction will give rise to misleading inferences in particular cases. In the absence of real knowledge about the distributions of the observations the analysis will, as previously emphasized (see §§ 4.2, 6.2, and 7.2), be in error to some unknown extent. *If y* were known to be normally distributed the methods of § 12.8 would be preferred to that now described.

To see whether the straight line defined by (12.6.7) fits the observations, the logarithms of the observations are tabulated in Table 12.6.4.

TABLE 12.6.5

Gaussian analysis of variance of $log_{10}y$

Source	d.f.	SSD	MS	F	P
Linear regression	1	4·2467	4·2467	874·1	≪0·001
Deviations from linearity	3	0·0337	0·0112	2·31	0·1 —0·2
Between times	4	4·2804	1·0701	220·3	≪0·001
Error	10	0·04858	0·004858		
Total	14	4·3290			

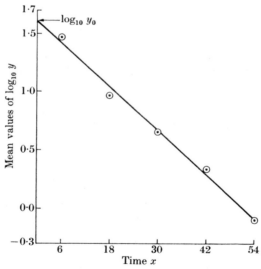

FIG. 12.6.2.   Same data as Fig. 12.6.1, but the mean value of the log_{10} adrenaline concentration (from Table 12.6.4) is plotted against time. The line is that found by the method of least squares, eqn (12.6.12).

The mean log concentrations are plotted against time in Fig. 12.6.2. The graph looks much straighter than Fig. 12.6.1. The analysis of variance of the log concentrations in Table 12.6.4 is now calculated in exactly the same way as the calculation of the analysis of variance of the concentrations themselves (Table 12.6.1). The result is Table 12.6.5. Compare Table 12.6.3.

The results in Table 12.6.5 show that almost all the variation of the log concentrations between times is accounted for by a straight line relationship between log y and time and the evidence against the null hypothesis that the true slope of this line, β, is zero is very strong. Deviations from linearity as large as, or larger than, those observed would be expected to occur in between 10 and 20 per cent of repeated experiments if the true population line were straight (if the assumptions made are correct). There is, therefore, no compelling reason to believe that the relation between log y and time is non-linear, i.e. the experiment provides no evidence that eqn. (12.6.7), and hence also eqn. (12.6.4), fit the observations inadequately. In other words there is no reason to believe that the concentration of adrenaline does not decay exponentially.

Having established that it is reasonable to fit a straight line to the log concentrations, the next step is to estimate the parameters (slope and intercept) of the line.

Fitting the straight line

If the log observations are denoted y', i.e.

$$y' = \log_{10} y, \tag{12.6.8}$$

the equation to be fitted (12.6.7) can be written as

$$\log_{10} Y \equiv Y' = a + b(x - \bar{x}), \tag{12.6.9}$$

which has the same form as in previous examples in this chapter.

Using (12.2.6) the estimate of a is

$$a = \bar{y}' = \frac{9{\cdot}9159}{15} = 0{\cdot}66106.$$

To estimate the slope, the sum of products is first found as described in the analysis of untransformed concentrations (eqns. (12.6.1) and (2.6.7))

$$\Sigma(y' - \bar{y}')(x - \bar{x}) = \Sigma xy' - \frac{(\Sigma x)(\Sigma y')}{N}$$

$$= (6 \times 4{\cdot}3883) + (18 \times 2{\cdot}8859) + \ldots + (54 \times -0{\cdot}3178) - \frac{(450)(9{\cdot}9159)}{15}$$

$$= -135{\cdot}446 \tag{12.6.10}$$

This is negative because y' decreases with x (see § 2.6). The sum of squares for x is found as in eqn. (12.6.2), and is 4320·000 as before. The slope is therefore estimated, by (12.2.8), to be

$$b = \frac{\Sigma(y'-\bar{y}')(x-\bar{x})}{\Sigma(x-x)^2} = \frac{-135\cdot466}{4320\cdot000} = -0\cdot03135. \quad (12.6.11)$$

Putting these values, and $\bar{x} = 450/15 = 30\cdot000$ as before, into (12.6.9) gives the least squares estimate of the straight line as

$$\log_{10} Y = 0\cdot66106 - 0\cdot03135\ (x-30)$$
$$= 1\cdot6016 - 0\cdot03135x. \quad (12.6.12)$$

Comparing this with (12.6.7) gives the estimates of the parameter as

$$\log_{10}y_0 = 1\cdot6016, \text{ so } y_0 = 39\cdot96\ \mu\text{g/ml} \quad (12.6.13)$$

and $\dfrac{-k}{2\cdot3026} = -0\cdot03135$ so $k = 0\cdot07219$ min^{-1} \quad (12.6.14)

In its original form (12.6.4) the estimated regression equation is thus

$$y = 39\cdot96e^{-0\cdot07219x} \quad (12.6.15)$$

The time constant (discussed above) for adrenaline catabolism is estimated to be

$$1/k = 1/0\cdot07219 = 13\cdot85 \text{ min}, \quad (12.6.16)$$

and from (12.6.6) the half-life of adrenaline is

$$x_{0.5} = \frac{0\cdot69315}{k} = 9\cdot602 \text{ min.} \quad (12.6.17)$$

It is shown in § A2.3 that $k^{-1} = 13\cdot85$ min can be interpreted as the mean lifetime of adrenaline molecules, and $x_{0.5} = 9\cdot602$ min can be interpreted as the median lifetime.

Confidence limits for the half-life. From the analysis of variance (Table 12.6.5), the variance of the log observations is estimated as var$[y'] = 0\cdot004858$ (the error mean square with 10 d.f.). Thus, using (12.4.2) and (12.6.2), the variance of the estimated slope (b in eqn. (12.6.9)) is var$[b] = \text{var}[y']/\Sigma(x-\bar{x})^2 = 0\cdot004858/4320\cdot000 = 1\cdot125 \times 10^{-6}$. The value of Student's t for $P = 0\cdot95$ and 10 d.f. (from tables, see § 4.4) is 2·228, so 95 per cent confidence limits for b follow from (12.4.3) as $b \pm t\sqrt{(\text{var}[b])} = -0\cdot03135 \pm 2\cdot228\sqrt{(1\cdot125 \times 10^{-6})} = -0\cdot03371$ to $-0\cdot02899$. The values for the half-life corresponding to these values of b are now found as above ($x_{0.5} = 0.69315/(-2.3026b)$). Because 2.3026 and 0.69315 are *constants*, not variables, no additional error enters in the conversion of b to $x_{0.5}$. The 95 per cent Gaussian confidence limits for the true half-life are thus 8.930 to 10.38 min. As usual these limits can be interpreted as in § 7.9 only *if* all the assumptions discussed in §§ 7.2, 11.2, and 12.2 are fulfilled. And, as usual, the limits are likely to be optimistic (see § 7.2).

A simplifying transformation of the x values

When, as in the present example, the x values are equally spaced *and* there is the same number of observations at each, the values of x can be transformed to make the arithmetic simpler. If x' is defined as $x/12 - 2\frac{1}{2}$ the scale becomes

x	6	18	30	42	54
$x/12$	$\frac{1}{2}$	$1\frac{1}{2}$	$2\frac{1}{2}$	$3\frac{1}{2}$	$4\frac{1}{2}$
x'	-2	-1	0	$+1$	$+2$

Thus $\Sigma x' = -2-2-2-1-1-1+0+0+0+1+1+1+2+2+2 = 0$, so $\bar{x}' = 0$. It follows that $\Sigma(x'-\bar{x}')^2 = \Sigma x'^2 = 3(2^2+1^2+0^2+1^2+2^2) = 30$ and $\Sigma(y'-\bar{y}')(x'-\bar{x}') \equiv \Sigma y'(x'-\bar{x}') = \Sigma y'x' = (-2\times 4.3883) + \ldots + (+2 \times -0.3187) = -11.2872$. These simplified calculations give, of course, the regression equation $Y' = a+b(x'-\bar{x}') = a+bx'$, the plot of log y against x'. The result is $Y' = 0.6611 - 0.3762x'$. Inserting the definition of x' gives $Y' = 0.6611 - 0.3762 \ (x/12 - 2\frac{1}{2}) = 1.602 - 0.03135x$, exactly as above (eqn. (12.6.12)).

12.7. Linearity, non-linearity, and the search for the optimum

In real life most graphs are *not* straight lines. Sometimes, as in § 12.6, they can be converted to lines that are near enough straight, but, as will be shown in § 12.8, this may be a hazardous process. Most elementary books do not discuss curves that are non-linear (in the sense to be defined in this section) because the mathematics is inconvenient to do by hand. Since most relationships that are based on some sort of physical model are non-linear, this is unfortunate. A simple computer method for fitting non-linear models will be given in § 12.8. Before this the principles of finding least squares estimates will be discussed, mainly in a pictorial way, and an attempt made to give an idea of the scope of linear (in the general sense) models.

Finding least squares solutions. The geometrical meaning of the algebra

In § 12.2 the least square estimates, $\hat{a}$ and $\hat{b}$, of the parameters, α and β, of the straight line (12.2.3) were found algebraically. (In this section, and in § 12.8, the symbols $\hat{a}$ and $\hat{b}$ will be used to distinguish least squares estimates from other possible estimates of the parameters.) It will be convenient to illustrate the approach to more complicated curves by first going into the case of the straighter line in greater detail.

The intention is to find the values of the parameter estimates that make the sum of the squares of the deviations of the observations (y) from the calculated values (Y), $S = \Sigma(y-Y)^2$ (eqns. (12.2.1) and (12.2.5)), as small as possible. *Notice that during the estimation procedure the experimental observations are treated as constants (the particular observations made) and various possible values of the parameters are considered.* The conventional way of finding a minimum, as in § 12.2, is to differentiate and equate to zero. How this works was illustrated in Fig. 12.2.2, in which S was plotted against various possible values for a(b being held constant). The slope of this graph (i.e. $\partial S/\partial a$) is zero at the minimum, and the corresponding value of a is taken as the least squares estimate of α. The curly ∂, indicating partial differentiation means that b is treated as a constant when differentiating (12.2.5) to obtain (12.2.6). This means that b is given a fixed value which is inserted, along with the experimental observations (from Table 12.7.1) into (12.2.5) so that S can be calculated for various values of a, giving the curve plotted in Fig. 12.2.2. It may occur to you to ask whether the value at which b is held constant makes any difference to the estimate of α. In fact it does not because the expression found for $\partial S/\partial a$ did not involve b, and similarly the expression for $\partial S/\partial a$ did not involve a. The geometrical meaning of this will be made clear using the data in Table 12.7.1.

Fitting a straight line in the form

$$Y = a+b(x-\bar{x}) \qquad (12.7.1)$$

gives $$Y = 8{\cdot}000+2{\cdot}107(x-1{\cdot}0), \qquad (12.7.2)$$

which is plotted in Fig. 12.7.1. The calculations and interpretation are the same as for the example in § 12.5. The corresponding analysis of variance in Table 12.7.2 shows that, if the true line were straight and the assumptions described in § 12.2 were true, then the slope of this line is

TABLE 12.7.1

x	y
-2	1
-1	4
0	6
1	9
2	11
3	10
4	15

	x	y
Total	7	56
Mean	1·0	8·0

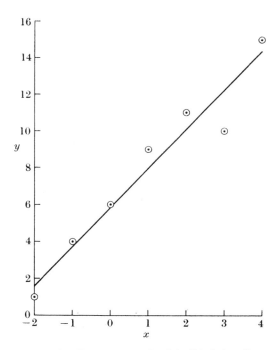

FIG. 12.7.1. A straight line (eqn. (12.7.2)) fitted by the method of least squares to the data in Table 12.7.1.

greater than could be reasonably expected if the population slope (β) were zero.

The least squares estimates given in (12.7.2) are $\hat{a} = \bar{y} = 8\cdot000$ and $\hat{b} = 2\cdot107$, calculated from eqns. (12.2.6) and (12.2.7). If the values of x and y from Table 12.7.1 are inserted in the expression for the sum of

TABLE 12.7.2

Source	d.f.	SS	MS	F	P
Linear regression	1	124·321	124·321	80·95	<0·001
Deviations from linearity	5	7·679	1·536		
Total	6	132·000			

squared deviations, S, (eqn. (12.2.5)), then S can be calculated for various possible values of a and b. There are three variables here so the results must be plotted as a three-dimensional graph. The most convenient way to represent this on two dimensional paper is to plot a contour map, the contours representing values of S (the height), i.e. the a- and b-axes are in the plane of the paper and the S-axis is sticking up perpendicular to the paper. The result, calculated for the results in Table 12.7.1 using eqn. (12.2.5), is shown in Fig. 12.7.2. The graph is seen to represent a valley with elliptical contours. The bottommost point of the valley corresponds to $\hat{a} = 8\cdot000$ and $\hat{b} = 2\cdot107$, i.e. the least squares estimates already found.

It can now be shown why the value of b used in constructing Fig. 12.2.2 did not matter. In Fig. 12.7.3 sections across the valley are shown for $b = \hat{b} = 2\cdot107$, $b = 3\cdot0$ and $b = 3\cdot4$. The lines along which these sections have been taken are shown in Fig. 12.7.2. It can be seen that wherever the section is taken (i.e. whatever value b is held constant at), the minimum in the curve occurs at the same place, viz. at $a = \hat{a} = 8\cdot00$. Similarly, if sections across the valley are taken at various fixed a values (i.e. at 90° to the sections illustrated), each section will give a plot of S against b, with slope $\partial S/\partial b$. The minima ($\partial S/\partial b = 0$) of these curves will clearly all be at $b = \hat{b} = 2\cdot107$ whatever the value of a. Clearly this independence of a and b arises because the axes of the ellipses in Fig. 12.7.2 are at right angles to the coordinates of the graph (the ellipses are said to be in *canonical* form).

The fundamental form of the straight-line equation is $\mu = \alpha' + \beta x$,

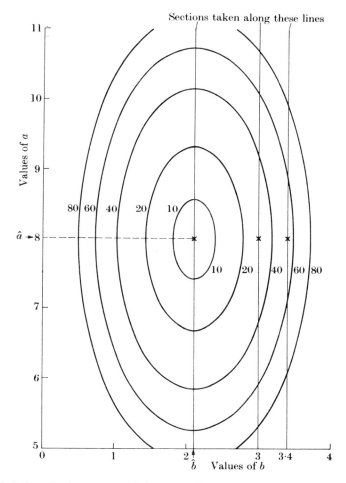

FIG. 12.7.2.   Contour map of the sum of squared deviations, S (on an axis perpendicular to the paper), plotted against various values of a and b using eqn. (12.2.5) and the data in Table 12.7.1 (plotted in Fig. 12.7.1). The contours for a straight line fitted in the form $Y = a+b(x-\bar{x})$ always have this appearance. Values of S are marked on the contours. The minimum value of S, at the bottom of the valley, gives the least squares estimates as $\hat{a} = 8 \cdot 000$ and $\hat{b} = 2 \cdot 107$. Sections across the valley, along the lines shown, are plotted in Fig. 12.7.3. The lowest point on each line (minima in Fig. 12.7.3) is marked ✕.

where α' is the intercept, β the slope, and μ the population value of y. Inserting the estimates of the parameters gives

$$Y = \hat{a}' + \hat{b}x \tag{12.7.3}$$

and, because (12.7.2) can be written as

$$Y = 5{\cdot}893 + 2{\cdot}107x, \tag{12.7.4}$$

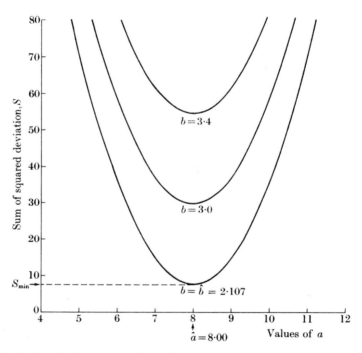

FIG. 12.7.3.   Sections across the valley along the lines indicated in Fig. 12.7.2. The slope of the line, $\partial S/\partial a$, is zero when S is at a minimum, as shown in Fig. 12.2.2. The value of S at the bottom of the valley is 7·679 as shown, and as found in Table 12.7.2.

it is seen that $\hat{a}' = 5{\cdot}893$ and $\hat{b} = 2{\cdot}107$. Comparison of (12.7.1) and (12.7.3) shows that in general, as in § 12.2,

$$\hat{a}' = \hat{a} - \hat{b}\bar{x}. \tag{12.7.5}$$

It may be asked why (12.7.4) was arrived at indirectly, through the seemingly more complicated form, (12.7.2). Why not apply the method of least squares directly to (12.7.3)? The answer to this will become clear when it is tried. The method of least squares will now be applied to the straight line in the form of (12.7.3), in just the same way as it was applied in § 12.2 to the straight line in the form of (12.7.1).

Denoting the observations y and the values calculated from (12.7.3) as Y, as in § 12.2, gives the sum of squared deviations, which is to be minimized, as

$$
\begin{aligned}
S &= \Sigma(y-Y)^2 = \Sigma(y-a'-bx)^2 \\
&= \Sigma(y^2+a'^2+b^2x^2-2a'y-2ybx+2a'bx) \\
&= \Sigma y^2+Na'^2+b^2\Sigma x^2-2a'\Sigma y-2b\Sigma yx+2a'b\Sigma x.
\end{aligned}
\qquad (12.7.6)
$$

This is analogous to (12.2.5), but notice that this time the last term is *not* zero. As in § 12.2, S is differentiated with respect to a', treating b as a constant, giving

$$
\frac{\partial S}{\partial a'} = 2Na'-2\Sigma y+2b\Sigma x,
\qquad (12.7.7)
$$

and equating this to zero to find the value of a' for which S is a minimum (see Fig. 12.7.5) gives

$$
Na'+b\Sigma x = \Sigma y.
\qquad (12.7.8)
$$

The value of a' for which S is a minimum is no longer independent of b, as shown by the presence of b in (12.7.8), the solution of which will depend on the value of b chosen.

Differentiating (12.7.6) with respect to b, holding a' constant, gives

$$
\frac{\partial S}{\partial b} = 2b\Sigma x^2-2\Sigma yx+2a'\Sigma x,
\qquad (12.7.9)
$$

and again equating to zero gives

$$
a'\Sigma x+b\Sigma x^2 = \Sigma yx.
\qquad (12.7.10)
$$

Again unlike the result in § 12.2, the estimate of b is seen to depend on the value of a'.

The required solution for $\hat{a}'$ and $\hat{b}'$ are those for which (12.7.8) and (12.7.10) are both true simultaneously. In fact, (12.7.8) and (12.7.10) are a pair of (linear) simultaneous equations (known, in regression analysis, as the *normal equations*), which can be solved for a' and b by school-book methods giving (with the values of x and y in Table 12.8) $\hat{a}' = 5\cdot893$ and $\hat{b} = 2\cdot107$ as found above.

What is the geometrical meaning of these results? If contours are plotted from (12.7.6) (using the data in Table 12.7.1) the results are as shown in Fig. 12.7.4.

The contours are still elliptical, but their axes are no longer parallel with the coordinates of the graph. When sections are made across the valley at the values of b shown in Fig. 12.7.4, the results are as shown in Fig. 12.7.5.

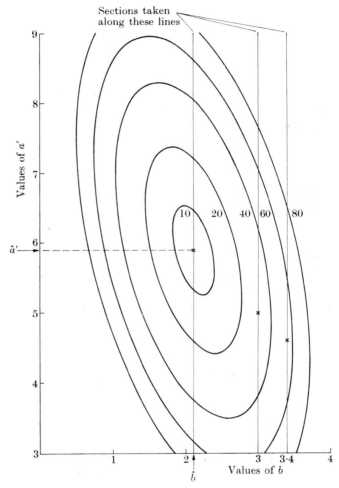

FIG. 12.7.4. Contour maps of S (values marked on the contours) for same data as Fig. 12.7.2, but straight line fitted in the form $Y = a' + bx$. Sections across the valley, along the lines shown, are plotted in Fig. 12.7.5. The lowest point along each line (minima in Fig. 12.7.5) is marked ×.

The value of a' for which S is a minimum is seen to depend on the value at which b was held constant when making the section across the valley, as expected from (12.7.8). Of course, the slope of the curves in Fig.

12.7.5, $\partial S/\partial a'$, is zero at the minimum of each curve. But the only point at which $\partial S/\partial b$ is *simultaneously* zero is at the bottommost point of the valley in Fig. 12.7.4 (hence the *simultaneous* equations). For example, on the curve for $b = 3\cdot4$, in Fig. 12.7.5, S is at a minimum (i.e. $\partial S/\partial a' = 0$) at the point $a' = 4\cdot6$. Inspection of Fig. 12.7.4 makes it clear that if a section is made across the valley (at 90° to the sections

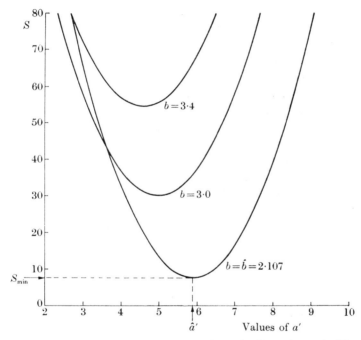

FIG. 12.7.5. Sections across the valley, along the lines shown in Fig. 12.7.4, when a straight line is fitted in the form $Y = a'+bx$. The value of S at the bottom of the valley is 7·679 as before.

in Fig. 12.7.5) at $a' = 4\cdot6$, giving a plot of S against b (with slope $= \partial S/\partial b$), the minimum will *not* be at $b = 3\cdot4$. That is to say, at the point $a' = 4\cdot6$, $b = 3\cdot4$, $\partial S/\partial a'$ is zero but $\partial S/\partial b$ is not.

It is now clear that the effect of writing the straight line in the form $Y = a+b(x-\bar{x})$, is to make the estimates $\hat{a}$ and $\hat{b}$ independent of each other so two simple independent equations (derived in § 12.2) can be used for their estimation. If the line is written in the form $Y = a'+bx$, then the estimates are no longer independent, but must be found by solving simultaneous equations.

What does linear mean?

The term *linear*, as usually based by statisticians, embraces more than the simple straight line. It includes any relationship of the form

$$Y = a + bx_1 + cx_2 + dx_3 + ..., \qquad (12.7.11)$$

where x_1, x_2, x_3, . . . are independent variables (see § 12.1; examples are given below), and a, b, c, d, . . . are estimates of parameters. This relationship includes, as a special case, the straight line ($Y = a + bx$), which has already been discussed at length. Equation (12.7.11) is described as a *multiple linear regression* equation (the 'linear' bit is, sad to say, often omitted). As well as describing straight line relationships for several variables (x_1, x_2, . . .), (12.7.11) also includes, for example, the *parabola* (or *second degree polynomial*, or quadratic), $Y = a + bx + cx^2$, as the special case in which x_2 is the square of x_1. (As discussed in § 12.1, an 'independent variable' in the regression sense is simply one the value of which can be fixed precisely by the experimenter; it does not matter that in this case x_1 and x_2 are not independent in the sense of §§ 2.4 and 2.7 since their covariance is not zero. All that is required is that the values of x_1, x_2, . . . be known precisely.) The parabola is not a straight line of course, but it is *linear in the sense that Y is a linear function* (p. 39) *of the parameters if the x values are regarded as constants* (they are fixed when the experiment is designed). This is the sense in which 'linear' is usually used by the statisticians. It turns out that for (12.7.11) in general (and therefore for the parabola), the estimates of the parameters are linear functions of the observations. This has already been shown in the case of the straight line for which $\hat{a} = \bar{y}$, and for which $\hat{b}$ has also been shown (eqn. (12.4.1)) to be a linear function of the observations. This means that the parameter estimates will be normally distributed if the observations are, and the standard deviations of the estimates can be found using (2.7.11). Also, if the parameter estimates are normally distributed, it is a simple matter to interpret their standard deviations in terms of significance tests or confidence limits. Furthermore, linear problems (including polynomials) give rise to linear simultaneous equations (like (12.7.8) and (12.7.10)) which are relatively easy to solve (cf. § 12.8). They can be handled by the very elegant branch of mathematics known as *matrix algebra*, or *linear algebra* (see, for example, Searle (1966), if you want to know more about this). It is doubtless partly the aesthetic pleasure to be found in deriving analytical solutions

in terms of matrix algebra that has accounted for the statistical literature being heavily dominated by empirical linear models with no
physical basis, and, much more dangerous, the widespread availability
of computer programs for fitting such models by people who do not
always understand their limitations (some of which are mentioned
below).

Polynomial curves

It does not change the nature of the problem if some x values in
(12.7.1) are powers of the others. Thus the general polynomial regression
equation

$$Y = a+bx+cx^2+dx^3+...$$ (12.7.12)

is still a linear statistical problem. Increasingly complex shapes can
be described by (12.7.12) by including higher powers of x. The highest
power of x is called the *degree* of the polynomial, so a straight line is a
first degree polynomial, the parabola is a second degree polynomial,
the cubic equation, $Y = a+bx^2+dx^3$, is a third degree polynomial,
and so on. Just as a straight line can always be found to pass exactly
through any two points, it can be shown that pth degree polynomial
can always be found that will pass exactly through any specified $p+1$
points. Because of the linear nature of the problem discussed above,
polynomials are relatively easy to fit (especially if the x values are
equally spaced). Methods are given in many textbooks (e.g. Snedecor
and Cochran (1967, pp. 349–58 and Chapter 13); Williams (1959,
Chapter 3); Goulden (1952, Chapter 10); Brownlee (1965, Chapter 13);
and Draper and Smith (1966)) and will therefore not be repeated here.
   Although polynomials are the only sort of curves described in most
elementary books, they are, unfortunately, not of much interest to
experimenters in most fields. In most cases the reason for fitting a
curve is to estimate the values of the parameters in an equation based
on a physical model for the process being studies (for example the
Michaelis–Menten equation in biochemistry, which is discussed in
§ 12.8). Very few physical models give rise to polynomials, which are
therefore mainly used in a completely empirical way. *In most situations
nothing more is learned by fitting an empirical curve (the parameters of
which have no physical meaning), than could be made obvious by drawing a
curve by eye.* One possible exception is when the line is to be used for
prediction, for example a calibration curve, and an estimate of error is

required for the prediction—see § 13.14. In this case a polynomial curve might be useful if the observed line was not straight.

Multiple linear regression

If, as is usually the case, the observation depends on several different variables, it might be thought desirable to find an equation to describe this dependence. For example if the response of an isolated tissue depended on the concentration of drug given (x_1, say), and also on the concentration of calcium (x_2, say) present in the medium in which the tissue was immersed, then the response, Y, might be described by a multiple linear regression equation like (12.7.11), i.e.

$$Y = a + bx_1 + cx_2. \tag{12.7.13}$$

This implies that the relationship between response and drug concentration is a straight line at any given calcium concentration, and the relationship between response and calcium concentration is a straight line at any given drug concentration (so the three-dimensional graph of Y against x_1 and x_2 is a flat plane). As already explained x_1 could be the log of the drug concentration, and x_2 could similarly be some transformation of the calcium concentration, the transformation being chosen so that (12.7.13) describes the observations with sufficient accuracy. Even so the linear nature of (12.7.12) is a considerable restriction on its usefulness. Furthermore all the assumptions described in § 12.2 are still necessary here. The process of fitting multiple linear regression equations is described, for example, by Snedecor and Cochran (1967, Chapter 13); Williams (1959, Chapter 3); Goulden (1952, Chapter 8); Brownlee (1965, Chapter 13); and Draper and Smith (1966).

The really serious hazards of multiple linear regression arise when the x values are not really independent variables in the regression sense (see § 12.1), i.e. when they are not fixed precisely by the experimenter, but are just observations of some variable thought to be related to Y, the variable of interest. Data of this sort always used to be analysed using the correlation methods described in § 12.9, but are now very often dealt with using multiple regression methods. There is much to be said for this as long as it is remembered that however the results are analysed it is impossible to infer causal relationships from them (see also §§ 1.2 and 12.9).

Consider the following example (which is inspired by one discussed by Mainland (1963, p. 322)). It is required to study the number of working days lost though illness per 1000 of population in various areas of a large city. Call this number y. It is thought that this may depend on the number of doctors per 1000 population (x_1) in the area and the level of prosperity (say mean income, x_2) of the area. Values of y, x_1, and x_2 are found by observations on a number of areas and an equation of the form of (12.7.13) is fitted to the results. Even supposing (and it is not a very plausible assumption) that such complex results can be described adequately by a linear relationship, and that the other assumptions (§ 12.2) are fulfilled, the result of such an exercize is very difficult to interpret. Suppose it were found that areas with *more* doctors (x_1) had *fewer* working days lost through illness (y). (If (12.7.13) were to fit the observations this would have to be true whatever the prosperity of the area.) This would imply that the coefficient b must be negative. Suppose it were also found that areas with *high* incomes had *few* working days lost through illness (whatever number of doctors were present in the area), so the coefficient c is also negative. Inserting the values of a, b, and c found from the data into (12.7.13) gives the required multiple regression equation. If x_1 in this equation is increased y will decrease (because b is negative). If x_2 is increased y will decrease (because c is negative). It might therefore be inferred (and often is) that if more doctors were induced to go to an area (increasing x_1), the number of working days lost (y) would decrease. This inference implies that it is believed that the presence of a large number of doctors is the *cause* of the low number of working days lost, and the data provide no evidence for this at all. Whatever happens in the equation, it is clear that in real life one still has no idea whatsoever what will happen if doctors go to an area. The number of working days lost might indeed decrease, but it might equally well increase. For example, it might be that doctors are attracted to areas of the city which are near to the large teaching hospitals, and that these areas also tend to be more prosperous. It is quite likely, then that most people in these areas will do office jobs which do not involve much health hazard, and this might be the real *cause* of the small number of working days lost in such areas. Conversely, less prosperous areas, away from teaching hospitals, where many people work at industrial jobs with a high health hazard, (and where, therefore, many working days are lost through illness) attract fewer doctors. If the occupational health hazard were the really important factor then inducing more doctors to go to an area might, far from decreasing the number of working days lost according to the naive interpretation of the regression equation, might

18

actually *increase* the number lost, because the occupational health hazards would be unchanged, and the larger number of doctors might increase the proportion of cases of occupational disease that were diagnosed. Similarly it cannot be predicted what effect in the change in the prosperity of an area will have on the number of working days lost. The regression equation describes (at most) only the static situation at the time the survey was made and says *nothing at all* about what would happen if the x values were changed.

Clearly a correlation or regression relationship based on static survey data of this sort, in which the x values are correlated with each other and with other variables that have not been included in the regression equation because they have not been thought of or cannot be measured, is the sort of thing for which it is possible to think up half-a-dozen plausible explanations before breakfast. The only use of the results, apart from describing, if you are lucky, the situation as it was when the survey was done, is to provide hints about what sort of proper experiments might be worth-while. The only way to find out what effect increasing the number of doctors in an area has is to increase the number. In a proper experiment various numbers of doctors would be allocated strictly at random (see § 2.3) to the areas being tested. This point has been discussed already in Chapter 1.

Further discussions will be found in § 12.9, and in Mainland (1963, p. 322). More quantitative descriptions of the hazards of multiple regression will be found in Tukey (1954), Snedecor and Cochran (1967, pp. 393–400), and Brownlee (1965, pp. 452–4).

Linear models and the analysis of variance

It is worth mentioning in passing that the analysis of variance can be written in the form of a multiple linear regression problem. Consider for example the comparison of two treatments on two independent samples of test objects (the problem discussed at length in Chapter 9). It was pointed out in § 11.2 that in doing the analysis based on Gaussian (normal) distribution (i.e. Student's t test in the case of two samples— see § 9.4) it is assumed that the ith observation on the first treatment can be represented as $y_{i1} = \mu + \tau_1 + e_{i1}$, and for the second treatment $y_{i2} = \mu + \tau_2 + e_{i2}$ (eqn (11.2.1)), where μ is a constant, and τ_1 and τ_2 are constants characteristic of the first and second treatments respectively. This model can be written in the form of a multiple linear regression equation

$$y_{ij} = \mu + \tau_1 x_1 + \tau_2 x_2 + e_{ij}, \qquad (12.7.14)$$

where x_1 is defined to have the value 1 for all responses to treatment 1 $(j = 1)$ and 0 for all responses to treatment $2(j = 2)$, and x_2 is 1 for responses to treatment 2, and 0 for responses to treatment 1. Inserting these values (12.7.14) reduces to $y_{i1} = \mu + \tau_1 + e_{i1}$ for treatment 1, and to $y_{i2} = \mu + \tau_2 + e_{i2}$ for treatment 2, exactly as in § 11.2 If the estimates of τ_1 and τ_2 from the data are called b and c, and estimate of μ is called a, the estimated value for ith response to the jth treatment becomes $Y = a + bx_1 + cx_2$, identical with (12.7.13). The estimation of treatment effects (values of τ) is the same problem as the estimation of the regression coefficients. An intermediate level discussion of this approach will be found in the first (1960) edition of Brownlee's (1965) book.

12.8. Non-linear curve fitting and the meaning of 'best' estimate

For the purposes of illustration, the problem of fitting the Michaelis–Menten hyperbola† (or Clark equation, or Langmuir isotherm) will be discussed. In biochemical terms the equation states that the velocity of an enzyme catalysed reaction is $\mathscr{V} x/(\mathscr{K} + x)$ where x is the concentration of substrate (the independent variable in the sense of § 12.1, and the parameters *true* or *population* values) of the equation are $\mathscr{V}$ (the maximum velocity, approached as $x \to \infty$), and $\mathscr{K}$, the Michaelis constant (the substrate concentration necessary for half-maximum velocity; if $\mathscr{K} = x$ the velocity is $\mathscr{V}/2$). The observed velocity, y say (the dependent variable, see § 12.1), will differ from this by some error. If V and K are estimates from experimental results, of $\mathscr{V}$ and $\mathscr{K}$, the estimated velocity of the reaction will be

$$Y = \frac{Vx}{K + x}.\qquad (12.8.1)$$

The shape of this curve is shown in Fig. 12.8.1.

Notice that the *parameters*, $\mathscr{V}$ and $\mathscr{K}$, are not linearly related to Y so this *is* a non-linear problem in the sense defined in § 12.7.

There are many ways of estimating $\mathscr{V}$ and $\mathscr{K}$. The relative merits of some of them will be considered below. First, the problem of finding least squares estimates for non-linear models will be discussed. The

---

† The general formula for a hyperbola is $(Y - c_1)(x - c_2) = $ constant, where the constants c_1 and c_2 are the asymptotes of the hyperbola. If $c_1 = V$ and $c_2 = -K$ rearranging (12.8.1) shows that $(Y - V)(x + K) = -VK = $ constant, which has the same form as the general formula.

problem of estimating the error of these estimates from the experimental results is important but complicated, and it will not be considered here (see Draper and Smith 1966). Oliver (1970) has given formulas for calculating the asymptotic variances of V and K from scatter of the observations $s(y)$. If there were several observations (y values) at each x then $s(y)$ would be estimated from the scatter of these values 'within

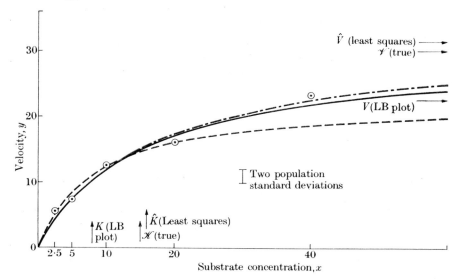

Fɪɢ. 12.8.1.   Fitting the Michaelis–Menten hyperbola.

⊙ 'Observed' values from Table 12.8.1.
—— True (population) hyperbola (known only because the 'observations' were computer-simulated, not real. See discussion on p. 268). The population standard deviation is $\sigma(y) = 1 \cdot 0$ at all x values.
— · — Least squares estimate of population line found from 'observed' values.
– – – Lineweaver–Burk (LB, or double reciprocal plot) estimate of population line found from the same 'observations'.

The true values, $\mathcal{V}$ and $\mathcal{K}$, and their values estimated by the two methods (from Table 12.8.4) are marked on the graph.

x values', but in the following example where there is only one observation at each x, the best that can be done is to assume the population curve follows (12.8.1), in which case the sum of squares of deviations from the fitted curve, $S_{\min}$ will be an estimate of $s^2(y)$. This is exactly like the situation for a straight line discussed in § 12.6. The formulas involve the population values $\mathcal{V}$ and $\mathcal{K}$ for which the experimental values V and K must be substituted. No allowance is made for the uncertainty resulting from the use of sample values V, K, and $s(y)$ in place of population values $\mathscr{V}$, $\mathscr{K}$, and $\sigma(y)$ so the formulas are to some extent optimistic. Using them is just like using the normal deviate u instead of Student's t (see §§ 4.3 and 4.4).

Least squares estimates for non-linear models

The approach is exactly as in § 12.7. It is required to find the estimates of the parameters that minimize the sum of the squares of deviations between observed (y) and calculated (Y) velocities, $S = \Sigma(y-Y)^2$. In this example these least squares estimates will be denoted $\hat{V}$ and $\hat{K}$, as in § 12.7.

From (12.8.1),

$$S = \Sigma(y-Y)^2 = \sum\left(y-\frac{Vx}{K+x}\right)^2$$

$$= \sum\left[y^2 - 2\frac{Vyx}{K+x} + \left(\frac{Vx}{K+x}\right)^2\right]$$

$$= \Sigma y^2 - 2V\sum\left(\frac{yx}{K+x}\right) + V^2\sum\left(\frac{x}{K+x}\right)^2. \tag{12.8.2}$$

If, as in § 12.7, this expression is differentiated first with respect to V holding K constant (giving $\partial S/\partial V$), and then with respect to K holding V constant (giving $\partial S/\partial K$), and the two derivatives equated to zero, the result is a pair of simultaneous equations (*the normal equations*) that can be solved for $\hat{V}$ and $\hat{K}$, just as (12.7.8) and (12.7.10) could be solved for $\hat{a}'$ and $\hat{b}$ in § 12.7. The only snag is that in this case they are non-linear simultaneous equations that cannot be solved by school-book methods. Another difficulty is that there may well be (as in this example) more than one set of solutions. The sort of difficulty that may be encountered can be illustrated using a numerical example. The figures in Table 12.8.1 represent the results of an enzyme kinetic 'experiment'.

Using these figures and eqn. (12.8.2), contours for various S values can be calculated and plotted against V and K as shown in Fig. 12.8.2 (a) and (b).

The contours are not simple ellipses like those found in § 12.7 (Fig. 12.7.2 and 12.7.4). The required solution is clearly the bottom-most point of the valley in Fig. 12.8.2(a) (where $\partial S/\partial V$ and $\partial S/\partial K$ are simultaneously zero, see § 12.7), and it can be seen that this point

TABLE 12.8.1

Results of an enzyme kinetic experiment. The population (true) velocities are also given. They are known only because the 'experiment' was not real, but was simulated on a computer, as discussed later in this section.

Substrate concentration (x)	'Observed' velocity (y)	Population mean velocity (μ)
2·5	5·576	4·2857
5·0	7·282	7·5000
10·0	12·521	12·0000
20·0	16·138	17·1429
40·0	23·219	21·8182

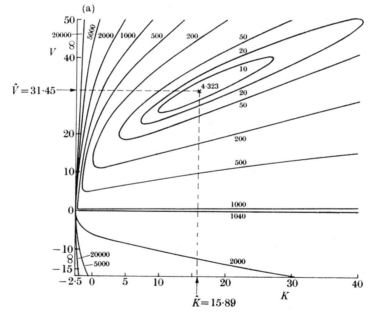

FIG. 12.8.2(a) Fitting the Michaelis–Menten hyperbola. Contour map of the sum of squared deviations, S (on an axis perpendicular to the paper), against various values of K and V. This figure is analogous to Figs. 12.7.2 and 12.7.4 which referred to the fitting of a straight line. The values of S, calculated from eqn. (12.8.2) using the observations in Table 12.8.1, are marked on the contours. (a) This covers the (physically important) positive values of V and K. The minimum value of S, 4·323, at the bottom of the valley corresponds to the least square estimates $\hat{V} = 31\cdot45$ and $\hat{K} = 15\cdot89$.

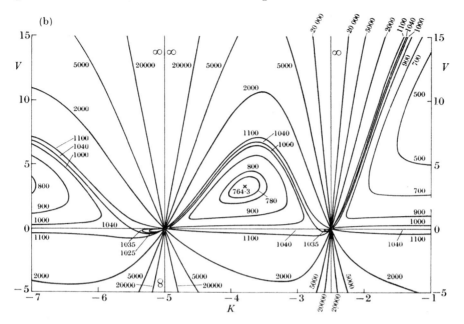

FIG. 12.8.2(b) This shows the contour map in the region of negative (physically impossible) K values. There is seen to be a subminimum at $V = 3{\cdot}244$ and $K = -3.793$, but this corresponds to $S = 764{\cdot}3$, a far worse fit than the lowest minimum, $S = 4.323$.

The contours marked 1040 are actually for $S = \Sigma y^2 = 1040{\cdot}4726$. For values of S equal to or greater than this, the contour lines behave curiously.

corresponds to least squares estimates of $\hat{V} = 31{\cdot}45$ and $\hat{K} = 15{\cdot}89$ at the minimum value, $S = 4{\cdot}323$. But the contours behave in a curious and complicated fashion in the region of negative K values shown in Fig. 12.8.2. There are infinitely high ridges at $K = -2{\cdot}5$, $K = -5$, etc., because at these points $K+x$ in eqn. (12.8.2) becomes zero. The astonishing behaviour of the contour lines for high values of $S(\geqslant 1040{\cdot}4726 = \Sigma y^2)$ can be seen in Fig. 12.8.2(b). The contours all cross each other, and cross the infinitely high ridges at $K = -2{\cdot}5$, $-5{\cdot}0$, etc. The points of intersection of the contours have curious properties. The height, i.e. the value of S, at these points depends on the direction from which the points are approached and, although anyone who has climbed a mountain will feel that this fact is not surprising, topographers might think that it was a warning against pushing the geographical analogy too far.

There are, in fact, several solutions to the simultaneous 'normal equations' in this case.† For example, there is another pit at the point $V = 3 \cdot 244$ and $K = -3 \cdot 793$, shown in Fig. 12.8.2(b). Although these values correspond to a minimum in S, the minimum is merely a hollow in the mountain side. The value of S at this minimum, $764 \cdot 3$, is far greater than the value of S at the least of all the minimums, $4 \cdot 323$, as shown at the bottom of the valley in Fig. 12.8.2(a). If there are several minimums that with the smallest S, i.e. the best fitting curve, corresponds to the least squares estimates. In this case (though not necessarily in all problems) all of the subminimums correspond to negative values of K that are physically impossible and can therefore be ruled out.

There are many methods of finding the least squares solutions (see, for example, Draper and Smith (1966, Chapter 10), Wilde (1964)). In almost all non-linear problems the solution involves successive approximations (iteration). The procedure is to make a guess at the solution and then to apply a method for correcting the guess to bring it nearer to the correct solution. The method is applied repeatedly until further corrections make no important difference. The final solution should, of course, be independent of the initial guess. Geometrically, the initial guess corresponds to some point on Fig. 12.8.2 (say $V = 10$, $K = 2$ for example). The mathematical procedure is intended to proceed by steps down the valley until it reaches (sufficiently nearly) the bottom, which corresponds to $V = \hat{V}$ and $K = \hat{K}$. One method, which sounds intuitively pleasing is to follow the *direction of steepest descent* (which is perpendicular to the contours) from the initial guess point to the minimum. However inspection of Fig. 12.8.2 shows that the direction of steepest descent often points nowhere near the minimum. Furthermore if the search for the minimum is started in the precipitous terrain shown in Fig. 12.8.2(b), or if this region is reached at some time during the search, the direction of steepest decent may be completely misleading. Although this and other sophisticated methods (see, e.g Draper and Smith (1966, Chapter 10)) have had much success, many people now favour simpler search methods which seem to be rather more robust (see Wilde 1964). One such method which has proved useful for curve fitting (Hooke and Jeeves 1961; Wilde 1964; Colquhoun, 1968, 1969) will now be described.

---

† There is also, in general, the possibility of a saddle point or mountain pass when a minimum in the plot of S against one parameter coincides with a maximum in the plot of S against the other. Such a point also satisfies the normal equations because both derivatives are zero.

Patternsearch minimization

In Table 12.8.2 a computer program (an Algol 60 procedure) is

TABLE 12.8.2

Patternsearch procedure (in Algol 60) written by M. Bell (University of London Institute of Computer Science) to whom I am grateful for permission to reproduce it

A Fortran IV version can be supplied on request.
For this procedure the following must be supplied:

k = number of variables on which the function to be minimized depends

$bp[1:k]$ = basepoint, the initial guesses for the values of each variable (parameter estimate)

$np[1:k]$ = newpoint, a **real array**

$step[1:k]$ = initial step size for altering each variable in search for better values

$redfact[1:k]$ = step reduction factor for each variable (usually between 0·1 and 0·5)

$critstep[1:k]$ = smallest permissible step size for each variable. This controls accuracy with which the minimum is located.

eps = half the smallest of the critsteps

$eval$ = number of evaluation, an integer variable

$evalim$ = maximum permissible number of evaluations of function

pat = patternfactor (usually 1·0, but other values may help in some cases)

min = a real variable

The function to be minimized is declared as

real procedure *function* (P); **real array** (P); (see Table 12.8.3 for an **example**)
On exit, after calling patternsearch,

min = minimum value of the function

np = values of the variables corresponding to the minimum (the least squares parameter estimates for example)

$eval$ = number of evaluations of the function during the search

procedure *patternsearch* (*function*, k, bp, np, $step$, $redfact$, $critstep$, eps, $eval$, $evalim$, pat, min); **integer** k, $eval$, $evalim$; **real** eps, pat, min; **real array** bp, np, $step$, $redfact$, $critstep$; **real procedure** *function*;
begin real array *move* $[1:k]$; **integer** i, $fails$; **real** $value$, $minstore$;
  procedure *explore*;
    begin real *home*; **integer** j;
      $fails := 0$
      for $i := 1$ **step** 1 **until** k **do**
      begin $home := np[i]$; $j := 1$;
        $ADD\ S: np[i] := home + step[i]$; $value := function\ (np)$; $eval := eval + 1$;
        if $value < min$ **then** $min := value$
        else begin
            if $j = 2$ **then begin** $np[i] := home$; $fails := fails + 1$ **end**
            else begin $step[i] := -step[i]$; $j := 2$; **goto** $ADD\ S$ **end**
          end
      end
    end *of explore*;

$min := function\ (bp);\ eval := 1;$
$GO\ ON:$ **for** $i := 1$ **step** 1 **until** k **do** $np[i] := bp[i];$
$TRY:\ explore;$
if $fails = k$ **then**
   begin for $i := 1$ **step** 1 **until** k **do**
         if $abs(step[i]) \geqslant critstep[i]$ **then goto** $CONT;$
      goto $EXIT;$
      $CONT:$ **for** $i := 1$ **step** 1 **until** k **do** $step\ [i] := redfact[i] \times step[i];$
      goto TRY
   end;
for $i := 1$ **step** 1 **until** k **do** $move[i] := np[i] - bp[i];$
$PATTERNING:$ **if** $eval > evalim$ **then goto** $EXIT;$
for $i := 1$ **step** 1 **until** k **do**
   begin $bp[i] := np[i];\ np[i] := bp[i] + pat \times move[i];$
     if $move[i] \times step[i] < 0$ **then** $step[i] := -step[i]$
   end;
$minstore := min;\ min := function\ (np);\ eval := eval + 1;$
$explore;$
if $min < minstore$ **then**
   begin for $i := 1$ **step** 1 **until** k **do** $move[i] := np[i] - bp[i];$
     for $i := 1$ **step** 1 **until** k **do if** $abs(move[i]) > eps$ **then goto** $PATTERNING$
   end;
$min := minstore;$ **goto** $GO\ ON;$
$EXIT:$ **end** *of patternsearch*;

given that can be used to minimize any function, i.e. that will find the values of the k variables (in the present example $k = 2$ variables, viz. K and V) required to make the function (in the present example, S given by (12.8.2)) a minimum. The procedure was written by Bell on the basis of the work of Hooke and Jeeves (1961). The procedure starts from the initial guess (*basepoint*) by trying steps (of specified size) in each variable to see whether the function is reduced. The size of the reduction is not taken into account. When a successful pattern of moves has been found it is repeated, the step size increasing while the moves are successful (i.e. while they reduce the function value). When the function cannot be decreased any further the step size is reduced (by a specified factor) and a further exploration carried out. When the steps fall below a specified size the search terminates on the assumptions that a minimum has been found. Further details are given by Wilde (1964).

Of course, if the surface has several pits *patternsearch* will locate only one of them, which one depending on the initial guess, step sizes, etc.

A typical procedure for calculating values of the function is shown in Table 12.8.3. It calculates the sum of square deviations (eqn. (12.8.2)) for fitting the Michaelis–Menten equation. It incorporates a simple device for preventing the search venturing into the craggy (and physically impossible) region of negative V and K values.

When the *patternsearch* program was used for fitting the Michaelis–Menten curve to the results in Table 12.8.1 a minimum of $S = 4\cdot32299$ was found at $\hat{V} = 31\cdot45004$ and $\hat{K} = 15\cdot89267$ after 215 evaluations of S (from Table 12.8.3) with various trial values of V and K. In this case the initial guesses, bp in Table 12.8.2, were set to $V = 2\cdot0$ $K = 50\cdot0$,

TABLE 12.8.3

An Algol 60 procedure for calculating the function to be minimized for fitting the Michaelis–Menten equation. The arrays containing the n observations, y $[1\!:\!n]$, and the n substrate concentrations, $x[1\!:\!n]$, are declared and read in before calling patternsearch. If the Boolean variable constrained is set to true the search is restricted to non-negative values of V and K

```
real procedure function (P); real array (P);
  begin integer j; real S, K, V, Ycalc;
    S: = 0;
    if constrained then for j: = 1, 2 do if P[j] < 0 then P[j]: = 0;
    V: = P[1]; K: = P[2];
    for j: = 1 step 1 until n do
    begin Ycalc: = V × x[j]/(K + x[j]);
      S: = S + (y[j] − Ycalc) ↑ 2
    end;
    function: = S
  end of function;
```

step sizes were $1\cdot0$ for both V and K, reduction factor was $0\cdot2$ for both V and K, and *critstep* was 10^{-5} for both V and K. Patternfactor was $2\cdot0$. In another run, the same except that patternfactor was set to $1\cdot0$ virtually the same point was reached ($S = 4\cdot32299$ at $\hat{V} = 31\cdot45019$ and $\hat{K} = 15\cdot89286$) after 228 evaluations of S—not quite as fast. If the initial guesses were $V = 1\cdot0$, $K = 2\cdot0$ then again the virtually same minimum ($S = 4\cdot32299$ at $\hat{V} = 31\cdot45018$ and $\hat{K} = 15\cdot89283$) was reached after 191 trial evaluations of S. On the other hand if the initial guesses are $V = 2\cdot5$, $K = -3\cdot8$ and the step sizes $0\cdot01$ then the program locates the subminimum ($S = 764\cdot299$ at $V = 3\cdot2443$ and $K = -3\cdot793$) shown in Fig. 12.8.2(b), if not constrained.

Other uses for patternsearch

The program in Table 12.8.2 can be used for any sort of minimization (or maximization) problem. It can, for instance, be used to solve any set of simultaneous equations (linear or non-linear). If the n equations are denoted $f_i(x_1,...,x_n) = 0$ $(i = 1,...,n)$ then the values of x corresponding to the minimum value of Σf_i^2 (which will be zero if the equations have an exact solution) are the required solutions.

The meaning of 'best' estimate

The method of least squares has been used throughout Chapter 12 (and implicitly, in earlier chapters). It was stated in § 12.1 that least squares (LS) estimates have certain desirable properties (unbiasedness and minimum variance; see below) in the case of linear (see § 12.7) problems. It cannot automatically be assumed that least squares estimates will be the best in the case of non-linear problems (and even if they are best, they may not be so much better than others that it is worth finding them if doing so is much more troublesome than the alternatives). *If* the distribution of the observations is normal then the method of least squares becomes the same as the method of maximum likelihood (see Chapter 1) and this method gives estimates that have some good properties. Maximum likelihood (ML) estimates, however, are often biased, as in the case of the variance for which the maximum likelihood estimate is $\Sigma(x-\bar{x})^2/N$, see § 2.6 and Appendix 1, eqn. (A1.3.4). And in general ML estimates have minimum variance only when the size of the sample is large (they are said to be *asymptotically efficient*, meaning that as the size of the sample tends to infinity the variance of the ML estimate is at least as small as that of any other estimate). Such results for large samples (*asymptotic* results) are often encountered but are not much help in practice because most experiments are done with more or less small samples. There are few published results about the relative merits of different sorts of estimates for non-linear models when the estimates are based on small experiments. Such knowledge as there is does not contradict the view that if the errors are roughly constant (homoscedastic) and roughly normally distributed, it is probably safest to prefer the LS estimates in the absence of real knowledge. The ideas involved will be illustrated by means of the Michaelis–Menten curve fitting problem discussed above.

As in all estimation problems, there are many ways of estimating the parameters ($\mathscr{V}$ and $\mathscr{K}$ of (12.8.1) in the present example), given some experimental results. And as usual all methods will, in general, give different estimates. The methods most widely used in the Michaelis–Menten case all depend on transformation of (12.8.1) to a straight line (cf. § 12.6). The most widely used (and worst) method is the double reciprocal plot (or Lineweaver–Burk plot). This depends on rearrangement of (12.8.1) into the form

$$\frac{1}{Y} = \frac{1}{V} + \frac{K}{V}\left(\frac{1}{x}\right),\qquad(12.8.3)$$

which shows that a plot of $1/y$ against $1/x$ should be straight with intercept $1/V$ and slope K/V. Such a plot is shown in Fig. 12.8.3. A straight line has been fitted to the results by the simple (unweighted) method of least squares described in § 12.5 (in laboratory practice

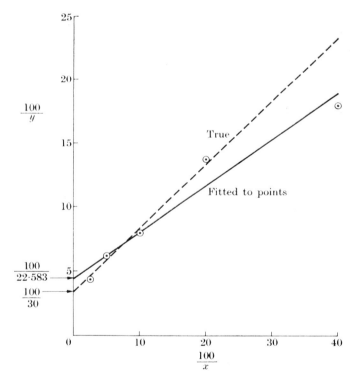

FIG. 12.8.3. Double reciprocal (or Lineweaver–Burk) plot ($1/y$ against $1/x$) for the 'observations' in Table 12.8.1. See also Table 12.8.4.

⊙ 'Observations'.
—— Straight line fitted (see text) to 'observations'.
    Intercept $= 100/V = 100/22\cdot58$.
    Slope $= K/V = 8\cdot16/22\cdot58$.
- - - - True line corresponding to population mean velocities in
    Table 12.8.1 (i.e. $\mathcal{V} = 30$ $\mathcal{K} = 15$, see Table 12.8.4).
    Intercept $= 100/30 = 3\cdot33$.
    Slope $= \mathcal{V}/\mathcal{K} = 0\cdot5$.

usually either this done, or a line is fitted by eye). From its slope and intercept the estimates of V and K are found to be $V = 22 \cdot 58$ and $K = 8 \cdot 16$.

Another method is based on the rearrangement of (12.8.1) in the form

$$Y = V - K\left(\frac{Y}{x}\right), \qquad (12.8.4)$$

from which it is seen that plot of y against y/x should be a straight line with slope $-K$ and intercept V. This plot is shown in Fig. 12.8.4. Again a straight line was fitted using the method of § 12.5 in spite of the fact that the abscissa, y/x, is not free of error as assumed in § 12.1 (because it now involves the observations, y). From the slope and intercept of this line the estimates are found to be $V = 25 \cdot 76$ and $K = 10 \cdot 13$.

The results of applying these various estimation methods to the observations in Table 12.8.1 are compared in Table 12.8.4. They are not very informative as they stand, but it will now be shown that they are not untypical.

TABLE 12.8.4

	V	K
True population value	30·00	15·00
Least squares estimate	31·45	15·89
Lineweaver–Burk estimate (eqn. (12.8.3))	22·58	8·16
y against y/x estimate (eqn. (12.8.4))	25·76	10·13

In fact the 'observations' in Table 12.8.1 were taken from a study in which simulated† experiments were used to investigate various methods of estimation under various conditions (Colquhoun 1969; cf. Dowd and Riggs 1965). An 'experiment' was performed by picking at random an observation from a normally distributed population known to have the mean (μ) given in Table 12.8.1 (and plotted in Fig. 12.8.1),

† The simulation method avoids the mathematical difficulties of finding the distribution of estimates, but the results are not very general. Fig. 12.8.5 would look different for different sorts of error, different distributions for the observations and different experimental designs (spacing and number of substrate concentrations, i.e. of x values).

and known to have a standard deviation $\sigma(y) = 1 \cdot 0$ at every concentration (i.e. the 'observations' were homoscedastic—see Fig. 12.2.3). The 'observations' were generated using computer methods. The observations are thus known to be unbiased (their population means, μ, are

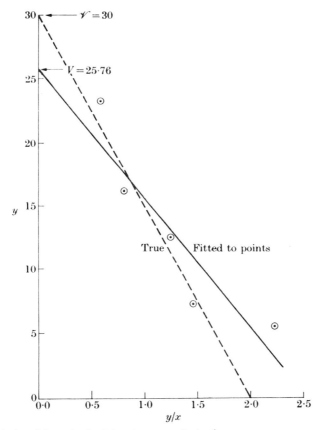

Fig. 12.8.4.   Linearized plot using y against y/x.

⊙ 'Observations' from Table 12.8.1.
—— Straight line fitted (see text) to 'observations'
  Intercept $= V = 25 \cdot 76$, slope $= -K = -10 \cdot 13$
  (see Table 12.8.4).
- - - - True line corresponding to population mean velocities in
  Table 12.8.1, i.e. intercept $= \mathscr{V} = 30$, slope $= -\mathscr{K}$
  $= -15$.

known to lie exactly on the calculated curve in Fig. 12.8.1) and, unlike what happens in any real experiment, their distribution and population means and standard deviations are known. Seven hundred and fifty such 'experiments' were performed, and from each 'experiment' estimates of V and K were calculated by five methods (three of which have been mentioned above). The resulting 750 estimates of V and K were grouped to form histograms. The distributions so obtained of the estimates of V are shown in Fig. 12.8.5 for three methods of estimation.

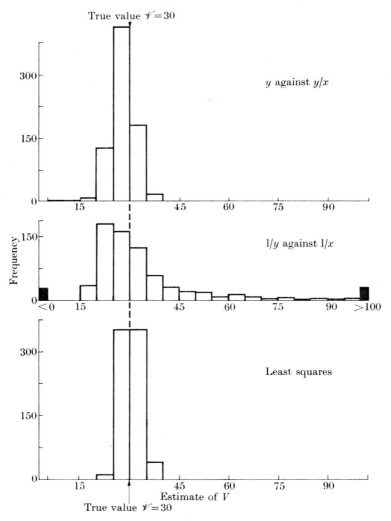

FIG. 12.8.5. Distributions of the 750 estimates of $\mathscr{V}$ ($= 30$) obtained, using three methods, in 750 simulated experiments. Top: estimates from plot of y against y/x (as shown for one 'experiment' in Fig. 12.8.4). Middle: double reciprocal (Lineweaver–Burk) plot (as shown for one 'experiment' in Fig. 12.8.3). Bottom: method of least squares.

The distributions of estimates of K are similar, which is expected in the light of the finding that the estimates of V and K are highly correlated, i.e. experiments that yield an estimate of V that is too high tend to give an estimate of K that is too high also, whichever method of estimation is used. Inspection of Fig. 12.8.5 shows that in this particular case (the μ values shown in Table 12.8.1 and Fig. 12.8.1, with normally distributed homoscedastic observations) the method of least squares is in fact the best of the three methods. The LS estimates are more closely grouped round the population value ($V = 30{\cdot}0$) than the estimates found by the other methods (i.e. they have the smallest variance), and the average value of the LS estimates (viz. 30.4) is close to the population value (i.e. they have little bias).

By comparison the Lineweaver–Burk method is clearly terrible— the scatter of estimates being very much greater (near infinite estimates will be obtained when the plot in Fig. 12.8.3 goes nearly through the origin giving $1/V \simeq 0$, and these distort the average estimate so much that no realistic estimate of the bias is possible).

The plot of y against y/x falls in between these extremes. In spite of breaking the rules for fitting straight lines by having error in the quantity (y/x) plotted along the abscissa, the estimates are obviously much less variable than those found by the Lineweaver–Burk method (their standard deviation is only about 28 per cent greater than that of the LS estimates in this case). The estimates from the y vs.y/x plot are, however clearly consistently too low—they have a negative bias. The average of all 750 estimates is 28·0, well below the population value of 30·0, and about 73 per cent of estimates are too low (i.e. below 30·0). This bias is purely a property of the method of estimation. In these simulated experiments the observations themselves were known to be completely unbiased (a similar situation was seen in the case of the standard deviation, see § 2.6 and Appendix 1). *In real life there would in addition be some unknown amount of bias in the observations themselves* (see §§ 1.2 and 7.2).

If, as is usually the case, experiments are repeated several times, bias would be considered a more serious problem than large variance. This is because the variance of an estimate can always be reduced by doing a large enough number of experiments, whereas bias remains however many experiments are averaged, and there is no way of detecting the presence of bias from the results of repeated experiments. These results are only valid for the particular conditions under which they were obtained. In fact different results are obtained if the errors are not constant or the observations not normally distributed (Dowd and Riggs 1965). For example, if the observations are normally distributed but heteroscedastic, i.e. they do not have the same standard deviation at each x value, then it is found, in the case when the coefficient of variation (standard deviation/mean) is the same at each x value, that linear transformations give *better* estimates of V and K than the least squares method (Colquhoun 1969). The only exception is the linear Lineweaver–Burk plot which is always awful.

Why are the Lineweaver–Burk estimates so bad?

The problem is mainly one of weighting. In fitting the straight line to the plot of $1/y$ against $1/x$, the dependent variable, $1/y$, has been treated as though it had constant variance (see §§ 12.1 and 12.2), and if the straight line is fitted by eye rather than by the method of § 12.5, the result is usually much the same. In fact, in this example y had constant variance ($= 1\cdot0$ at every x value). The variance of $1/y$ is therefore, from (2.7.14), approximately proportional to $1/\mu^4$—very far from constant. Inspection of Fig. 12.8.3 shows that in the particular experiment illustrated the poor estimates were mainly the result of the error in the top point of graph ($1/x = 0\cdot4$, $x = 2\cdot5$). This observation was somewhat too high (see Table 12.8.1), so $1/y$ is too low, and this point has been given far too much weight in plotting the straight line in Fig. 12.8.3. It has pulled the line down distorting the parameter estimates. From (2.7.14), and the values of μ in Table 12.8.1, it is seen that the variance of $1/y$ at $x = 2\cdot5$ is approximately $var(y)/\mu^4 = 1\cdot0/(4\cdot2857)^4$, and the variance of $1/y$ at the highest substrate concentration ($x = 40$) is approximately $1\cdot0/(21\cdot8182)^4$—far more precise. Each point should really have a weight inversely proportional to its variance (see § 2.5) so the point for $x = 40$ ($1/x = 0\cdot025$) should have $(21\cdot8182)^4/(4\cdot2857)^4 \simeq 670$ times the weight of the point for $x = 2\cdot5$ ($1/x = 0\cdot4$), not the equal weight it was given in Fig. 12.8.3. The impression that the point for $x = 2\cdot5$ has been given far too much importance in the Lineweaver–Burk plot is confirmed. The correctly weighted Lineweaver–Burk plot is quite satisfactory, but in real life the weights (population variances) would not be known so fitting it would be no less arithmetically inconvenient than finding the LS estimates.

12.9. Correlation and the problem of causality

So far in this chapter it has been assumed that the x variable (or variables) can be fixed precisely by the experimenter. In many cases, especially in social and behavioral sciences, when often it is not possible, or thought not to be possible, to do proper experiments (see Chapter 1), two (or more) variables are measured, neither (or none) of which can be fixed by the experimenter, or assigned by him to particular individuals. Results of this sort are far more difficult to interpret, and therefore far less satisfactory, than the results of proper experiments as discussed in Chapter 1, but they are sometimes unavoidable.

Examples of the sort of questions usually treated by correlation methods are (a) do people with good scores in school exams also have

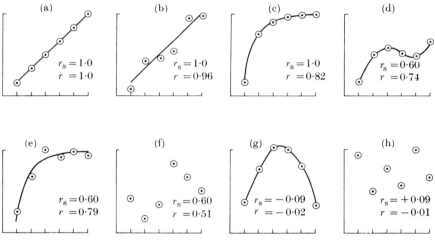

FIG. 12.9.1. Behaviour of the Spearman rank correlation coefficient r_S, and the product moment correlation coefficient, r, on various sorts of data. Clearly non-linearity can result in coefficients of almost any value even when there is a perfectly smooth relationship between x and y. In these small samples it can be seen from Table 12.9.2 that there is no evidence against the null hypothesis that the population value of r_S is zero in figures (d)–(h).

high scores in university exams? (b) are people who smoke a lot of cigarettes more likely to die of lung cancer than those who smoke few? (c) do parts of the country that have a large number of doctors per 1000 of population have more or fewer working days lost because of illness than less well supplied areas? and so on. In each of these cases there are two sets of figures (e.g. school and university exam scores for a number of people) which can be plotted on a graph or scatter diagram like those in Fig. 12.9.1. The tendency of one variable to increase (or decrease) as the other variable increases can be measured by a *correlation coefficient.* There are many different sorts of correlation coefficient, of which two will be described briefly. For detailed descriptions of correlation methods see, for example, Guilford (1954).

If a correlation is observed between two variables (*A* and *B* say), and if it is large enough for it to be unlikely that it arose by chance, then it can be concluded that

either (1) *A* causes *B*,
  or (2) *B* causes *A*,
  or (3) some other factor, directly or indirectly, causes both *A* and *B*,
  or (4) an unlikely event has happened and a large correlation has arisen by chance from an uncorrelated population (see § 6.1).

Usually there is no reason, other than the observer's prejudice, for preferring one of these explanations to the others. As explained in Chapter 1, the only way to choose between (1), (2), and (3) is to do a proper experiment. For example, using the example already discussed in § 12.7, if it were found that areas with more doctors (*x*) had fewer working days (*y*) lost through illness, the relationship might be presented in the form of a correlation coefficient, which would be negative, between *x* and *y*, or by fitting a curve to the graph of *y* against *x*. If a straight line, $Y = a + bx$, was an adequate representation of the observations the slope, *b*, would be negative. However, as mentioned in § 12.2, the least squares estimate (*b*) of the slope found by minimizing $\Sigma(y - Y)^2$ will not be quite the same as the estimate found by using the horizontal deviations from the line in Fig. 12.2.1 (i.e. treating *x* as the dependent variable and minimizing $\Sigma(x - X)^2$). Since there is no independent variable in this case it is not obvious which line to fit. This problem is avoided with correlation coefficients, into which *x* and *y* enter in a symmetrical way. The interpretation of the relationship, however it is presented, is clearly very difficult because chosen numbers of doctors were not allocated at random to selected areas. This has already been discussed at length in § 12.7. As stated there, and in Chapter 1, the only way out of the difficulty is to do a proper experiment.

Correlation based on ranks. Spearman's coefficient, r_s

This coefficient, like other methods based on ranks, does not depend on assumptions about normal distributions or the straightness of lines. And, like other correlation coefficients, a value of $+1$ corresponds to perfect correlation between *x* and *y*, a value of 0 corresponds to no correlation and a value of -1 corresponds to perfect negative correlation (y decreasing as x increases). However, what is meant by 'perfect correlation' is not the same for different coefficients (see Fig. 12.9.1). In the case of the Spearman coefficient it means that the ranking of individuals is the same for both criteria. As an example take the $N = 6$ pairs of observations shown in Table 12.5.1. These were analysed by regression methods in § 12.5. They are reproduced in Table 12.9.1, in which the ranks of the x and of the y values are given, and also d_i = difference between ranks for the ith pair of observations. In this case one variable might be a measure of the rarity of doctors in the ith area, and the other variable a measure of the number of working days lost through illness in that area.

TABLE 12.9.1

pair no. (i)	x_i	y_i	rank of x_i	rank of y_i	d_i	d_i^2
1	160	59	1	2	-1	1
2	165	54	2	1	$+1$	1
3	169	64	3	3	0	0
4	175	67	4	4	0	0
5	180	85	5	6	-1	1
6	188	78	6	5	$+1$	1
Total	1037	407	21	21	0	4

The Spearman rank correlation coefficient, r_S, is estimated using the same formula (eqn. (12.9.3)) as used for the Pearson coefficient (see below), but using the ranks rather than x and y themselves. It can be shown (e.g. Siegel 1956) that the same anwer is found more easily from

$$r_S = 1 - \frac{6\sum_{i=1}^{N} d_i^2}{N(N^2-1)}, \qquad (12.9.1)$$

where Σd^2 is the sum of the squares of the differences in rank for each pair of observations (as shown in Table 12.5.1) and N = number of pairs. From Table 12.5.1, $N = 6$ and $\Sigma d^2 = 4$ so

$$r_S = 1 - \frac{6\times4}{6(36-1)} = 0\cdot886.$$

This is a less than perfect positive correlation, as expected. If the ranks for y had been exactly the same as those for x, all the differences,

d_i, would have been zero, so it is obvious from (12.9.1) that r_S would have been $+1$. If the ranks for y had been in exactly the opposite order to the ranks for y then r_S would have been -1. And that is about all that can be said. In no sense does a correlation coefficient (of any sort) of 0·886 mean '88·6 per cent perfect correlation', and clearly r_S does not measure the slope of the line when the observations (or the ranks) are plotted against each other as shown in Fig. 12.5.1, as r_S can only vary between $+1$ and -1. Some examples of the Spearman and Pearson (see below) correlation coefficients calculated from particular sets of observations are shown in Fig. 12.9.1. to give an idea of their properties. It is obvious from this figure that far more information is to be gained from plotting the graph than from calculating a correlation coefficient.

Ties. Small numbers of ties can be given average ranks as in Chapters 8–10. For a description of the corrections necessary when there are many ties see, for example, Siegel (1956).

Is it unreasonable to suppose that the observed correlation arose by chance?

As usual this, put more precisely, means 'what is the probability that a correlation coefficient differing from zero by as much, or more than the observed value would be found by random sampling from an uncorrelated population?' (see § 6.1). The exact probability can be found in just the same sort of way as was used in Chapters 8–10. If the observations are from an uncorrelated population each of the N! possible rankings of y (permutations of the numbers 1 to N) would have an equal chance of being observed in combination with a given ranking of x. The probability of any particular ranking would therefore be $1/N$! so a correlation of $+1$ or -1 (when no more extreme values are possible) will have $P = 1/N$! (one tail) or $2/N$! (two tail, see Chapter 6). P can always be found by enumerating all N! possibilities and seeing how many give r_S equal to or larger than the observed value (cf. Chapters 8, 9–10). To save trouble, tables have been constructed giving the critical values of r_S corresponding to P (two tail) not more than 0·1, 0·05, and 0·01. For samples up to $N = 8$ the values are shown in Table 12.9.2.

In the present example $N = 6$ and $r_S = 0·886$ so $P = 0·05$ (from Table 12.9.2). For larger samples than 8 it is close enough to calculate

$$t = r\sqrt{\left(\frac{N-2}{1-r^2}\right)} \qquad (12.9.2)$$

and refer the value of t found to tables (described in § 4.4) of Student's t distribution with $N-2$ degrees of freedom. Equivalently, when $N > 8$, r_S can be referred to tables (e.g. Fisher and Yates, 1963, Table VII) of

TABLE 12.9.2

Critical values of r_S. If the observed r_S (taken as positive) is equal to or larger than the tabulated value then P(two tail) is not more than the specified value. Reproduced from Mainland (1963), by permission of author and publisher.

Number of pairs, N	P (two tail)		
	0·1	0·05	0·01
4	1·000		
5	0·900	1·000	
6	0·829	0·886	1·000
7	0·714	0·786	0·929
8	0·643	0·738	0·881

critical values of Pearson's correlation coefficient. In this case $t = 0·886$ $\sqrt{[(6-2)/(1-0·886^2)]} = 3·82$ with $6-2 = 4$ degrees of freedom. Reference to tables of t (see § 4.4) gives $P \simeq 0·02$, not a very good approximation to the exact value (0·05) when N is as small as 6.

Linear correlation. Pearson's product moment correlation coefficient (r)

If x and y are both normally distributed† (see Chapter 4) the closeness with which points cluster round a *straight* line is measured by Pearson's product moment correlation coefficient, r. This measure has been met already in § 10.7. The population value of r is estimated by

$$r = \frac{\Sigma(y-\bar{y})(x-\bar{x})}{\sqrt{[\Sigma(y-\bar{y})^2.\Sigma(x-\bar{x})^2]}}$$

$$= \frac{\mathrm{cov}(y,x)}{\sqrt{[\mathrm{var}(y).\mathrm{var}(x)]}}. \tag{12.9.3}$$

The second form follows from the definition of variance and covariance ((2.6.2) and (2.6.6)). It was shown in § 2.6 that the covariance measures the extent to which y increases as x increases. Pearson's r will be 1

† It is actually assumed that x and y follows a bivariate normal distribution (see, for example, Mood and Graybill (1963), p. 198).

(or -1) only if the points lie exactly on a *straight* line as shown in Fig. 12.9.1. The relationship between x and y may be perfectly predictable and yet have a low correlation coefficient if the relation is not a straight line, as illustrated in Fig. 12.9.1 (c), (d), and (g). The information to be gained from r is therefore limited.

Using the results in Table 12.5.1 and Table 12.9.1 as an example once again, r can be estimated easily because the sums of squares and products have already been calculated in § 12.5. Inserting their values in (12.9.3) gives

$$r = \frac{511 \cdot 833}{\sqrt{(526 \cdot 833 \times 682 \cdot 833)}} = 0 \cdot 853$$

a fairly large positive correlation. Its interpretation has been discussed above.

To find what the probability of observing a Pearson correlation coefficient as large or larger than $0 \cdot 852$ would be, if the observations were randomly selected from normal population with zero correlation, the procedure is to calculate t using (12.9.2). The value of t is referred to the tables of Student's t distribution (described in § 4.4) with $N-2$ degrees of freedom where N is the number of pairs of observations. In the present example $N = 6$ so

$$t = 0 \cdot 853 \sqrt{\left[\frac{6-2}{(1 - 0 \cdot 853^2)} \right]} = 3 \cdot 27.$$

Consulting the tables with $6-2 = 4$ degrees of freedom shows that the required probability is between $P = 0 \cdot 05$ (corresponding to $t = 2 \cdot 776$) and $P = 0 \cdot 02$ (corresponding to $t = 3 \cdot 747$). This is low enough to make one a little suspicious of the null hypothesis that the population correlation is zero. (The inference was, for practical purposes, the same when Spearman's coefficient was calculated using ranks.) The same result can be obtained, without calculation, from tables of critical values of r (e.g. Fisher and Yates (1963, Table VII)).

A little bit of algebra shows that the test of the hypothesis that the population correlation coefficient is zero is identical with test (in § 12.5) that the population slope (regression coefficient) is zero. The value of t just found is the same as that found at the end of § 12.5, and $t^2 = 3 \cdot 27^2 = 10 \cdot 7$ is the value of F found in the analysis of variance of the observations shown in Table 12.5.2.

13. Assays and calibration curves

'Il est vrai que certaines paroles et certaines cérémonies suffisent pour faire périr un tropeau de moutons, pourvu qu'on y ajoute de l'arsenic.'†

VOLTAIRE 1771

(Questions sur *l'Encyclopédie:* 'Enchantement')

† 'Incantations will destroy a flock of sheep if administered with a certain quantity of arsenic.'

(Translation: GEORGE ELIOT, *Middlemarch*, Chap. 17)

13.1. Methods for estimating an unknown concentration or potency

THE process of estimating an unknown concentration will be referred to as an *assay*. All biological assays and most chemical assays depend on comparison of the unknown substance with a standard so the principles involved in both chemical and biological assays are the same. The objects are to obtain (a) the 'best' (usually least squares, see §§ 12.1, 12.2, 12.7, and 12.8) estimate of the unknown concentration, (b) confidence limits for its true value, and (c) to test as many as possible of the assumptions involved in the assay. Unfortunately almost all the methods used involve the assumption of a Gaussian (normal) distribution (see § 6.2). As usual it is no exaggeration to say that there is rarely any reason to believe that this assumption is correct so the results must be interpreted with caution as indicated in §§ 4.2, 4.6 and 7.2. A detailed account of biological assay will be found in Finney (1964), whose notation has been used in most places to make this standard reference book as accessible as possible.

This chapter is pretty solid and it may help to go through the numerical examples in §§ 13.11–13.15 before looking at the theory in §§ 13.2–13.10. The object of the theoretical part is to derive the formulas used in parallel line assays using simple algebra only. This means putting all the steps in, avoiding 'evidently' and 'it is obvious that'. One result is that the theoretical part is rather long and, by mathematicians' standards, inelegant. Another result, I hope, is that the basis of the analysis of parallel line assays is made available to those who, like me, prefer to have the argument laid out in words of one syllable.

The experimental designs according to which the various concentrations of standard and unknown substance can be tested are discussed at the end of this section.

All the methods to be discussed involve the assumption, which may be tested, that the relationship between the measurement (y, e.g. response) and the concentration (x) is a straight line. Some transformation of either the dependent variable, y, or the independent variable, x, may be used to make the line straight. The effects of such transformations are discussed in § 12.2. In biological assay the transformed response is called the *response metameter* (i.e. the measure of response used for calculations) and the transformed concentration or dose is called the *dose metameter*. Of course the response metameter *may* be the response itself, when, as is often the case, no transformation is used.

Furthermore, all the methods to be discussed assume that the standard and unknown behave as though they were identical, apart from the concentration of the substance being assayed. Such assays are called *analytical dilution assays*. When this condition is not fulfilled the assay is called a *comparative assay*. Comparative assays occur when, for example, the concentration of one protein is estimated using a different protein as the standard, or when the potency of a new drug relative to a different standard drug is wanted. (Relative potency means the ratio of the concentrations or doses required to produce the same response.) One difficulty with comparative assays is that the estimate of relative concentration or potency may not be a constant, i.e. independent of the response level chosen for the comparison, so when a log dose scale is used the lines will not be parallel (see below).

Calibration curves

Chemical assays are often done by constructing a calibration curve, plotting response metameter (e.g. optical density) against concentration of standard. The concentration corresponding to the optical density (or whatever) of the unknown solution is then read off from the calibration curve. This sort of assay is discussed in § 13.14.

Continuous (or graded) and discontinuous (or quantal) responses

In chemical assays the 'response' is nearly always a continuous variable (see §§ 3.1 and 4.1), for example volume of sodium hydroxide or optical density. In biological assays this is often the case too.

For example the tension developed by a muscle, or the fall in blood pressure, is measured in response to various concentrations of the standard and unknown preparations. Assays based on continuous responses are discussed in this chapter. Sometimes, however, the proportion of individuals, out of a group of n individuals, that produced a fixed response is measured. For example 10 animals might be given a dose of drug and the number dying within 2 hours counted. This response is a discontinuous variable—it can only take the values 0, 1, 2, . . ., 10. The method of dealing with such responses is considered in Chapter 14, together with closely related *direct assay* in which the dose required to produce a fixed response is measured.

One of the assumptions involved in fitting a straight line by the methods of Chapter 12, discussed in § 12.2, is the assumption that the response metameter has the same scatter at each x value, i.e. is homoscedastic (see Fig. 12.2.3). This is usually assumed to be fulfilled for assays based on continuous responses (it should be tested as described in § 11.2). In the case of discontinuous (quantal) responses there is reason (see Chapter 14) to believe that the homoscedasticity assumption will not be fulfilled, and this makes the calculations more complicated.

Parallel line and slope ratio assays

In the case of the calibration curve described in § 13.14 the abscissa is measured in concentration (e.g. mg/ml or molarity). It is usual in biological assays to express the abscissa in terms of ml of solution (or mg of solid) administered. In this way the unknown and standard can both be expressed in the same units. The aim is to find the ratio of the concentrations of the unknown and standard, i.e. the potency ratio R.

$$R = \frac{\text{concentration of unknown}}{\text{concentration of standard}}$$

$$= \frac{\text{amount of standard for given effect } (z_S)}{\text{amount of unknown for same effect } (z_U)}. \tag{13.1.1}$$

For example, if the unknown is twice as concentrated as the standard only half as much, measured in ml or mg, will be needed to produce the same effect, i.e. to contain the same amount of active material. See also § 13.11.

Suppose it is found that the response metameter y, when plotted against the amount or dose, in ml or mg, gives a straight line. Obviously the response should be the same (zero, or control level) when the dose of either the standard or unknown preparation is zero. The straight line for standard can be written $Y_S = a + b_S z_S$, where b_S is the slope, z_S the dose (amount) of standard, and a the response to zero dose ($z_S = 0$); similarly for the unknown $Y_U = a + b_U z_U$, the response to zero dose being a, as for the standard. When $Y_S = Y_U$ it follows that $a + b_S z_S = a + b_U z_U$ so the potency ratio, from (13.1.1), is $R = z_S/z_U = b_U/b_S$, the ratio of the slopes of the lines, as illustrated in Fig. 13.1.1(a). An assay in which the abscissa is the dose or amount of substance is therefore called a *slope ratio assay* (c.f. § 13.14). This sort of assay is described in detail by Finney (1964).

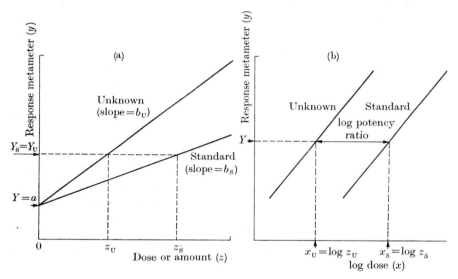

FIG. 13.1.   (a) Slope ratio assay. Response metameter plotted against dose. (b) Parallel line assay. Response metameter plotted against log dose. See text for discussion.

Consider now what happens if it is found empirically that, in order to obtain straight lines, the response metameter must be plotted against the logarithm of the dose, $x = \log z$ say. The ratio of doses required to produce any arbitrary constant effect Y, in Fig. 13.1.1(b) is again the potency ratio z_S/z_U from (13.1.1). Now from Fig. 13.1.1(b) the horizontal distance between the two lines is $x_S - x_U = \log z_S - \log z_U = \log(z_S/z_U) = \log R$. So the horizontal distance is the log of the potency ratio, and because (for analytical dilution assays, see above) the potency ratio (R) is a constant, the horizontal distance between the lines ($\log R$)

must also be a constant. This will be so whether or not the lines are straight (the argument has not involved the assumption that they are), but when they are straight it implies that they will be parallel. Assays in which the abscissa is on a logarithmic scale are therefore called *parallel line assays*. The *reason* for using a logarithmic dose scale is to produce a straight line. Parallelism is a *consequence* of using the logarithmic scale (see § 12.2 also). Another *consequence* of using the logarithmic dose scale is that the ratio between doses is usually kept constant so that the *interval* between the log doses will be constant. The spacing of the doses is, of course, a consequence of using a logarithmic scale, and not a reason for using it as is sometimes implied. Furthermore, the range covered by the doses has nothing to do with scale chosen. A wide range can be accommodated just as easily on an arithmetic scale as on a logarithmic scale.

A similar situation arises in pharmacological studies when the log dose–response curve is plotted in the presence and absence of a drug antagonist. The parallelism of the lines can be tested as described in the following sections. If they are parallel the potency ratio can be estimated. In this context the potency ratio is the ratio of the doses of drug required to produce the same response in the presence and absence of antagonist, and is called the *dose ratio*.

The rest of this chapter, except for § 13.14, will deal with parallel line assays with continuous responses. Sections 13.2–13.10 deal with the theory and numerical examples are worked out in §§ 13.11–13.15.

Types of parallel line assays

In biological assays, when the response, y, is plotted against log dose, x, the line is usually found to be sigmoid rather straight. But it is often sufficiently nearly straight over a central portion for the assumption to produce negligible error.

It is convenient to classify assays according to the number of dose levels of each preparation used. If k_S dose levels of the standard preparation are used, and k_U of unknown, the assay is described as a $(k_S + k_U)$ dose assay. The properties of various types are, briefly, as shown in Table 13.1.1.

The tests of validity possible in a $(2+2)$ dose assay will now be considered in slightly more detail before starting on the theory of parallel line assays. It is intuitively plausible that the following tests can be done (see § 13.7 for details).

(1) For slope (i.e. due to linear regression, see § 12.3). The null

hypothesis that the slope of the response-log dose curve is zero is tested. Obviously the assay is invalid unless it can be rejected. Possible reasons for an increase in dose not causing an increase in response are (a) insensitive test object, (b) doses not sufficiently widely spaced, or (c) responses all supramaximal.

(2) For difference between standard and unknown preparations, i.e. is the average response to the standard different from that to the

TABLE 13.1.1

Number of doses of		
Std (k_S)	Unknown (k_U)	
1	1	The responses to the two doses must be exactly matched. If not too much exactness is demanded this may be possible to achieve once, but a single match would allow no estimate of error. If the doses were given several times it is most improbable that the means would match and so no result could be obtained. This *matching assay* is therefore unsatisfactory.
2	1	The response-log dose line for the standard can be drawn with two points, if it is already known to be straight, and the dose of standard needed to produce the same response as the unknown can be *interpolated*. Error can be estimated. The assumption that the slope of the line is not zero can be tested but the assumptions of linearity and parallelism cannot. (See § 13.15.)
2	2	The (2+2) dose assay is better because, in addition to being able to test the slope of the dose response lines, their parallelism can be tested (see Fig. 12.1.2). It is still necessary to assume that they are straight.
3	3	With a (3+3) dose assay the assumptions of non-zero slope, parallelism, and linearity can all be tested.

test preparation? This is not usually of great interest in itself though it helps precision if $\bar{y}_S$ and $\bar{y}_U$ are not too different (see § 13.6). It will be seen later that this test emerges as a side effect of doing tests (1) and (3).

(3) Deviations from parallelism. The null hypothesis that the true (population) slopes for standard and test preparation are equal is

tested. If this hypothesis is rejected the assay must be considered invalid. In an analytical dilution assay the most probable cause of nonparallelism is that one of the preparations is off the linear part of the log dose-response curve. This is shown in Fig. 13.1.2.

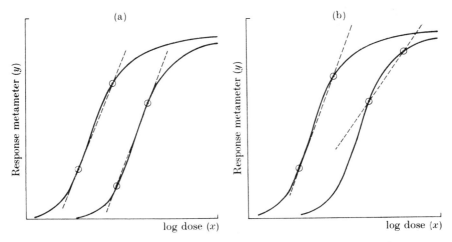

FIG. 13.1.2. Apparent deviations from parallelism can result when some doses are not on the straight part of the dose response curve, as shown in (b), even when the horizontal distance between the two curves *is* constant.

      ⊙ Observations.
    – – – Straight line fitted to observations.
    ——— True response–log dose curve.

Symmetrical parallel line assays

In the following section it will become obvious that the calculations can be very greatly simplified when the assay is symmetrical. In the context symmetry means that the assay has (a) the same number of dose levels of each preparation—either $(2+2)$ or $(3+3)$ usually, (b) each dose is administered the same number of times, (c) the *ratios* between all doses are equal, and the same for both standard and unknown, i.e. the *intervals* between doses are equal on the logarithmic scale. These conditions are summarized precisely in eqns. (13.8.1).

Designs for the administration of standard and unknown

Any of the usual experimental designs, some of which were described in Chapter 11, may be used. The various concentrations of standard and unknown are the *treatments*. See also § 13.8.

For example in a (3+3) dose assay there are 6 different solutions, each of which is to be tested several (say n) times. The $6n$ tests may be done in a *completely random* fashion as described in § 11.4. If each dose is tested on a separate animal this means allocating the $6n$ doses to $6n$ animals strictly at random (see §§ 2.3 and 11.4). Often all observations are made on the same individual (e.g. the same spectrophotometer or the same animal). In this case the *order* in which the $6n$ tests are done must be strictly random (see § 2.3), and, in addition, the size of a response must not be influenced by the size of previous responses (see discussion of single subject assays below).

If, for example, all $6n$ responses could not be obtained on the same animal, it might be possible to obtain 6 responses from each of n animals, the animals being *blocks* as described in § 11.6. Examples of assays based on *randomized block designs* (see § 11.6) are given in §§ 13.11 and 13.12. A second source of error could be eliminated by using a 6×6 *Latin square* design (this would force one to use $n = 6$ replicate observations on each of the 6 treatments). However it is safer to avoid small Latin squares (see § 11.8).

If the natural blocks were not large enough to accommodate all the treatments (for example, if the animals survived long enough to receive only 2 of the 6 treatments), the *balanced incomplete block* design could be used. References to examples are given in § 11.8 (p. 207).

The analysis of assays based on all of these designs is done using Gaussian methods. Many untested assumptions are made in the analysis and the results must therefore be treated with caution, as described in §§ 4.2, 4.6, 7.2, 11.2, and 12.2. In particular, the estimate of the error of the result is likely to be too small (see § 7.2).

Single subject assays

Assays in which all the doses are given, in random order, to a single animal or preparation (e.g. in the example in § 13.11, a single rat diaphragm) are particularly open to the danger that the size of a response will be affected by the size of the preceding responses(s). Contrary to what is sometimes said, the fact that responses are evoked in random order does not eliminate the requirement that they be independent of each other. Special designs have been worked out to make the allowance for the effect of one response on the next, but it is necessary to assume an arbitrary mathematical model for the interaction so it is much better to arrange the assay so as to prevent the effect of one dose on the next (see, for example, Colquhoun and Tattersall

(1969). If the doses have to be well separated in time to prevent interaction it may not be possible to give all the treatments to one subject,
so an incomplete block design may have to be used (see § 11.8 and, for
example, Colquhoun (1963)). The problem is discussed by Finney
(1964, p. 291).

13.2. The theory of parallel line assays. The response and dose metameters

Response metameter (y)

The object is to transform the response so that it becomes normally
distributed, homoscedastic, and produces a straight line when plotted
against log dose (see §§ 11.2, 12.2, p. 221, and 13.1). In many cases the
response itself is used. A linear transformation of the response, of the
form $y = c_1 + c_2 y$ where c_1 and c_2 are constants, may be used to simplify
the arithmetic. This will not affect the distribution, scedasticity, or
linearity. For example, in § 11.4 each observation was reduced by 100
to make the numbers smaller. For tests of normality, see § 4.6.

The dose metameter (x)

For parallel line assays this of course, by definition (see § 13.1),
the logarithm of the dose. The dose (measured in volume or weight) is
denoted z, as in § 13.1. Thus

$$x = \log z \qquad (13.2.1)$$

Usually logarithms to base 10 (common logs) will be used because the
tables are the most convenient; but it will be shown that for parallel
line assays which are symmetrical, as defined in § 13.1 and eqn. (13.8.1),
it will make the calculations much simpler to use a different base for
the logarithms. This will not, of course, affect the linearity or parallelism
of the lines. At this stage this only looks like an additional complication,
but the simplification will become apparent later. Numerical examples
are worked out in §§ 13.11, 13.12, 13.13, and 13.15.

The symmetrical (2+2) dose assay

Suppose that the ratio between the high and low doses is D, both for
the standard and for the unknown. Suppose further that each dose is
given n times so the total number of observations is $N = 4n$. Of these
$n_S = 2n = \frac{1}{2}N$ are standards and $n_U = \frac{1}{2}N$ are unknowns.

If the low doses of standard and unknown preparations are z_{LS} and z_{LU} then, by definition of D, the high doses will be

$$z_{\text{HS}} = Dz_{\text{LS}}, \text{ and } z_{\text{HU}} = Dz_{\text{LU}}. \qquad (13.2.2)$$

The most convenient base for the logarithms is $\sqrt{D}$. This looks most improbable at first sight, but the reason why it is so will now be shown. Taking the logarithms to the base $\sqrt{D}$ of the doses (remembering that $\log_{\sqrt{D}} D = 2$ whatever the value of D, because the log is defined as the power to which the base must be raised to give the argument) gives, from (13.2.1) and (13.2.2),

$$x_{\text{LS}} = \log_{\sqrt{D}} z_{\text{LS}} \qquad (13.2.3)$$
$$\begin{aligned} x_{\text{HS}} = \log_{\sqrt{D}} z_{\text{HS}} &= \log_{\sqrt{D}} (Dz_{\text{LS}}) \\ &= \log_{\sqrt{D}} D + \log_{\sqrt{D}} z_{\text{LS}} \\ &= 2 + x_{\text{LS}}. \end{aligned} \qquad (13.2.4)$$

Similarly, for the unknown, $x_{\text{LU}} = \log_{\sqrt{D}} z_{\text{LU}}$, $x_{\text{HU}} = 2 + x_{\text{LU}}$.

The mean value of the log dose for the standard preparation, if the high and low doses are given an equal number of times (n), will be, using (13.2.4)

$$\left. \begin{aligned} \bar{x}_{\text{S}} &= \frac{n x_{\text{LS}} + n x_{\text{HS}}}{2n} = \frac{x_{\text{LS}} + x_{\text{HS}}}{2} = 1 + x_{\text{LS}} \\ \text{and} \qquad \bar{x}_{\text{U}} &= 1 + x_{\text{LU}}. \end{aligned} \right\} (13.2.5)$$

Combining these results with (13.2.4) gives

$$\left. \begin{aligned} (x_{\text{HS}} - \bar{x}_{\text{S}}) &= +1, \\ (x_{\text{LS}} - \bar{x}_{\text{S}}) &= -1, \\[4pt] (x_{\text{HU}} - \bar{x}_{\text{U}}) &= +1, \\ (x_{\text{LU}} - \bar{x}_{\text{U}}) &= -1. \end{aligned} \right\} (13.2.6)$$

and similarly

Using logs to the base $\sqrt{D}$ has made $(x - \bar{x})$ takes the value $+1$ for the high doses (of both standard and unknown), and -1 for the low doses. This means that $(x - \bar{x})^2 = 1$ for every dose; and since there are $\frac{1}{2}N$ doses of standard and $\frac{1}{2}N$ doses of unknown, it follows from (2.1.7) that

$$\Sigma(x_{\text{S}} - \bar{x}_{\text{S}})^2 = \Sigma(x_{\text{U}} - \bar{x}_{\text{U}})^2 = \tfrac{1}{2}N, \qquad (13.2.7)$$

where the summations are over all $\frac{1}{2}N$ doses of standard (or unknown). Thus the total sum of squares for x, pooling standard and unknown, is

$$\sum_{\text{S,U}} \Sigma(x - \bar{x})^2 \equiv \Sigma(x_{\text{S}} - \bar{x}_{\text{S}})^2 + \Sigma(x_{\text{U}} - \bar{x}_{\text{U}})^2 = N, \qquad (13.2.8)$$

where the symbol $\sum_{S,U}$ means 'add the value of the following for the standard to its value for the unknown' (as shown in the central expression of (13.2.8)). The sums of squares are greatly simplified by using logs to the base $\sqrt{D}$.

The symmetrical (3+3) dose assay

The most convenient base for the logarithms in this case is D rather than $\sqrt{D}$. The low, middle, and high doses will be indicated by the subscripts S1, S2, and S3 for standard, and U1, U2, and U3 for unknown. The ratio between each dose and the one below it is D, as before. Thus

$$z_{S2} = Dz_{S1},$$
$$z_{S3} = Dz_{S2} = D^2 z_{S1} \tag{13.2.9}$$

Taking logarithms to the base D (remembering $\log_D D = 1$ and $\log_D D^2 = 2$, whatever the value of D) gives

$$\left. \begin{aligned} x_{S1} &= \log_D z_{S1}, \\ x_{S2} &= \log_D z_{S2} = \log_D(Dz_{S1}) = \log_D D + \log_D z_{S1} = 1 + x_{S1}, \\ x_{S3} &= \log_D z_{S3} = \log_D D^2 + \log_D z_{S1} = 2 + x_{S1}. \end{aligned} \right\} \tag{13.2.10}$$

The mean standard dose, if each dose level is given the same number of times (n), will be, using (13.2.10),

$$\left. \begin{aligned} \bar{x}_S &= \frac{nx_{S1} + nx_{S2} + nx_{S3}}{3n} = 1 + x_{S1} \\ \text{and} \qquad \bar{x}_U &= 1 + x_{U1}. \end{aligned} \right\} \tag{13.2.11}$$

Combining this with (13.2.10) gives, for the standard

$$\left. \begin{aligned} (x_{S1} - \bar{x}_S) &= -1 \\ (x_{S2} - \bar{x}_S) &= 0 \\ (x_{S3} - \bar{x}_S) &= +1 \end{aligned} \right\} \tag{13.2.12}$$

and similarly $(x_U - \bar{x}_U) = -1, 0, +1$ for low, middle, and high doses of unknown.

Because the assay is symmetrical (see §§ 13.1 and 13.8.1) each dose is given the same number of times, n. The total number of observations is $N = 6n$ and the number of standards is $n_S = 3n = \frac{1}{2}N$, and of unknowns $n_U = 3n = \frac{1}{2}N$. Now $(x - \bar{x})^2 = +1$ for all high and low doses, and 0 for all middle doses so

$$\Sigma(x_S - \bar{x}_S)^2 = \Sigma(x_U - \bar{x}_U)^2 = \frac{1}{3}N \tag{13.2.13}$$

the summations being over all $\frac{1}{2}N$ doses (n low, n middle, and n high) of standard or unknown. The total sum of squares for x, pooling standard and unknown, is

$$\underset{\text{S,U}}{\sum\sum}(x - \bar{x})^2 \equiv \Sigma(x_\text{S} - \bar{x}_\text{S})^2 + \Sigma(x_\text{U} - \bar{x}_\text{U})^2 = \tfrac{2}{3}N \quad (13.2.14)$$

where $\underset{\text{S,U}}{\sum}$ means sum over preparations as in (13.2.8).

13.3. The theory of parallel line assays. The potency ratio

This discussion applies to any parallel line assay, symmetrical (see §§ 13.1 and 13.8) or not. Numerical examples are given in §§ 13.11, 13.12, 13.13, and 13.15.

According to (13.1.1) the ratio of the concentration of unknown to concentration of standard is

$$R = \frac{\text{concentration of unknown}}{\text{concentration of standard}}$$

$$= \frac{\text{amount of standard for given effect}}{\text{amount of unknown for same effect}} = \frac{z'_\text{S}}{z'_\text{U}}, \quad (13.3.1)$$

where the prime indicates doses estimated to produce identical responses. As in § 13.1 the dose z will be measured in the same units (e.g. volume of solution, or weight of solid) for both standard and unknown (what happens when the units are different is explained in the numerical example in § 13.11).

The conventional symbol for the log of the potency ratio is M so, from (13.3.1),

$$M \equiv \log R = \log z'_\text{S} - \log z'_\text{U} = x'_\text{S} - x'_\text{U}. \quad (13.3.2)$$

As in § 13.1, $M = x'_\text{S} - x'_\text{U}$ the difference between the logs of equi-effective doses, is the horizontal distance between the parallel lines as shown in Fig. 13.3.1. The least squares estimate of this quantity will now be derived.

When straight lines are fitted to the standard and unknown responses the lines are constrained to be parallel, i.e. an average of the observed slopes for S and U is used for both (see § 13.4). If this common slope is called b, the linear regression equations (see §§ 12.1 and 12.7, and eqn. (12.3.1)) are written

$$Y_\text{S} = \bar{y}_\text{S} + b(x_\text{S} - \bar{x}_\text{S}),$$
$$Y_\text{U} = \bar{y}_\text{U} + b(x_\text{U} - \bar{x}_\text{U}). \quad (13.3.3)$$

When the response is the same for standard and unknown $Y_S = Y_U$, so these can be equated giving

$$\bar{y}_S + b(x_S' - \bar{x}_S) = \bar{y}_U + b(x_U' - \bar{x}_U),$$

where x_S' and x_U' are the log doses giving equal responses as above. Rearranging this to give $M = x_S' - x_U'$, from (13.3.2), gives the result

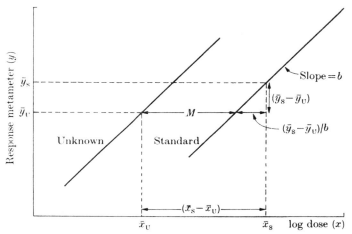

F IG . 13.3.1. Geometrical meaning of the equation (derived in the text) for the log potency ratio ($M = \log R$) in any parallel line assay.

$$M = \log R = x_S' - x_U' = (\bar{x}_S - \bar{x}_U) + \frac{(\bar{y}_U - \bar{y}_S)}{b}. \qquad (13.3.4)$$

The geometrical meaning of the right-hand side is illustrated in Fig. 13.3.1. To find the potency ratio, R, the antilog of M can be found from tables if common logarithms (to base 10) have been used. However in symmetrical assays it has been shown that it is better to use logarithms to a different base, say base r in general (it was shown in § 13.2 that $r = \sqrt{D}$ is best for 2+2 dose assays and $r = D$ for 3+3 dose assays). Since antilog tables are available only for base 10 logarithms it will be necessary to convert to base 10 before looking up antilogs. The general formula† for changing the base of logarithms from a to b is

$$\log_a z = \log_b z . \log_a b \qquad (13.3.5)$$

from which it follows that

$$\log_{10} R = \log_r R . \log_{10} r = M . \log_{10} r. \qquad (13.3.6)$$

† *Proof.* From the definition of logs, antilog$_b$ $x \equiv b^x$ and so $b^{\log_b z} = z$. Also, in general, $n \log x = \log x^n$. Thus $\log_b z . \log_a b = \log_a (b^{\log_b z}) = \log_a z$.

Therefore, multiplying (13.3.4) by the conversion factor, $\log_{10} r$, gives

$$\log_{10} R = \left[(\bar{x}_S - \bar{x}_U) + \frac{(\bar{y}_U - \bar{y}_S)}{b} \right] . \log_{10} r. \qquad (13.3.7)$$

This is a perfectly general formula whatever sort of logarithms are used. If common logs were used, $r = 10$ so the conversion factor is $\log_{10} 10 = 1$.

For symmetrical assays (as defined in (13.8.1) and at the end of § 13.1) this expression can be simplified, as shown in § 13.10.

13.4. The theory of parallel line assays. The best average slope

For estimation of the potency ratio it is essential that the response-log dose lines be parallel (see §§ 13.1, 13.3). Inevitably the line fitted to the observations on the standard will not have *exactly* the same slope as the line fitted to the unknown; but, if the deviation from parallelism is not greater than might reasonably be expected on the basis of experimental error, the observed slope for the standard (b_S) is averaged with that for the unknown (b_U) to give a pooled estimate of the presumed common value.

By 'best' is meant, as usual, least squares (see §§ 12.1, 12.2, 12.7, and 12.8). A weighted average of the slopes for standard and unknown is found using (2.5.1). Calling the average slope b, this gives

$$b = \frac{\sum\limits_{S,U} wb}{\sum\limits_{S,U} w} \equiv \frac{w_S b_S + w_U b_U}{w_S + w_U}, \qquad (13.4.1)$$

where the weights are the reciprocals of the variances of the slopes (see § 2.5). The estimated variances of the individual slopes, by (12.4.2), are

$$\text{var}[b_S] = \frac{s^2[y]}{\Sigma(x_S - \bar{x}_S)^2} \equiv \frac{1}{w_S},$$

$$\text{var}[b_U] = \frac{s^2[y]}{\Sigma(x_U - \bar{x}_U)^2} \equiv \frac{1}{w_U}, \qquad (13.4.2)$$

where $s^2[y]$ is, as usual, the estimated error variance of the observations (the error mean square from the analysis of variance).

Now in general the variance of the weighted mean $\bar{y} = \Sigma w_i y_i / \Sigma w_i$ will be given by (2.7.12) as

$$\text{var}[\bar{y}] = \frac{1}{\Sigma w_i}. \tag{13.4.3}$$

Taking $w_{\text{S}} = 1/\text{var}[b_{\text{S}}]$ and $w_{\text{U}} = 1/\text{var}[b_{\text{U}}]$ from (13.4.2), and inserting the estimate of the slope from (12.2.7) gives

$$w_{\text{S}} b_{\text{S}} = \frac{\Sigma(x_{\text{S}} - \bar{x}_{\text{S}})^2}{s^2[y]} \cdot \frac{\Sigma y_{\text{S}}(x_{\text{S}} - \bar{x}_{\text{S}})}{\Sigma(x_{\text{S}} - \bar{x}_{\text{S}})^2} = \frac{\Sigma y_{\text{S}}(x_{\text{S}} - \bar{x}_{\text{S}})}{s^2[y]} \tag{13.4.4}$$

and similarly for unknown. Inserting these results in (13.4.1) gives the weighted average slope

$$b = \frac{\Sigma y_{\text{S}}(x_{\text{S}} - \bar{x}_{\text{S}}) + \Sigma y_{\text{U}}(x_{\text{U}} - \bar{x}_{\text{U}})}{\Sigma(x_{\text{S}} - \bar{x}_{\text{S}})^2 + \Sigma(x_{\text{U}} - \bar{x}_{\text{U}})^2} \equiv \frac{\displaystyle\sum_{\text{S,U}}\Sigma y(x - \bar{x})}{\displaystyle\sum_{\text{S,U}}\Sigma(x - \bar{x})^2} \tag{13.4.5}$$

where the symbol $\displaystyle\sum_{\text{S,U}}$ means, as before, 'add the value of the following quantity for the standard to its value for the unknown'. In other words, the average slope is simply (pooled sum of products for S and U)/ (pooled sum of squares of x).

For symmetrical assays it was shown in § 13.2 that $\Sigma(x - \bar{x})^2$ is the same for standard and unknown so, from (13.4.2), the weights are equal and the two slopes (b_{S} and b_{U}) are simply averaged.

From (13.4.2) and (13.4.3) it follows that the variance of the average slope is, in general, estimated as

$$\text{var}[b] = \frac{1}{\Sigma w} = \frac{s^2[y]}{\Sigma(x_{\text{S}} - \bar{x}_{\text{S}})^2 + \Sigma(x_{\text{U}} - \bar{x}_{\text{U}})^2} \equiv \frac{s^2[y]}{\displaystyle\sum_{\text{S,U}}\Sigma(x - \bar{x})^2} \tag{13.4.6}$$

(compare this with (12.4.2)). It is, of course being assumed that the variance of the observations, $s^2[y]$, is the same for standard and unknown as well as for each dose level—see §§ 11.2, 12.2, and 13.1.

13.5. Confidence limits for the ratio of two normally distributed variables: derivation of Fieller's theorem

The solution to the problem posed in § 7.5 will now be given. The result, in its general form, looks rather complicated; but the numerical examples in §§ 13.11–13.14 show how easy it is to use.

Although the sum or difference (or any linear combination, see p. 39) of two normally distributed (Gaussian) variables is itself normally distributed, their *ratio* is not. Therefore, as discussed in § 7.5, the methods discussed so far cannot provide confidence limits for the ratio. A solution of the problem will now be described.

The simplest application of the result is to find the confidence limits for the ratio (= *m*, say) of two means (see § 14.1), the problem discussed in § 7.5. It is shown below that if *g* (eqn (13.5.8)) is very small compared with one, so $(1-g) \simeq 1$, the result of using Fieller's theorem is the same as the approximate result, $m \pm t\sqrt{[\text{var}(m)]}$, where var(*m*) is given, approximately, by (2.7.16).

The theorem is needed to find confidence limits for the value of the independent variable (*x*) necessary to produce a given value of the dependent variable (*y*) as discussed in § 12.4. A numerical example of this 'calibration curve problem' is given in § 13.14. The confidence limits for a potency ratio are also found using Fieller's theorem.

Before considering a ratio, the argument of § 7.4 leading to confidence limits for a single Gaussian variable, *y*, will be repeated in a rather more helpful form. If *y* is normally distributed with population mean μ and estimated variance s^2 then $y-\mu$ is normal with population mean $= \mu-\mu = 0$ and variance s^2, so, as in § 4.4, $t = (y-\mu)/s^2$. As in § 7.4 the 100 α per cent confidence limits for the value of μ are based on Student's *t* distribution (§ 4.4) which implies

$$P[-ts \leqslant (y-\mu) \leqslant +ts] = \alpha \qquad (13.5.1)$$

or, in other words (see § 11.3, p. 182),

$$P[(y-\mu)^2 \leqslant t^2 s^2] = \alpha. \qquad (13.5.2)$$

The deviation $(y-\mu)$ will border on significance when it is equal to $-ts$ or $+ts$, i.e. when

$$(y-\mu)^2 = t^2 s^2.$$

This is a quadratic equation in μ and solving it for μ using the usual formula† gives as the two solutions $\mu = y-ts$ and $\mu = y+ts$, the confidence limits for μ found in § 7.4. This seems a long way round to get the same answer as before, but the approach will now be used to find the confidence limits for a ratio.

Consider, in general, any ratio $\mu = \alpha/\beta$. The estimate (*m*) of the population value (μ) from the observations will be written $m = a/b$ where *a* is the estimate of α, and *b* is the estimate of β. The case of interest, or, at any rate, the case to be dealt with, is when *a* and *b* are

---

† In general, if $ax^2+bx+c = 0$ then $x = [-b \pm \sqrt{(b^2-4ac)}]/2a$.

normally distributed variables (with population means α and β). The variances of a and b must be specified and this will be done using a new notation. This notation is based on the fact that not only the variances but also the covariances (in analysis of variance problems that are linear in the general sense discussed in § 12.7) can be expressed as multiples of the error variance of the observations, $s^2[y]$ (as usual this is the error mean square from the analysis of variance). For example, the variance of a mean $\bar{y}$, is, by (2.7.8), $1/n$ times the error variance. Similarly the variance of a slope, b, is, by (12.4.2), $1/\Sigma(x-\bar{x})^2$ times the error variance. If these multiplying factors are symbolized v then one can define

$$\left.\begin{array}{l} \text{var}[a] = v_{11}s^2, \\ \text{var}[b] = v_{22}s^2, \\ \text{cov}[a,\, b] = v_{12}s^2, \end{array}\right\} \quad (13.5.3)$$

where s^2 is written for $s^2[y]$. The subscripts distinguishing the *variance multipliers*, v, are arbitrary (cf. § 2.1), but the notation used emerges naturally from a more advanced treatment, and is that used in Finney (1964), who discusses Fieller's theorem and two of its extensions. For example, if a was a mean, $\bar{y}$, then $v_{11} = 1/n$ as above.

Since a and b are normally distributed and μ is a constant, the variable $(a-\mu b)$ is a linear function of normal variables, and is therefore normally distributed. The population mean of $(a-\mu b)$ will be $\alpha-\mu\beta = 0$ and its estimated variance will be, using (13.5.3), (2.7.2), (2.7.5), and (2.7.6),

$$\begin{aligned} \text{var}[(a-\mu b)] &= \text{var}[a]+\text{var}[\mu b]-2\,\text{cov}[a,\mu b] \\ &= \text{var}[a]+\mu^2\,\text{var}[b]-2\mu\,\text{cov}[a,b] \\ &= s^2(v_{11}+\mu^2 v_{22}-2\mu v_{12}). \end{aligned} \quad (13.5.4)$$

Now, by direct analogy with (13.5.2), it follows from the definition of Student's t that

$$P[(a-\mu b)^2 \leqslant t^2 s^2(v_{11}+\mu^2 v_{22}-2\mu v_{12})] = \alpha. \quad (13.5.5)$$

And, again by analogy, the 100 α per cent confidence limits for μ are found by solving for μ the equation

$$(a-\mu b)^2 = t^2 s^2(v_{11}+\mu^2 v_{22}-2\mu v_{12}). \quad (13.5.6)$$

This is again a quadratic equation in μ and when solved for μ by the usual formula (see above) the two solutions are the required confidence limits for μ. They are

$$\frac{1}{(1-g)}\left[m-\frac{gv_{12}}{v_{22}}\pm\frac{ts}{b}\sqrt{\left\{v_{11}-2mv_{12}+m^2v_{22}-g\left(v_{11}-\frac{v_{12}^2}{v_{22}}\right)\right\}}\right] \quad (13.5.7)$$

where

$$g = \frac{t^2s^2v_{22}}{b^2}. \quad (13.5.8)$$

Simplifications of Fieller's theorem in special cases

If a and b are independent, i.e. $v_{12} = 0$, the result simplifies considerably, giving the confidence limits for μ as

$$\frac{m}{(1-g)}\pm\frac{ts}{b(1-g)}\sqrt{\left[(1-g)v_{11}+m^2v_{22}\right]}. \quad (13.5.9)$$

The quantity g defined in (13.5.8) can be considered an index of the significance of the difference of b (the denominator of the ratio) from zero. This is clearly important because if the denominator could be zero, the ratio could be infinite. The effect of the $(1-g)$ in front of the $\pm$ sign is to raise both upper and lower limits, i.e. unless g is very small the limits are not symmetrical about m. Since $\text{var}(b) = v_{22}s^2$ by (13.5.3) it follows that if $b^2 < t^2s^2v_{22}$, i.e. if $g > 1$, then b would be judged 'not significantly different' from zero at the level of significance fixed by the value of t chosen, and useful limits could not be found. In other words $1/g$ is the square of the ratio of b to t times the standard deviation of b.

If g is very small, as it will be in good experiments, then the general formula, (13.5.7), simplifies giving the (symmetrical) confidence limits for μ as

$$m\pm\frac{ts}{b}\sqrt{(v_{11}-2mv_{12}+m^2v_{22})}. \quad (13.5.10)$$

This is the result that would be obtained by treating m as roughly normally distributed and calculating $m\pm t\sqrt{[\text{var}(m)]}$, as in § 7.4, using the approximate formula, (2.7.18), together with (13.5.3) to give

$$\text{var}(m) \simeq m^2\left(\frac{\text{var}(a)}{a^2}+\frac{\text{var}(b)}{b^2}-\frac{2\,\text{cov}(a,b)}{ab}\right)$$

$$= \frac{s^2}{b^2}(v_{11}-2mv_{12}+m^2v_{22}). \quad (13.5.11)$$

If a and b are uncorrelated ($v_{12} = 0$), as well $g \ll 1$, then the confidence limits for μ can again be found as $m \pm t\sqrt{[\text{var}(m)]}$, the approximate expression for var(m), (13.5.11), simplifying even further to

$$\text{var}(m) \simeq m^2\left(\frac{\text{var}(a)}{a^2}+\frac{\text{var}(b)}{b^2}\right)$$

$$= \frac{s^2}{b^2}(v_{11}+m^2 v_{22}). \qquad (13.5.12)$$

This is the variance given by the approximate formula, (2.7.16) (because $C^2(m) = \text{var}(m)/m^2$, etc., from the definition (2.6.4)).

Examples of the use of the results in this section will occur in §§ 13.6 and 13.10–13.14.

13.6. The theory of parallel line assays. Confidence limits for the potency ratio and the optimum design of assays

This discussion applies to any parallel line assay. The simplifications possible in the case of symmetrical assays (see § 13.1) are given later in § 13.10, and numerical examples in § 13.11 onwards.

The logarithm of the potency ratio (R) is $M = \log R$ as in § 13.3. It will be convenient to rearrange the formula for the potency ratio, (13.3.4), to give

$$M-(\bar{x}_{\text{S}}-\bar{x}_{\text{U}}) = \frac{(\bar{y}_{\text{U}}-\bar{y}_{\text{S}})}{b}. \qquad (13.6.1)$$

The term $(\bar{x}_{\text{S}}-\bar{x}_{\text{U}})$ has zero variance because x is supposed to be measured with negligible error (see §§ 12.1, 12.2, and 12.4), and so can be treated as a constant. The approach is therefore to find confidence limits for the population value of $(\bar{y}_{\text{U}}-\bar{y}_{\text{S}})/b$ and then add the constant $(\bar{x}_{\text{S}}-\bar{x}_{\text{U}})$ to the results. Now if the observations are normally distributed then so are $(\bar{y}_{\text{U}}-\bar{y}_{\text{S}})$ and (as explained in § 12.4) the average slope, b. The right-hand side of (13.6.1) is therefore the ratio of two normally distributed variables, and confidence limits for it can be found using Fieller's theorem (§ 13.5).

The variance multipliers defined in (13.5.3) are required first. From (2.7.3) and (2.7.8) it follows that $\text{var}[\bar{y}_{\text{U}}-\bar{y}_{\text{S}}] = s^2[y]/n_{\text{U}}+s^2[y]/n_{\text{S}}$ and therefore

$$v_{11} = \left(\frac{1}{n_{\text{U}}}+\frac{1}{n_{\text{S}}}\right) \qquad (13.6.2)$$

where n_U and n_S are the numbers of responses to the unknown and standard preparations. The variance of the average slope, b, in the denominator, is, from (13.4.6), $\text{var}[b] = s^2[y]/\sum_{S,U}\sum(x-\bar{x})^2$ so

$$v_{22} = \frac{1}{\sum_{S,U}\sum(x-\bar{x})^2}, \tag{13.6.3}$$

the notation being explained in § 13.4. Because it can be shown (see § 13.10) that $(\bar{y}_U-\bar{y}_S)$ and b are uncorrelated, i.e. have zero covariance, it follows that $v_{12} = 0$ (see also § 12.7). Thus the simplified form of Fieller's theorem, eqn. (13.5.9), can be used to find confidence limits for the ratio $(\bar{y}_U-\bar{y}_S)/b$. Using (13.6.2) and (13.6.3) these are

$$\frac{(\bar{y}_U-\bar{y}_S)/b}{(1-g)} \pm \frac{st}{b(1-g)}\sqrt{\left[(1-g)\left(\frac{1}{n_U}+\frac{1}{n_S}\right)+\frac{(\bar{y}_U-\bar{y}_S)^2/b^2}{\sum_{S,U}\sum(x-\bar{x})^2}\right]}, \tag{13.6.4}$$

where from (13.5.8) and (13.6.3),

$$g = \frac{t^2s^2}{b^2\sum_{S,U}\sum(x-\bar{x})^2}. \tag{13.6.5}$$

From (13.6.1) it follows that (13.6.4) gives the confidence limits for $M-(\bar{x}_S-\bar{x}_U)$, so the confidence limits for the log potency ratio, M, are $(\bar{x}_S-\bar{x}_U)+[13.6.4]$.

To find the confidence limits for the potency ratio itself the anti-logarithms of these limits are required. Now, as discussed in § 13.3, the calculations are often carried out not with logarithms to base 10 of the dose, but with some other convenient base, say r. In this case $M = \log_r R$, and, as explained in § 13.3, it is necessary to multiply by $\log_{10} r$, to convert to logarithms to base 10, before looking up the anti-logarithms. The confidence limits for true value of $\log_{10} R$ are thus

$$\left[(\bar{x}_S-\bar{x}_U)+\frac{(\bar{y}_U-\bar{y}_S)/b}{(1-g)}\pm\frac{st}{b(1-g)}\sqrt{\left\{(1-g)\left(\frac{1}{n_U}+\frac{1}{n_S}\right)+\frac{(\bar{y}_U-\bar{y}_S)^2/b^2}{\sum_{S,U}\sum(x-\bar{x})^2}\right\}}\right]\log_{10}r.$$

$$\tag{13.6.6}$$

A numerical example of the use of this general equation occurs in § 13.13.

Simplification of the calculation for good assays

If the slope of the log dose-response line, b, is large compared with the experimental error then g will be small (see § 13.5), so $(1-g) \simeq 1$. Inserting this into (13.6.6), together with the definition of $M = \log_r R$ from (13.3.4), gives the confidence limits for $\log_{10} R$ as approximately

$$\left[M \pm \frac{st}{b} \middle/ \sqrt{\left\{ \left(\frac{1}{n_U} + \frac{1}{n_S} \right) + \frac{(\bar{y}_U - \bar{y}_S)^2/b^2}{\sum\limits_{S,U}\sum(x - \bar{x})^2} \right\}} \right] \log_{10} r. \qquad (13.6.7)$$

This is equivalent to treating the log potency ratio, M, as approximately normally distributed and calculating the limits as $M \pm t\sqrt{[\text{var}(M)]}$, as in § 7.4, with

$$\text{var}(M) \simeq \frac{s^2}{b^2} \left(\frac{1}{n_U} + \frac{1}{n_S} + \frac{(\bar{y}_U - \bar{y}_S)^2/b^2}{\sum\limits_{S,U}\sum(x - \bar{x})^2} \right), \qquad (13.6.8)$$

which can alternatively be inferred directly from (13.5.12

The optimum design of assays

The aim is to make the confidence limits for the potency ratio as narrow as possible, i.e. the result of the assay as precise as possible. Ways of doing this can be inferred from (13.6.6).

(1) g should be small. In other words (see discussion at the end of § 13.5) the slope of the response–log dose lines should be as large as possible relative to its standard deviation. If g is large (approaching 1) the limits for the log potency ratio will become wide because of the term involving g after the $\pm$ sign in (13.6.6), and also unsymmetrical about M because of the g in the term before the $\pm$ sign *both* upper *and* lower limits are raised when g is large.

(2) s, the error standard deviation should be small. That is the responses should be as reproducible as possible; and the error variance reduced, if possible, by giving the doses in designs such as randomized blocks, as described in § 13.1 and illustrated in §§ 13.11 and 13.12.

(3) b should be large, to minimize the term after the $\pm$ sign in (13.6.6). A steep slope will also minimize g.

(4) $(1/n_U + 1/n_S)$ should be small. That is, as many responses as possible should be obtained. For a fixed total number of responses $(1/n_U + 1/n_S)$ is at a minimum when $n_U = n_S$ so a symmetrical design (see § 13.1) is preferable.

(5) $(\bar{y}_U - \bar{y}_S)$ should be small because it occurs after the $\pm$ sign in § 13.6. That is, the size of the responses to standard and unknown should be as similar as possible. The assay will be more precise if a good guess at its result is made beforehand.

(6) $\Sigma\Sigma(x - \bar{x})^2$ should be large. That is, the doses should be as far apart as possible, making $(x - \bar{x})$ large; but the responses must, of course, remain of the straight part of the response–log dose curve.

13.7. The theory of parallel line assays. Testing for non-validity

This discussion is perfectly general for any parallel line assay with at least 2 dose levels for both standard and unknown, i.e. (2+2) and larger assays. The simplifications possible in the case of symmetrical assays are described in §§ 13.8 and 13.9 and numerical examples, are worked out in §§ 13.11–13.13. The $(k+1)$, e.g. (2+1), dose assay is discussed in § 13.15.

In the discussion in § 13.1 it was pointed out that it will be required to test whether the slope of the response–log dose lines differs from zero ('linear regression' as in § 12.5), and whether there is reason to believe that the lines are not parallel. If more than 2 dose levels are used for either standard or unknown it will also be possible to test the hypothesis that the lines are really straight. The method of doing these tests will now be outlined.

Each dose level gives rise to a group of comparable observations and these can be analysed using an analysis of variance appropriate to the design of the assay, the dose levels being the 'treatments' of Chapter 11, as discussed in §§ 12.6 and 13.1. For example, for a (2+2) dose assay there 4(= k, say) 'treatments' (high and low standard, and high and low unknown), and a (3+4) dose assay has $k = 7$ 'treatments'. The number of degrees of freedom for the 'between treatments sum of squares' will be one less than the number of 'treatments' (or 'doses'), i.e. at least 3 as this section deals only with (2+2) dose or larger assays (cf. § 11.4). Now the 'between treatments' (or 'between doses') sum of squares can be subdivided into components just as in § 12.6. This *partition* can be done in many different ways (see § 13.8 and, for example, Mather (1951), and Brownlee (1965, p. 517)), but only one of these ways is of interest. Each component must be un-correlated with all others and this will be demonstrated, in the case of symmetrical assays, in § 13.8. Three components, each with one degree of freedom, can always be separated; (a) linear regression, (b) deviations from parallelism, and (c) difference between standard and unknown responses, as described in § 13.1. If there are more than 3 degrees of freedom (i.e. more than 4 'treatments') the remainder can be lumped together as 'deviations from linearity' (cf. § 12.6), as in Table 13.1, or further subdivided as in §§ 13.10 and 13.12. The analysis thus has the appearence of Table 13.1 if there are k dose levels ('treatments') and N responses altogether.

TABLE 13.7.1

Source of variation	Degrees of freedom	Sum of squares
Linear regression	1	A
Deviations from parallelism	1	B
Between standard and unknown	1	C
Deviations from linearity	$k-4$	$D-(A+B+C)$
Between 'treatments' or dose levels	$k-1$	D
Error (within 'treatments')	$N-k$	
Total	$N-1$	

The bottom part of the analysis would look like Table 11.6.1 or Table 11.8.2 if a randomized block or Latin square design (respectively) were used.

(1) *Linear regression.* To test whether the population value of the slope differs from zero, the appropriate sum of squares (SSD) is, from (13.4.5) by analogy with (12.3.4),

SSD for linear regression =

$$\frac{[\Sigma y_S(x_S-\bar{x}_S)+\Sigma y_U(x_U-\bar{x}_U)]^2}{\Sigma(x_S-\bar{x}_S)^2+\Sigma(x_U-\bar{x}_U)^2}. \tag{13.7.1}$$

(2) *Deviations from parallelism.* To test whether the lines are parallel it seems reasonable to calculate the difference between (a) the total sum of squares for linear regression for lines fitted separately to standard and unknown (from 12.3.4), and (b) the sum of squares for linear regression when the slopes are averaged (i.e. (13.7.1)), because this difference will be zero if the lines are parallel. Thus

SSD for deviations from parallelism $=$

$$\frac{[\Sigma y_S(x_S - \bar{x}_S)]^2}{\Sigma(x_S - \bar{x}_S)^2} + \frac{[\Sigma y_U(x_U - \bar{x}_U)]^2}{\Sigma(x_U - \bar{x}_U)^2} - \text{SSD for linear regression.} \quad (13.7.2)$$

(3) *Between standard and unknown.* This is found directly from (11.4.5) as

$$\text{SSD between S and U} = \frac{(\Sigma y_S)^2}{n_S} + \frac{(\Sigma y_U)^2}{n_U} - \frac{G^2}{N}. \quad (13.7.3)$$

(4) *Deviations from linearity.* This is found as the difference between the sum of squares between 'treatments' (from (11.4.5)), as in Table 11.4.3, and the total of the above 3 components. It must be zero for a $(2+2)$ dose assay $(k = 4)$ when the sum of (13.7.1), (13.7.2), and (13.7.3) can be shown to add up to the sum of squares between treatments.

A numerical example of the use of these relations is worked out in § 13.13.

13.8. The theory of symmetrical parallel line assays. Use of orthogonal contrasts to test for non-validity

Numerical examples are given in §§ 13.11 and 13.12. Symmetrical (in this context) means, summarizing the definition in § 13.1,

$n =$ number of responses at each of the k dose levels
    (see § 13.7), same for all
$N = kn =$ total number of responses,
$k_S = k_U = \frac{1}{2}k =$ number of dose levels for standard and
         to unknown (same for both),   (13.8.1)

$D =$ ratio between each dose and the one below it. The same for all doses, and for standard and unknown (see also §§ 13.1 and 13.2 and Fig. 13.8.1), so the doses are equally spaced (by $\log D$) on the logarithmic scale.

The symmetrical $(2+2)$ dose assay. Contrasts

There are $k = 4$ dose levels, low standard (LS), high standard (HS), low unknown (LU) and high unknown (HU). The $k-1 = 3$ degrees of freedom between dose levels can be separated into 3 components as described in §§ 13.1 and 13.7 (Table 13.7.1). A simpler approach than that in § 13.7 is possible.

As usual a hypothesis is formulated. Then the probability that observations *would* be made, deviating from the hypothesis by as much as, or more than, the experimental results do, *if* the hypothesis were in fact true, is calculated (cf. § 6.1).

(1) *Linear regression.* From Fig. 13.8.1 it is clear that *if* the null hypothesis (that the true value, β, of the average slope, see § 13.4, is zero) were true *then*, in the long run, the responses to the high doses

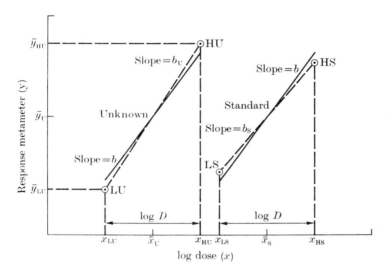

FIG. 13.8.1. The symmetrical 2+2 dose parallel line assay.

    ⊙ Mean of n observed responses (e.g. $\bar{y}_{HU}$ is the mean of the n responses to x_{HU}).

    – – – Straight lines between observed points, with slope b_U for unknown and b_S for standard.

    ——— Best fitting parallel lines with slope b (= average of b_S and b_U, see § 13.4).

would be the same as those to low doses, i.e. $\bar{y}_{HU}+\bar{y}_{HS}=\bar{y}_{LU}+\bar{y}_{LS}$. It follows that if the *regression contrast*, L_1, is defined as

$$L_1 = -\Sigma y_{LS}+\Sigma y_{HS}-\Sigma y_{LU}+\Sigma y_{HU} \qquad (13.8.2)$$

(a linear combination of the responses), it will be a measure of departure from the null hypothesis. If the null hypothesis were true the population mean value of L_1 would be zero (as long as each dose level is given the same number of times so the total responses can be used in place of the mean responses). In a small experiment L_1 would not be exactly zero even if the null hypothesis were true, and it is shown § 13.9 how to judge whether L_1 is large enough for rejection of the null hypothesis.

(2) *Deviations from parallelism.* From Fig. 13.8.1 it is clear that *if* the null hypothesis (that the population lines are parallel, $\beta_S = \beta_U$, see § 13.4) were true then, in the long run, $\bar{y}_{HU} - \bar{y}_{LU} = \bar{y}_{HS} - \bar{y}_{LS}$. Therefore deviations from parallelism are measured, as above, by a *deviations from parallelism contrast, L_1'* defined as

$$L_1' = \Sigma y_{LS} - \Sigma y_{HS} - \Sigma y_{LU} + \Sigma_{HU}. \tag{13.8.3}$$

Again the population value of L_1' will be zero if the null hypothesis is true.

(3) *Between standard and unknown preparations.* If the null hypothesis that the population mean response to standard is the same as that for unknown were true then, in the long run, $\bar{y}_{LS} + \bar{y}_{HS} = \bar{y}_{LU} + \bar{y}_{HU}$. Departure from the null hypothesis is therefore measured by the *between S and U* (or *between preparations*) *contrast, L_p,* defined as

$$L_p = -\Sigma y_{LS} - \Sigma y_{HS} + \Sigma y_{LU} + \Sigma y_{HU}, \tag{13.8.4}$$

which will have a population mean of zero if the null hypothesis is true.

These contrasts are used for calculation of the analysis of variance and potency ratio, as described below and in §§ 13.9 and 13.10.

The subdivision of a sum of squares (cf. § 13.7) using contrasts is quite a general process described, for example, by Mather (1951) and Brownlee (1965, p. 517). The set of contrasts used must satisfy two conditions.

(1) The sum of the coefficients of the contrast must be zero. In Table 13.8.1 the coefficients (which will be denoted α) of the response totals for the contrasts defined in (13.8.2), (13.8.3), and (13.8.4) are summarized. In each case $\Sigma\alpha = 0$ as required. This means that the population mean value of the contrast will be zero when the null hypothesis is true.†

(2) Each contrast must be independent of every other. A set of mutually independent contrasts is described as a set of *orthogonal contrasts*. It is easily shown (e.g. Brownlee (1965, p. 518)) that two contrasts will be uncorrelated (and therefore, because a normal distribution is assumed, independent, see § 2.4) when the sum of products of corresponding coefficients for the two contrasts is zero. All results

---

† In the language of Appendix 1, $E[L] = E[\Sigma\alpha T_j]$ where T_j is the total of the n responses of the jth treatment (dose). If all the observations were from a single population, all $E[T_j] = n\mu$ where $E[y] = \mu$. Thus $E[L] = n\mu\Sigma\alpha = 0$ if $\Sigma\alpha = 0$.

necessary for the proof have been given in § 2.7. It is shown in the lower part of Table 13.8.1 that this condition is fulfilled for all three possible pairs of contrasts.

TABLE 13.8.1

The upper part summarizes the coefficients (α) of the response totals for the validity tests for the symmetrical $(2+2)$ dose assay. The lower part demonstrates the orthogonality (i.e. independence) of the contrasts

	Σy_{LS}	Σy_{HS}	Σy_{LU}	Σy_{HU}	$\Sigma \alpha$	$\Sigma \alpha^2$
Linear regression L_1	-1	$+1$	-1	$+1$	0	4
Parallelism L_1'	$+1$	-1	-1	$+1$	0	4
Preparations (S and U)						
L_p	-1	-1	$+1$	$+1$	0	4

					Total
$\alpha_{L_1} \times \alpha_{L_1'}$	-1	-1	$+1$	$+1$	0
$\alpha_{L_1} \times \alpha_{Lp}$	$+1$	-1	-1	$+1$	0
$\alpha_{L_1'} \times \alpha_{Lp}$	-1	$+1$	-1	$+1$	0

These conditions mean that if two contrasts are defined, to measure linear regression and deviations from parallelism, say, there is no choice about the third, which happens to measure the difference between $\bar{y}_S$ and $\bar{y}_U$.

The symmetrical $(3+3)$ dose assay

There are $k = 6$ levels, say S1, S2, S3, U1, U2, and U3, where S1 is the lowest, S2 the middle, and S3 the highest standard dose. There are $k-1 = 5$ degrees of freedom between dose levels so after separating components for linear regression, deviations from parallelism and between S and U, there are two degrees of freedom left for deviations from linearity (see Table 13.7.1). The first three contrasts are constructed from response totals by the same sort of reasoning as for the $(2+2)$ assay and the coefficients (α) are given in Table 13.8.2. Deviations from linearity can be further divided into two components each with one degree of freedom. If the average curve for S and U is straight then (for a symmetrical assay) the responses to the middle doses will be equal to the mean of the responses to the high and low doses, i.e., in the long run $(\bar{y}_{S1} + \bar{y}_{S3} + \bar{y}_{U1} + \bar{y}_{U3})/4 = (\bar{y}_{S2} + \bar{y}_{U2})/2$.

Therefore a *deviation from linearity contrast*, L_2, measuring departure from the hypothesis of straightness, can be defined as

$$L_2 = \Sigma y_{\text{S1}} - 2\Sigma y_{\text{S2}} + \Sigma y_{\text{S3}} + \Sigma y_{\text{U1}} - 2\Sigma y_{\text{U2}} + \Sigma y_{\text{U3}} \qquad (13.8.5)$$

and this will be zero in the long run if the average line is straight. The fifth contrast is dictated by the conditions mentioned above. It is called L_2', and inspection of the coefficients in Table 13.8.2 shows that

TABLE 13.8.2

Coefficients (α) *of the response totals for the orthogonal contrasts in a symmetrical* (3+3) *dose assay*

Contrast	Response totals						$\Sigma\alpha$	$\Sigma\alpha^2$
	Σy_{S1}	Σy_{S2}	Σy_{S3}	Σy_{U1}	Σy_{U2}	Σy_{U3}		
L_1	-1	0	1	-1	0	1	0	4
L_1'	1	0	-1	-1	0	1	0	4
L_p	-1	-1	-1	1	1	1	0	6
L_2	1	-2	1	1	-2	1	0	12
L_2'	-1	2	-1	1	-2	1	0	12

It can easily be checked that, as in Table 13.8.1, the sum of the products of the coefficients of corresponding totals is zero for all possible pairs of contrasts, so all pairs of contrasts are orthogonal.

it is a measure of the extent to which deviations from linearity are the same for the standard and unknown. It is therefore called the *difference of curvature contrast* (cf. L_1' which measures the extent to which the *linear* regressions differ between S and U, i.e. deviations from parallelism).

13.9. The theory of symmetrical parallel line assays. Use of contrasts in the analysis of variance

The notation used is defined in § 2.1 and (13.8.1). In conformity with the usual approach in the analysis of variance, it is required to calculate from each contrast a quantity (the mean square) that will be an estimate of the error variance, $\sigma^2[y]$, if the appropriate null hypothesis is true. These estimates will then be compared with the error variance (which estimates σ^2 whether or not the null hypotheses are true), in the usual way (see § 11.4). Numerical examples are given in §§ 13.11 and 13.12.

The first step is to estimate the variance of a contrast. If T_j is used to stand for the total of the n responses to the jth dose level, then the contrasts defined in § 13.8 all have the form

$$L = \sum_{j=1}^{k} \alpha_j T_j. \tag{13.9.1}$$

The variance of this, from (2.7.10), is $\Sigma \alpha_j^2 \mathrm{var}(T_j)$ and from (2.7.4) it follows that $\mathrm{var}(T_j) = ns^2[y]$ where $s^2[y]$ is the estimated variance of the observations and n is the number of observations in each total. Thus

$$\mathrm{var}[L] = ns^2 \Sigma \alpha^2. \tag{13.9.2}$$

The values of $\Sigma \alpha^2$ are worked out in Tables 13.8.1 and 13.8.2.

It might be supposed that it is not possible to estimate the variance of L directly from the observed scatter of values of L, because there is only a single experimentally observed value for each contrast. However it is what happens when the null hypothesis is true that is of interest, and it was shown in § 13.8 that when it is true the population mean value of each L will be zero. Now it was pointed out in § 2.6 (eqn. (2.6.3)) that if there are N observations of y, and if the population mean value (μ) of y is known, then the estimate of the variance of y is $\Sigma(y-\mu)^2/N$ (the divisor is only $N-1$ when the sample mean is used in place of μ). For a single value of L it follows that, on the null hypothesis,

$$\mathrm{var}[L] = (L-0)^2/1 = L^2. \tag{13.9.3}$$

Equating (13.9.2) and (13.9.3) shows that when the null hypothesis (that the population value of L is zero) is true, an estimate of the error variance is provided by

$$s^2 = \frac{L^2}{n\Sigma \alpha^2}, \tag{13.9.4}$$

and this expression also gives the sum of squares required for the analysis of variance, because each sum of squares has one degree of freedom—see §§ 13.7, 13.8, 13.11, and 13.12—so the sum of squares is the same as the mean square.

It is not difficult to show (try it) that, when the appropriate base is used for the logarithms giving (13.2.6), (13.2.7), (13.2.12), and (13.2.13), the sums of squares for testing validity given by the general formulas (13.7.1), (13.7.2), and (13.7.3) are the same as those given by (13.9.4), using the definitions of the contrasts in § 13.8. The demonstrations follow the lines used in the next section.

13.10. The theory of symmetrical parallel line assays. Simplified calculation of the potency ratio and its confidence limits

The general results in §§ 13.3, 13.4, and 13.6 can be simplified when the appropriate dose metameter is used (see § 13.2). The notation, and the definition of *symmetrical*, are given in (13.8.1). Numerical examples are given in §§ 13.11 and 13.12. Try to suspend your belief that this is a very complicated sort of simplification until you have compared the calculations for symmetrical assays (in §§ 13.11 and 13.12) with those for an unsymmetrical assay (§ 13.13).

The symmetrical (2+2) *dose assay*

The best dose metameter in this case was shown in § 13.2 to be $x = \log_r z$ where $z = $ dose and $r = \sqrt{D}$. The consequences of using this base for the logarithms, derived in § 13.2, can be used to simplify the ratio $(\bar{y}_U - \bar{y}_S)/b$ which occurs in the potency ratio and its confidence limits. Taking the numerator first, $(\bar{y}_U - \bar{y}_S)$ is, as expected, simply related to the between-preparations contrast, L_p. Thus, from (13.8.1) and (13.8.4),

$$(\bar{y}_U - \bar{y}_S) = \frac{\Sigma y_U}{n_U} - \frac{\Sigma y_S}{n_S} = \frac{\Sigma y_{LU} + \Sigma y_{HU} - \Sigma y_{LS} - \Sigma y_{HS}}{\frac{1}{2}N}$$

$$= \frac{L_p}{\frac{1}{2}N}. \tag{13.10.1}$$

The average slope, b (see § 13.4) is related to the regression contrast, L_1, as expected. From (13.4.5), (13.2.6), (13.2.8), and (13.8.2) it follows that

$$b = \frac{\Sigma y_S(x_S - \bar{x}_S) + \Sigma y_U(x_U - \bar{x}_U)}{\Sigma(x_S - \bar{x}_S)^2 + \Sigma(x_U - \bar{x}_U)^2}$$

$$= \frac{(x_{LS} - \bar{x}_S)\Sigma y_{LS} + (x_{HS} - \bar{x}_S)\Sigma y_{HS} + (x_{LU} - \bar{x}_U)\Sigma y_{LU} + (x_{HU} - \bar{x}_U)\Sigma y_{HU}}{\displaystyle\sum_{S,U}\sum (x - \bar{x})^2}$$

$$= \frac{-\Sigma y_{LS} + \Sigma y_{HS} - \Sigma y_{LU} + \Sigma y_{HU}}{N} = \frac{L_1}{N}. \tag{13.10.2}$$

Combining (13.10.1) and (13.10.2) gives

$$(\bar{y}_U - \bar{y}_S)/b = 2L_p/L_1. \tag{13.10.3}$$

Furthermore, from (13.2.5),

$$\bar{x}_{\mathrm{S}} - \bar{x}_{\mathrm{U}} = x_{\mathrm{LS}} - x_{\mathrm{LU}} = \log_r z_{\mathrm{LS}} - \log_r z_{\mathrm{LU}} = \log_r(z_{\mathrm{LS}}/z_{\mathrm{LU}})$$
(13.10.4)

and from (13.3.6)

$$(\bar{x}_{\mathrm{S}} - \bar{x}_{\mathrm{U}}) \log_{10} r = \log_{10}(z_{\mathrm{LS}}/z_{\mathrm{LU}}).$$
(13.10.5)

The potency ratio

Substituting (13.10.3) and (13.10.5) into the general formula for the log potency ratio, (13.3.7), gives

$$\log_{10} R = \log_{10}\left(\frac{z_{\mathrm{LS}}}{z_{\mathrm{LU}}}\right) + \frac{2L_{\mathrm{p}}}{L_1}\log_{10} r.$$

Putting $r = \sqrt{D}$ (remembering that $\log \sqrt{D} = \log D^{\frac{1}{2}} = \frac{1}{2}\log D$), taking antilogarithms gives

$$R = \left(\frac{z_{\mathrm{LS}}}{z_{\mathrm{LU}}}\right)\mathrm{antilog}_{10}\left[\frac{L_{\mathrm{p}}}{L_1}\log_{10} D\right].$$
(13.10.6)

The confidence limits

It was mentioned in § 13.6 that $(\bar{y}_{\mathrm{U}} - \bar{y}_{\mathrm{S}})$ is not correlated with b and so $v_{12} = 0$. This follows (using (13.10.1) and (13.10.2)) from the fact that L_{p} and L_1 were shown in § 13.8 to be uncorrelated. From (13.6.2) and (13.8.1),

$$v_{11} = (1/n_{\mathrm{U}} + 1/n_{\mathrm{S}}) = (1/\tfrac{1}{2}N + 1/\tfrac{1}{2}N) = 4/N.$$
(13.10.7)

And from (13.6.3) and (13.2.8),

$$v_{22} = 1/\sum_{\mathrm{S,U}}\sum(x - \bar{x})^2 = 1/N.$$
(13.10.8)

Substituting (13.10.2), (13.10.3), (13.10.5), (13.10.7), (13.10.8) and $r = \sqrt{D}$ (again $\log \sqrt{D} = \frac{1}{2}\log D$) into the general formula for the confidence limits for $\log_{10} R$ (13.6.6), and taking antilogarithms, gives the confidence limits for the population value of R, the potency ratio in a symmetrical (2+2) dose assay, as

$$\left(\frac{z_{\mathrm{LS}}}{z_{\mathrm{LU}}}\right).\mathrm{antilog}_{10}\left[\left(\frac{L_{\mathrm{p}}}{L_1(1-g)} \pm \frac{st}{L_1(1-g)}\sqrt{\left\{N(1-g) + N\left(\frac{L_{\mathrm{p}}}{L_1}\right)^2\right\}}\right).\log_{10} D\right]$$
(13.10.9)

where, from (13.6.5), (13.10.2), and (13.10.8),

$$g = \frac{N s^2 t^2}{L_1^2}.$$ (13.10.10)

If g is very small so $(1-g) \simeq 1$, then (13.10.9) can be further simplified. As explained in § 13.6, this equivalent to treating $\log_{10} R$ as approximately normally distributed, and calculating confidence limits for its population value as $\log_{10} R \pm t\, s[\log_{10} R]$ as in § 7.4, where the approximate standard deviation of $\log_{10} R$, from (13.10.9) (or from (13.6.8)), is

$$s[\log_{10} R] \simeq \frac{\log_{10} D}{L_1} \cdot \sqrt{[s^2 N\{1 + (L_p/L_1)^2\}]}.$$ (13.10.11)

A numerical example of the use of (13.10.9) and (13.10.11) is given in § 13.11.

The symmetrical (3+3) dose assay

The simplifications follow exactly the same lines as those just described. (From (13.8.1) and the definitions of the contrasts in Table 13.8.2,

$$(\bar{y}_U - \bar{y}_S) = \frac{\Sigma y_U}{n_U} - \frac{\Sigma y_S}{n_S} = \frac{\Sigma y_{U1} + \Sigma y_{U2} + \Sigma_{U3} - \Sigma y_{S1} - \Sigma y_{S2} - \Sigma y_{S3}}{\tfrac{1}{2}N}$$

$$= \frac{L_p}{\tfrac{1}{2}N},$$ (13.10.12)

and, from the general definition of the slope b, (13.4.5), using (13.2.12), (13.2.14), and the definition of L_1 in Table 13.8.2,

$$b = \frac{1}{\sum_{S,U} \Sigma (x - \bar{x})^2}[(x_{S1} - \bar{x}_S)\Sigma y_{S1} + (x_{S2} - \bar{x}_S)\Sigma y_{S2} + (x_{S3} - \bar{x}_S)\Sigma y_{S3} +$$

$$+ (x_{U1} - \bar{x}_U)\Sigma y_{U1} + (x_{U2} - \bar{x}_U)\Sigma y_{U2} + (x_{U3} - \bar{x}_U)\Sigma y_{U3}]$$

$$= \frac{-\Sigma y_{S1} + \Sigma y_{S3} - \Sigma y_{U1} + \Sigma_{U3}}{\tfrac{2}{3}N} = \frac{L_1}{\tfrac{2}{3}N}.$$ (13.10.13)

Combining (13.10.12) and (13.10.13) gives

$$(\bar{y}_U - \bar{y}_S)/b = 4L_p/3L_1.$$ (13.10.14)

Furthermore, from (13.2.11)

$$(\bar{x}_S - \bar{x}_U) = (x_{S1} - x_{U1}) = \log_r z_{S1} - \log_r z_{U1} = \log_r(z_{S1}/z_{U1}) \tag{13.10.15}$$

so, from (13.3.6),

$$(\bar{x}_S - \bar{x}_U).\log_{10} r = \log_{10}(z_{S1}/z_{U1}). \tag{13.10.16}$$

Substituting these results, together with $r = D$ (see § 13.2), into the general formulas, as above, gives the potency ratio, from (13.3.7), as

$$R = \left(\frac{z_{S1}}{z_{U1}}\right).\text{antilog}_{10}\left(\frac{4L_p}{3L_1}.\log_{10} D\right). \tag{13.10.17}$$

Confidence limits for the population of R from (13.6.6), (with $v_{11} = 4/N$, $v_{22} = 1/\frac{2}{3}N$) are

$$\left(\frac{z_{S1}}{z_{U1}}\right).\text{antilog}_{10}\left[\left(\frac{4L_p}{3L_1(1-g)} \pm \frac{4st}{3L_1(1-g)}\sqrt{\left\{N(1-g) + \frac{2N}{3}\left(\frac{L_p}{L_1}\right)^2\right\}}\right).\log_{10} D\right],$$
$$\tag{13.10.18}$$

where

$$g = \frac{2Ns^2t^2}{3L_1^2}. \tag{13.10.19}$$

Again if, g is small, so $1 - g \simeq 1$, further simplification is possible. As explained above and in § 13.6 the confidence limits for the population value of $\log_{10} R$ can be found, as in § 7.4, from $\log_{10} R \pm ts[\log_{10} R]$ where the approximate standard deviation of $\log_{10} R$, from (13.10.18) or (13.6.8), can be written in the form

$$s[\log_{10} R] \simeq \frac{4\log_{10} D}{3L_1} \cdot \sqrt{\left[s^2 N\left\{1 + \frac{2}{3}\left(\frac{L_p}{L_1}\right)^2\right\}\right]}. \tag{13.10.20}$$

A numerical example of the use of (13.10.18) and (13.10.20) is given in § 13.12.

13.11. A numerical example of a symmetrical (2 + 2) dose parallel line assay

The results of a symmetrical assay of (+)-tubocurarine based on a randomized block design (see § 11.6) are shown in Table 13.11.1. The mean responses are plotted in Fig. 13.5. The response, y, was the percentage inhibition, caused by each dose, of the contraction (induced

by stimulation of the phrenic nerve) of the isolated rat diaphragm.
The four doses (or 'treatments') were allotted arbitrarily to the numbers
0, 1, 2, 3 as described in § 2.3:

Dose 0 ≡ LU = 0·28 ml of unknown solution,
    1 ≡ HU = 0·32 ml of unknown solution,
    2 ≡ HS = 16·0 μg of pure (+)-tubocurarine,
    3 ≡ LS = 14·0 μg of pure (+)-tubocurarine.

Each dose was given four times, so sixteen doses were given altogether.
The doses were given in sequence to the same tissue (see § 13.1, p. 286),
the blocks, in this case, corresponding to periods of time. The analysis

TABLE 13.11.1

*Responses to (+)-tubocurarine. The doses were given in random order in
each block (time period) as specified in the text, not in the order shown in
the table*

		LS	HS	LU	HU	Totals
	1	43	62	41	61	207
Block	2	48	62	48	68	226
	3	53	66	53	70	242
	4	52	70	56	72	250
Totals		196	260	198	271	925

will therefore help to eliminate error due to changes of sensitivity of
the tissue with time (which occurred in this experiment). However it
seems most unlikely that the responses in one block (period of time)
will differ from the responses in another by a constant amount, as
specified in the model (eqn. (11.2.2)) on which the analysis is based
(see §§ 11.2 and 11.6), so the analysis should be regarded as only an
approximation. The four doses were given in strictly random order
(see § 2.3) in each time period, the random number tables producing
the sequence: first block: 1, 0, 2, 3; second block 3, 1, 2, 0; third block,
0, 3, 1, 2; fourth block, 1, 0, 3, 2.

The assumptions involved in the analysis (normal distribution of
errors, equal scatter in all groups, size of response not affected by previous responses, additivity etc.) have been discussed in §§ 11.2 and
13.1, p. 279. The analysis is the same as that for randomized block

experiments (§ 11.6), with the addition that the between treatment sum of squares can be split into components as described in § 13.7. Because this assay is symmetrical (see (13.8.1)) the arithmetic can be simplified using the results in §§ 13.8, 13.9 and 13.10. Remember that

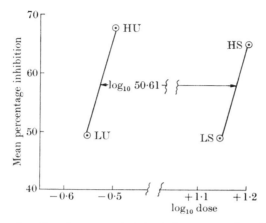

FIG. 13.11.1.   Results of symmetrical 2+2 dose assay from Table 13.11.1.

⊙ Observed mean responses.
—— Least squares lines constrained to be parallel (i.e. with mean slope, see §§ 13.4, 13.10 and calculations at end of this section).
Notice break on abscissa. The question of the units for the potency ratio, 50·61, is discussed later in this section.

the assumptions discussed in §§ 4.2, 7.2, 11.2, and 12.2 have not been tested, so the results are more uncertain than they appear.

The analysis of variance of the response (y)

The figures in Table 13.11.1 are actually identical with those in Table 11.6.2 which were used to illustrate the randomized block analysis. The lower part of the analysis of variance (Table 13.11.2) is therefore identical with Table 11.6.3. The calculations were described on p. 199. (A similar example is worked out in § 13.12.) All that remains to be done is to partition the between-treatment sum of squares, 1188·6875, using the simplifications made possible by the symmetry of the assay.

(a) *Linear regression.* The linear regression contrast defined in (13.8.2) (or by the coefficients in Table 13.8.1) is found, using the response totals from Table 13.11.1, to be

$$L_1 = -196 + 260 - 198 + 271 = 137.$$

The sum of squared deviations (SSD) for linear regression is found using eqn. (13.9.4):

$$\text{SSD} = \frac{L_1^2}{n\Sigma\alpha^2} = \frac{137^2}{4\times4} = 1173{\cdot}0625.$$

In this expression n is the number of responses in each total used in the calculation of L (see (13.8.1)), and $\Sigma\alpha^2$ is the sum of the squares of the coefficients of the response totals in L (given in Table 13.8.1; in the particular case of the $(2+2)$ dose assay it is 4 for all 3 contrasts).

(b) *Deviations from parallelism.* The deviations from parallelism contrast defined in (13.8.3) is

$$L_1' = 196 - 260 - 198 + 271 = 9{\cdot}0$$

The sum of squares is found, as above, from (13.9.4)

$$\text{SSD} = \frac{L_1'^2}{n\Sigma\alpha^2} = \frac{9^2}{4\times4} = 5{\cdot}0625.$$

(c) *Between standard and unknown preparations.* The contrast, defined (in 13.8.4), is

$$L_p = -196 - 260 + 198 + 271 = 13{\cdot}0,$$

and the sum of squares, using (13.9.4) as above, is

$$\text{SSD} = \frac{13^2}{4\times4} = 10{\cdot}5625.$$

(d) *Check on arithmetical accuracy.* The sum of the three components is $1173{\cdot}0625 + 5{\cdot}0625 + 10{\cdot}5625 = 1188{\cdot}6875$, the same as the between treatments sum of squares which was calculated independently.

These results are assembled in the analysis of variance, Table 13.11.2.

Interpretation of the analysis of variance

Dividing each mean square by the error mean square, in the usual way, gives the variance ratios F. As usual all the mean squares would be an estimate of the same variance σ^2 if all 16 observations were randomly selected from a single population with variance σ^2. This is the basic all-embracing form of the null hypothesis because if it were true there would obviously be no differences between treatments, blocks, preparations, etc. In fact, when the variance ratio for linear regression, $F = 319{\cdot}3$ with $f_1 = 1$ and $f_2 = 9$ degrees of freedom, is looked up in tables of the distribution of the variance ratio, as described in § 11.3, it is found that a value of $F(1,9)$ as large as, or larger than, 319·3 would be very rarely ($P \leqslant 0·001$) observed if both 1173·0625 and 3·674 were estimates of the same variance (σ^2), i.e. if there were in fact no tendency for the high doses to give larger responses than low doses (see §§ 13.8 and 13.9). It is therefore preferred to reject the null hypothesis in favour of the alternative hypothesis that response *does*

TABLE 13.11.2

Analysis of variance of responses for symmetrical (2+2) *dose assay of* (+) *tubocurarine. The lower part of the analysis is identical with* Table *11.6.3 which was calculated using the same figures*

Source of variation	d.f.	SSD	MS	F	P
Linear regression	1	1173·0625	1173·0625	319·3	$\leqslant 0·001$
Deviations from parallelism	1	5·0625	5·0625	1·38	$> 0·2$
Between preps. (S and U)	1	10·5625	10·5625	2·87	0·1 —0·2
Between treatments	3	1188·6875	396·229	107·86	$\leqslant 0·001$
Between blocks	3	270·6875	90·229	24·56	$< 0·001$
Error	9	33·0625	3·674		
Total	15	1492·4375			

change with dose (i.e. that β, the population value of b, is *not* zero, cf. § 12.5). The logical reason for this preference was discussed in § 6.1.

Proceeding similarly for the other variance ratios shows that deviations from parallelism such as those observed would be quite common if the true (population) lines were parallel. The same (or larger deviations from parallelism would be expected in more than 20 per cent of repeated experiments if the population lines had the same slope ($\beta_S = \beta_U$). There is therefore no evidence against the hypothesis of parallelism.

Similarly there is little evidence that the average responses are different for standard and unknown. Of course it is most unlikely that they are *exactly* equipotent, but differences as large as, or larger than those observed would not be very uncommon if they were (see p. 93).

There appears to be a real difference between blocks. Differences as large as, or larger than, those observed would be expected in less than 1 in 1000 experiments if the population block means were equal; cf.

§ 11.6. Inspection of the results reveals a tendency for the responses to get larger with time, and the analysis suggests that this cannot be attributed to experimental error. The arrangement in blocks has therefore helped to decrease the experimental error.

All these inferences depend on the assumption of §§ 4.2, 7.2, 11.2, and 12.2 being sufficiently nearly true. If they were, the conclusion would be that there is no evidence that the assay is invalid so it is not unreasonable to carry on and calculate the potency ratio and its confidence limits.

The potency ratio and the question of units

The simplified result for the symmetrical (2+2) dose parallel line assay, eqn. (13.10.6), gives the least squares estimate of the potency as

$$R = \left(\frac{z_{LS}}{z_{LU}}\right) . \text{antilog}_{10}\left[\frac{L_p}{L_1}\text{log}_{10}D\right]$$

$$= \left(\frac{14.0}{0.28}\right) . \text{antilog}_{10}\left[\frac{13.0}{137.0} . \text{antilog}_{10}(1.14286)\right] = 50.61 \ \mu\text{g/ml}.$$

D is the ratio between high and low doses (see (13.8.1)), i.e. $D = 0.32/0.28 = 16.0/14.0 = 1.14286$, and the contrasts, L_p and L_1 have been already calculated. In § 13.3 and later sections it was assumed that all doses were expressed in the same units. This means that z_{LS}/z_{LU}, and hence R, is a dimensionless ratio. In this case the dose of standard was given in μg, and that of unknown in ml, so $z_{LS}/z_{LU} = 14.0 \ \mu\text{g}/0.28 \ \text{ml} = 50.0 \ \mu\text{g/ml}$. If these units are used z_{LS}/z_{LU}, and hence R, will have the units μg/ml, suggesting that, if these units are used, R is actually the potency (concentration in μg/ml) of the unknown, rather than a potency *ratio*. It can easily be seen that this is so by converting standard and unknown to the same units. For example the doses of standard could be assumed to be 16.0 ml and 14.0 ml of a 1.0 μg/ml standard solution of (+)-tubocurarine (the fact that they are more likely in reality, to have been 0.16 ml and 0.14 ml of a 100 μg/ml solution does not alter the dose given). This would give $z_{LS}/z_{LU} = 14.0 \ \text{ml}/0.28 \ \text{ml} = 50$ (a dimensionless ratio). The potency *ratio* would therefore be 50.61, as above, also a dimensionless ratio. The concentration of the unknown is, from the definition of the potency ratio (13.3.1), $R \times$ concentration of standard $= 50.61 \times 1.0 \ \mu\text{g/ml} = 50.61 \ \mu\text{g/ml}$, as found above.

Confidence limits for the potency ratio

The simplified form of Fieller's theorem appropriate to this assay is eqn. (13.10.9), which gives confidence limits for the population value of the potency ratio as

$$\left(\frac{z_{\text{LS}}}{z_{\text{LU}}}\right) . \text{antilog}_{10}\left[\left(\frac{L_{\text{p}}}{L_1(1-g)}\pm\frac{st}{L_1(1-g)}\sqrt{\left\{N(1-g)+N\left(\frac{L_{\text{p}}}{L_1}\right)^2\right\}}\right) . \log_{10}D\right]$$

where $g = Ns^2t^2/L_1^2$, according to (13.10.10).

If the doses are expressed in their original units the equation will give confidence limits for the concentration of unknown, rather than for the potency ratio, for exactly the reasons explained above in connection with the potency ratio calculation. In this example:

$z_{\text{LS}}/z_{\text{LU}} = 14\cdot0/0\cdot28 = 50\cdot0 \ \mu g/ml$ as above;

$L_{\text{p}}/L_1 = 13\cdot0/137\cdot0 = 0\cdot09489$;

$\log_{10}D = \log (0\cdot32/0\cdot28) = 0\cdot05529$;

$s^2[y] = 3\cdot674$, the error variance (from Table 13.11.2) with 9 degrees of freedom;

$t = 2\cdot262$ for $P = 0\cdot05$ (as 95 per cent confidence limits are wanted) with 9 d.f. (from tables of Student's t, see §§ 4.4 and 7.4); thus

$g = 16\times3\cdot674\times2\cdot262^2/137\cdot0^2 = 0\cdot0160$, from equation (13.10.10), and $(1-g) = 1-0\cdot016 = 0\cdot984$.

The fact that g is considerably less than one implies that the slope, b, is much larger than its standard deviation (as inferred from the large variance ratio for linear regression in Table 13.11.2). This means that it is safe to use an approximate equation, based on (2.7.16), for the variance of the log potency ratio (as discussed in §§ 13.5, 13.6, and 13.10, and illustrated below). However, it is very little trouble to use the full equation. Substituting the above quantities into the equation for limits gives

$$50.\text{antilog}_{10}\left[\left(\frac{0\cdot09489}{0\cdot984}\pm\frac{1\cdot917\times2\cdot262}{137\cdot0\times0\cdot984}\sqrt{\left\{(16\times0\cdot984)+16(0\cdot09489)^2\right\}}\right)\right.$$

$$\left.\times0\cdot05529\right]$$

$= 50 \ \text{antilog}_{10} \ [-0\cdot001757 \ \text{and} \ +0\cdot01242]\dagger$

$= 49\cdot80 \ \mu g/ml$ and $51\cdot45 \ \mu g/ml$.

† If necessary, see p. 325 for a footnote describing how to find the antilog of a negative number.

Approximate confidence limits

Because g is much less than 1 the approximate formula for the limits, eqn. (13.10.11), can be used (see §§ 13.5, 13.6, and 13.10). Substituting the quantities already calculated into (13.10.11) gives the estimated standard deviation of $\log_{10} R$ as

$$s[\log_{10} R] \simeq \frac{0\cdot05529}{137\cdot0}\sqrt{[3\cdot674 \times 16(1+0\cdot09489^2)]} = 0\cdot003108.$$

The approximate confidence limits are therefore, as in § 7.4,

$$\log_{10} R \pm ts[\log_{10} R] = \log_{10} 50\cdot61 \pm (2\cdot262 \times 0\cdot003108) = 1\cdot6972$$
$$\text{and } 1\cdot7113.$$

Taking antilogs gives the approximate 95 per cent Gaussian confidence limits for the true value of R as 49·80 and 51·44 μg/ml, not much different from the values found from the full equation, which are themselves, of course, only approximate as explained in § 7.2.

Summary of the result

There is no evidence that the assay is invalid, and the estimated potency of the unknown tubocurarine solution is 50·61 μg/ml, with 95 per cent Gaussian confidence limits 49·9 μg/ml to 51·45 μg/ml. These conclusions are based on the assumptions discussed in §§ 4.2, 7.2, 11.2, and 12.2. The confidence limits are, as usual, likely to be too narrow (see § 7.2). Notice that the confidence limits for R are not equally spaced on each side of R, unlike the limits encountered in Chapter 7. In fact even the limits for $\log R$ are not equally spaced on each side of $\log R$ unless g is small (see §§ 13.5, 13.6, and 13.10).

How to plot results. Conversion to convenient units

When the results of the assay is plotted, as in Fig. 13.11.1, it will be preferable to plot the least squares lines. The calculated average slope, b, has been found using logs to base $\sqrt{D}$ (see § 13.2) so these must be used in plotting the graph (they can be found from logs to base 10 using (13.3.5)). Alternatively, if the graph is plotted with $\log_{10}$ (dose) along the abscissa, as in Fig. 13.11.1, the calculated slope must be converted to the correct units. In this example $b = L_1/N = 137\cdot0/16 = 8\cdot5625$ (from eqn. (13.10.2)).

To convert from logs to base $\sqrt{D}$ to logs to base 10, it is necessary, using (13.3.5), to multiply the former by $\log_{10} \sqrt{D}$ as in § 13.3. Because dose occurs in the denominator of the slope, b must be *divided*† by $\log_{10}\sqrt{D}$, i.e. by $\frac{1}{2}\log_{10}D = \frac{1}{2}\log_{10}(0\cdot32/0\cdot28) = 0\cdot0290$. The required slope is therefore $b' = 8\cdot5625/0\cdot0290 = 295\cdot3$. The dose response curves have the eqns. (13.3.3),

$$Y_{\mathrm{S}} = \bar{y}_{\mathrm{S}}+b'(x_{\mathrm{S}}-\bar{x}_{\mathrm{S}}),$$
$$Y_{\mathrm{U}} = \bar{y}_{\mathrm{U}}+b'(x_{\mathrm{U}}-\bar{x}_{\mathrm{U}}),$$

where x is now being used to stand for $\log_{10}(\text{dose})$, the abscissa of Fig. 13.11.1. The response means are, from Table 13.11.1, $\bar{y}_{\mathrm{S}} = (196 +260)/8 = 57\cdot0$ and $\bar{y}_{\mathrm{U}} = (198+271)/8 = 58\cdot625$. The dose means have not been needed explicitly because of the simplifications resulting from the choice of dose metameter. For the standard, $\log_{10}16\cdot0 = 1\cdot2041$ and $\log_{10}14\cdot0 = 1\cdot1461$ so $\bar{x}_{\mathrm{S}} = (4\times1\cdot2041+4\times1\cdot1461)/8 = 1\cdot1751$ (each dose occurs four times remember). Similarly $\log_{10}z_{\mathrm{HU}} = \log_{10}0\cdot32 = -0\cdot4949$ and $\log_{10}z_{\mathrm{LU}} = \log_{10}0\cdot28 = -0\cdot5528$, so $\bar{x}_{\mathrm{U}} = (4\times-0\cdot4949+4\times-0\cdot5528)/8 = -0\cdot5238$.

Substituting these results gives the lines plotted in Fig. 13.11.1 as $Y_{\mathrm{S}} = 57\cdot0+295\cdot3(x_{\mathrm{S}}-1\cdot1751)$, $Y_{\mathrm{U}} = 58\cdot625+295\cdot3(x_{\mathrm{U}}+0\cdot5238)$.

13.12. A numerical example of a symmetrical (3 +3) dose parallel line assay

The results in Table 13.12.1, which are plotted in Fig. 13.12.1, are measures of the tension developed by the isolated guinea pig ileum in response to pure histamine (standard), and to a solution of a histamine preparation containing various impurities as well as an unknown amount of histamine. Five replicates of each of the six doses were given, all to the same tissue, so there is a danger that one response may affect the size of the next, contrary to the necessary assumption that this does not happen (see discussion in § 13.1, p. 286). The doses were arranged into five random blocks. The purpose of this arrangement is the same as in § 13.11, and, as in that example, the order in which the six doses were arranged in each block was decided strictly at random using random number tables (see § 2.3).

This is a symmetrical assay as defined in (13.8.1), there being $n = 5$ responses at each of the $k = 6$ dose levels; $k_{\mathrm{S}} = k_{\mathrm{U}} = 3$ dose levels for standard and for unknown; $n_{\mathrm{S}} = n_{\mathrm{U}} = 15$ responses for standard

† More rigorously, the slope using $\log_{10}$ (dose) is

$$\frac{\mathrm{d}y}{\mathrm{d}\log_{10}z} = \frac{\mathrm{d}y}{\mathrm{d}(\log_{\sqrt{D}}z.\log_{10}\sqrt{D})} = \frac{1}{\log_{10}\sqrt{D}}\cdot\frac{\mathrm{d}y}{\mathrm{d}\log_{\sqrt{D}}z} = \frac{b}{\log_{10}\sqrt{D}}.$$

TABLE 13.12.1

Responses of the isolated ileum. The doses were given in random order (see text) in each block (time period), not in the order shown in the table

| | Standard histamine dose | | | Unknown dose | | | |
Block	S1 4 ng/ml	S2 8 ng/ml	S3 16 ngl/ml	U1 8 ng/ml	U2 16 ng/ml	U3 32 ng/ml	Total
1	20·5	27·0	38·0	18·5	30·0	35·0	169·0
2	18·5	31·5	44·0	15·0	24·0	34·5	167·5
3	20·0	26·0	35·5	13·0	26·0	38·0	158·5
4	18·0	23·5	41·5	13·5	26·0	35·0	157·5
5	20·0	25·0	38·5	12·0	25·0	32·0	152·5
Total	97·0	133·0	197·5	72·0	131·0	174·5	805·0

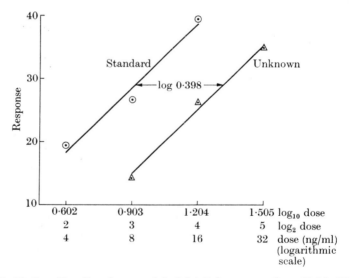

FIG. 13.12.1.   Results of symmetrical 3+3 dose assay from Table 13.12.1.

⊙ Observed mean responses to standard.
△ Observed mean responses to unknown.
— Least squares lines constrained to be parallel (see §§ 13.4, 13.10 and end of this section).

The analysis indicates that these straight lines may well not fit the observations adequately. The abscissa shows three equivalent ways of plotting the log dose. Note that the ordinate does not start at zero.

and for unknown. The ratio between each dose and the one below it is $D = 2$ throughout. The first stage is to perform an analysis of variance on the responses to test the assay for non-validity. As for all assays, this is a Gaussian analysis of variance, and the assumptions that must be made have been discussed in §§ 4.2, 7.2, 11.2, and 12.2, which should be read. Uncertainty about the assumptions means, as usual, that the results are more uncertain than they appear.

Analysis of variance of the response (y)

The first thing to be done is, as in § 13.11, a conventional Gaussian analysis of variance for randomized blocks. Proceeding as in § 11.6,

(1) correction factor $\dfrac{G^2}{N} = \dfrac{805^2}{30} = 21600 \cdot 8333$;

(2) sum of squares between doses (treatments), with $k-1 = 5$ degrees of freedom, from (11.4.5),
$$= \frac{97 \cdot 0^2}{5} + \frac{133 \cdot 0^2}{5} + \ldots + \frac{174 \cdot 5^2}{5} - 21600 \cdot 8333 = 2179 \cdot 0667;$$

(3) sum of squares between blocks, with $n-1 = 4$ degrees of freedom, from (11.6.1),
$$= \frac{169 \cdot 0^2}{6} + \ldots + \frac{152 \cdot 5^2}{6} - 21600 \cdot 8333 = 32 \cdot 8333;$$

(4) total sum of squares, from (2.6.5) or (11.4.3),
$$= \Sigma(y - \bar{y})^2 = 20 \cdot 5^2 + 18 \cdot 5^2 + \ldots + 35 \cdot 0^2 + 32 \cdot 0^2 - 21600 \cdot 8333 = 2328 \cdot 6667;$$

(5) error (or residual) sum of squares, by difference,
$$= 2328 \cdot 6667 - (2179 \cdot 0667 + 32 \cdot 8333) = 116 \cdot 7667$$
with $29 - 5 = 6(5 - 1) = 24$ degrees of freedom.

These results can now be entered in the analysis of variance table, Table 13.12.2. The next stage is to account for the differences observed between the responses to the six doses, i.e. to partition the between doses sum of squares into components representing different sources of variability, as described in § 13.7. The simplified method described in § 13.8 can be used because the assay is symmetrical. The coefficients, α, for construction of the contrasts are given in Table 13.8.2.

(a) *Linear regression.* From the coefficients in Table 13.8.2, and the response totals in Table 13.12.1, the linear regression contrast is

$$L_1 = -97 \cdot 0 + 197 \cdot 5 - 72 \cdot 0 + 174 \cdot 5 = 203 \cdot 0.$$

The corresponding sum of squares for linear regression is found, using (13.9.4), to be

$$\text{SSD} = \frac{L_1^2}{n\Sigma\alpha^2} = \frac{203^2}{5\times 4} = 2060\cdot45.$$

In this expression $n = 5$ is the number of responses at each dose level (i.e. in each total), and $\Sigma\alpha^2$, the sum of squares of the coefficients, is given in Table 13.8.2.

(b) *Deviations from parallelism.* The deviations from parallelism contrast, from Table 13.8.2, is

$$L_1' = 97\cdot0-197\cdot5-72\cdot0+174\cdot5 = 2\cdot0.$$

The corresponding sum of squares is

$$\text{SSD} = \frac{L_1'^2}{n\Sigma\alpha^2} = \frac{2\cdot0^2}{5\times 4} = 0\cdot20.$$

(c) *Between standard and unknown preparations.* The contrast, from Table 13.8.2, is

$$L_p = -97\cdot0-133\cdot0-197\cdot5+72\cdot0+131\cdot0+174\cdot5 = -50\cdot0.$$

The sum of squares, from (13.9.4) (using $\Sigma\alpha^2 = 6$ from Table 13.8.2), is

$$\text{SSD} = \frac{L_p^2}{n\Sigma\alpha^2} = \frac{50\cdot0^2}{5\times 6} = 83\cdot33.$$

(d) *Deviations from linearity.* The contrast, from Table 13.8.2, is

$$L_2 = 97\cdot0-(2\times 133\cdot0)+197\cdot5+72\cdot0-(2\times 131\cdot0)+174\cdot5 = 13\cdot0$$

and the corresponding sum of squares, as before, is

$$\text{SSD} = \frac{L_2^2}{n\Sigma\alpha^2} = \frac{13\cdot0^2}{5\times 12} = 2\cdot82.$$

(e) *Difference of curvature.* The contrast, from Table 13.8.2, is

$$L_2' = -97\cdot0+(2\times 133\cdot0)-197\cdot5+72\cdot0-(2\times 131\cdot0)+174\cdot5 = -44\cdot0,$$

and the corresponding sum of squares

$$\text{SSD} = \frac{(L_2')^2}{n\Sigma\alpha^2} = \frac{(-44\cdot0)^2}{5\times 12} = 32\cdot27.$$

(*f*) *Check on arithmetical accuracy.* The total of the five sums of squares just calculated is

$$2060 \cdot 45 + 0 \cdot 20 + 83 \cdot 33 + 2 \cdot 82 + 32 \cdot 27 = 2179 \cdot 07$$

agreeing, as it should, with the sum of squares between doses which was calculated independently above.

All these results are now assembled in an analysis of variance table, Table 13.12.2, which is completed as usual (cf. §§ 11.6 and 13.7). Divide each sum of squares by its number of degrees of freedom to find

TABLE 13.12.2

The P value marked † is found from reciprocal F = 5·838/0·2,
see text

Source	d.f.	SSD	MS	F	P
Lin regression	1	2060·45	2060·45	352·9	≪0·001
Deviation from parallelism	1	0·20	0·20	0·034	0·8 −0·9†
Between S and U	1	83·33	83·33	14·27	≃0·001
Deviations from linearity	1	2·82	2·82	0·48	>0·2
Difference of curvature	1	32·27	32·27	5·53	<0·05
Between doses	5	2179·07	435·813	74·65	≪0·001
Between blocks	4	32·83	8·208	1·41	>0·2
Error	20	116·77	5·838		
Total	29				

the mean squares. Then divide each mean square by the error mean square to find the variance ratios. The value of P is found from tables of the distribution of the variance ratio as described in § 11.3. As usual P is the probability of seeing a variance ratio equal to or greater than the observed value *if* the null hypothesis (that all 30 observations were randomly selected from a single population) were true.

Interpretation of the analysis of variance

The interpretation of analyses of variance has been discussed in §§ 6.1, 11.3, and 11.6 and in the preceding example, § 13.11. As usual it is conditional on the assumptions being sufficiently nearly true, and must be regarded as optimistic (see §§ 7.2, 11.2, and 12.2). There is no evidence for differences between blocks, so little or nothing was gained, and some degrees of freedom were lost, by using the block arrangement in this particular case (cf. § 13.11). The average slope of the dose response curves, shown in Fig. 13.12.1, is clearly not likely to be zero because if it were, a value of $F \geqslant 352 \cdot 9$ would be exceedingly rare. The question of parallelism is interesting, especially as the standard and unknown were not identical substances. The variance ratio, $F(1,20) = 0 \cdot 2/5 \cdot 838 = 0 \cdot 034$, is very small so there is no hint of deviations from parallelism. To find the P value for $F < 1$ the method described in § 11.3 can be used. Looking up $F(20,1) = 5 \cdot 838/0 \cdot 2 = 29 \cdot 2$ in tables of the variance ratio gives the probability of observing an F value of $29 \cdot 2$ or larger as something between $0 \cdot 1$ and $0 \cdot 2$. Therefore the probability of observing $F(1,20) \leqslant 0 \cdot 034$ is $0 \cdot 1 - 0 \cdot 2$,—not so rare that the lines must be considered as more *nearly* parallel than would be expected on the basis of the observed experimental error. Another way of stating the result is that in 80–90 per cent of repeated experiments the F value for deviations from parallelism would be predicted to be *greater* than $0 \cdot 034$ if the population lines were parallel.

Though neither the standard nor the unknown observations lie on straight lines, as seen in Fig. 13.12.1, the analysis of variance gives no hint of deviations from linearity. This is because the average of the two lines (to which the analysis refers) *is* very nearly straight. The observations lie on lines that curve in opposite directions so the curvatures cancel when the slopes are averaged. In fact an F value corresponding to a difference in curvature as large as, or larger than, the observed one would be expected to occur, as a result of experimental error, in rather less than 5 per cent of repeated experiments. This cannot be explained further without doing more experiments. There could be a real difference in curvature as a result of the impurities in the unknown solution. In intuitive pharmacological grounds this does not seem very likely so perhaps there is no real difference in curvature and a rarish (rather less than 1 in 20) chance has come off (see § 6.1). More experiments would be needed to tell.

If the possibility of a real difference in curvature were not considered to invalidate the assay, the potency ratio and its confidence limits would be calculated as follows.

The potency ratio

In this example the doses of both standard and unknown are expressed in the same units (ng/ml), so the units problem discussed in

§ 13.11 does not arise. The least squares estimate of the potency ratio, from (13.10.17), is

$$R = \left(\frac{z_{S1}}{z_{U1}}\right) . \text{antilog}_{10}\left(\frac{4L_p}{3L_1} . \log_{10} D\right)$$

$$= \left(\frac{4}{8}\right) . \text{antilog}_{10}\left(\frac{4\times(-50)}{3\times 203} . \log_{10} 2\right)$$

$$= 0.5\ \text{antilog}_{10}(-0.09885) = 0.5\times 0.7964 = 0.398\dagger$$

From the definition of the potency ratio, (13.3.1), concentration of unknown $= R\times$ concentration of standard. The unknown preparation is thus estimated to contain 39·8 per cent by weight of histamine, assuming that the impurities in it do not interfere with the assay.

Confidence limits for the potency ratio

The simplified form of Fieller's theorem for the (3+3) dose symmetrical assay is (13.10.18), which gives confidence limits for the population value of the potency ratio as

$$\left(\frac{z_{S1}}{z_{U1}}\right) . \text{antilog}_{10}\left[\left(\frac{4L_p}{3L_1(1-g)}\pm\frac{4st}{3L_1(1-g)}\right.\right.$$

$$\left.\left.\sqrt{\left\{N(1-g)+\frac{2N}{3}\left(\frac{L_p}{L_1}\right)^2\right\}}\right) . \log_{10} D\right]$$

where $g = 2Ns^2t^2/3L_1^2$ according to (13.10.19). For this example

$z_{S1}/z_{U1} = 4/8 = 0.5$,
$L_p/L_1 = -50/203 = -0.2463$,
$\log_{10} D = \log_{10} 2 = 0.3010$,
$s^2[y] = 5.838$, the error variance (from Table 13.12.2) with 20 degrees of freedom,
$s = \sqrt{(5.838)} = 2.416$,
$t = 2.086$ for $P = 0.05$ (for 95 per cent confidence limits) and 20 d.f. (from tables of Student's t, see §§ 4.4 and 7.4).

---

† To find the antilog of a negative number write it as the sum of a negative integer and a positive part between 0 and 1. Thus, to find antilog (-0.09885), write -0.09885 in the form $-1+0.9011$, which is conventionally written $\bar{1}.9011$. Look up antilog 0.9011 $= 7.964$, and move the decimal point one place to the left (because of the $\bar{1}$) giving antilog $(-0.09885) = 0.7964$. Working from first principles, $\text{antilog}_{10}(-0.09885)$ $= 10^{-0.09885}$, from the definition of logarithms, and $10^{-0.09885} = 10^{-1}10^{+0.9011} = 10^{-1}$ antilog (0.9011).

Thus $g = 2 \times 30 \times 5 \cdot 838 \times 2 \cdot 086^2/(3 \times 203^2) = 0 \cdot 01233$
and $(1-g) = 0 \cdot 9877$.

As in the last example, g is small so the approximate formula for the limits could be used, but before doing this the full equation given above will be used to make sure that the approximation is adequate. Substituting the above quantities into the general formula gives

$$0 \cdot 5 \text{ antilog}_{10}\left[\left(\frac{4 \times (-0 \cdot 2463)}{3 \times 0 \cdot 9877}\right) \pm \right.$$

$$\left. \frac{4 \times 2 \cdot 416 \times 2 \cdot 086}{3 \times 203 \times 0 \cdot 9877}\sqrt{\left\{(30 \times 0 \cdot 9877) + \frac{4 \times 30}{3}(-0 \cdot 2463)^2\right\}}\right)0 \cdot 3010\right]$$

$$= 0 \cdot 5 \text{ antilog } (-0 \cdot 1561, -0 \cdot 04405) = 0 \cdot 349 \text{ to } 0 \cdot 452\dagger.$$

Approximate confidence limits

Because g is much less than 1, the approximate formula for the confidence limits (see §§ 13.5, 13.6, and 13.10) can be used, as in the last example. Substituting into (13.10.20) gives the estimated standard deviation of $\log_{10} R$ as

$$s[\log_{10}R] \simeq \frac{4 \times 0 \cdot 3010}{3 \times 203}\sqrt{\left[5 \cdot 838 \times 30\left(1 + \frac{2}{3}(-0 \cdot 2463)^2\right)\right]} = 0 \cdot 02669.$$

The approximate 95 per cent confidence limits are therefore, as in § 7.4,

$$\log_{10} R \pm ts[\log_{10}R] = \log_{10} 0 \cdot 3982 \pm (2 \cdot 086 \times 0 \cdot 02669)$$
$$= -0 \cdot 3999 \pm 0 \cdot 05568 = -0 \cdot 4556 \text{ and } -0 \cdot 3442.$$

Taking antilogs$\dagger$ gives the confidence limits as $0 \cdot 350$ and $0 \cdot 453$, similar to the values found from the full equation.

Summary of the result

The assay may have been invalid because of a difference in curvature between the standard and unknown logdose–response curves. If this difference were attributed to (a rather unlikely) chance the estimated potency ratio would be $0 \cdot 398$, with 95 per cent Gaussian confidence limits of $0 \cdot 349$ to $0 \cdot 452$. As usual, these confidence limits must be regarded as optimistic (see § 7.2).

$\dagger$ See footnote p. 325.

Plotting the results

The slope of the response-log dose lines, from (13.10.13), is $b = 203/20 = 10\cdot15$. This is the slope using $x = \log_D (\text{dose})$ (see § 13.2). It must be divided by $\log_{10} D = 0\cdot3010$, giving $b' = 33\cdot72$, the slope of the response against $\log_{10} (\text{dose})$ lines, which are plotted in Fig. 13.12.1. The full argument is similar to that for the $2+2$ dose assay given in detail in § 13.11.

13.13. A numerical example of an unsymmetrical (3 +2) dose parallel line assay

The general method of analysis for parallel line assays, when none of the simplifications resulting from symmetry (defined in (13.8.1)) can be

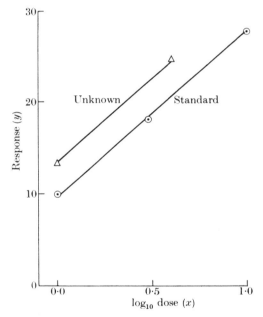

FIG. 13.13.1. Results of an unsymmetrical 3+2 dose assay from Table 13.13.1.

⊙ Observed mean responses to standard.
△ Observed mean responses to unknown.
—— Least squares lines constrained to be parallel (see § 13.4 and this section).

used, will be illustrated using the results shown in Table 13.13.1 and plotted in Fig. 13.13.1. The figures are not from a real experiment—in real life a symmetrical design would be preferred. The 15 doses should be allocated strictly at random (see § 2.3) so a one way analysis of variance (see § 11.4) is appropriate (given the assumptions described in § 11.2).

TABLE 13.13.1

Results of a 3 + 2 dose assay

	Standard doses			Unknown doses		
Dose (z)	1·0	3·0	10·0	1·0	4·0	
$\log_{10}$ dose ($x = \log_{10}z$)	0·0	0·4771	1·0	0·0	0·6021	Total
Responses(y)	9·4	18·0	27·7	13·6	25·1	
	10·8	18·8	28·1	12·8	25·0	
	10·1	17·9	28·2		24·0	
		18·1				
n	3	4	3	2	3	15
Mean	10·1	18·2	28·0	13·2	24·7	
Total	30·3	72·8	84·0	26·4	74·1	
Total		187·1			100·5	287·6

The analysis of variance of the responses

The one way analysis of variance is exactly as in § 11.4.

(1) Correction factor $\dfrac{G^2}{N} = \dfrac{287\cdot6^2}{15} = 5514\cdot25066.$

(2) Total sum of squares (from (2.6.5) or (11.4.3)), with $N-1 = 14$ degrees of freedom,
$$= 9\cdot4^2 + 10\cdot8^2 + \ldots + 24\cdot0^2 - 5514\cdot25066$$
$$= 650\cdot16933$$

(3) Sum of squares between doses (from (11.4.5)) with $5-1 = 4$ degrees of freedom,
$$= \frac{30\cdot3^2}{3} + \frac{72\cdot8^2}{4} + \ldots + \frac{74\cdot1^2}{3} - 5514\cdot25066$$
$$= 647\cdot48933.$$

(4) Error sum of squares, by difference,
$$= 650\cdot16933 - 647\cdot48933 = 2\cdot6800$$
with $14-4 = 10$ degrees of freedom.

The next stage is to divide up the sum of squares between doses, as described in § 13.7. It will be convenient first to calculate various quantities from the results.

For the standard the figures in Table 13.13.1 give

$n_S = 10$,
$\Sigma x_S = (0 \times 3) + (0 \cdot 4771 \times 4) + (1 \cdot 0 \times 3) = 4 \cdot 9084$
    (remember that each dose occurs several times; cf. Table 12.6.2),
$\bar{x}_S = 4 \cdot 9084/10 = 0 \cdot 49084$,
$\Sigma y_S = 30 \cdot 3 + 72 \cdot 8 + 84 \cdot 0 = 187 \cdot 1$,
$\bar{y}_S = 187 \cdot 1/10 = 18 \cdot 71$,
$\Sigma(x_S - \bar{x}_S)^2 = (0^2 \times 3) + (0 \cdot 4771^2 \times 4) + (1 \cdot 0^2 \times 3) - 4 \cdot 9084^2/10$
                $= 1 \cdot 50126$ (from (2.6.5); again each x occurs several
                times),
$\Sigma y_S(x_S - \bar{x}_S) = (0 \times 9 \cdot 4) + (0 \times 10 \cdot 8) + \dots + (1 \cdot 0 \times 28 \cdot 2) -$
                $(4.9084 \times 187 \cdot 1)/10$
                $= 26 \cdot 89672$ (found from (2.6.9) and (12.2.9), as in
                (12.6.1)).

Similarly, for the unknown preparation,

$n_U = 5$,
$\Sigma x_U = (0 \times 2) + (0 \cdot 6021 \times 3) = 1 \cdot 80630$,
$\bar{x}_U = 1 \cdot 80630/5 = 0 \cdot 36126$,
$\Sigma y_U = 26 \cdot 4 + 74 \cdot 1 = 100 \cdot 5$,
$\bar{y}_U = 100 \cdot 5/5 = 20 \cdot 10$,
$\Sigma(x_U - \bar{x}_U)^2 = (0^2 \times 2) + (0 \cdot 6021^2 \times 3) - 1 \cdot 8063^2/5 = 0 \cdot 43503$,
$\Sigma y_U(x_U - \bar{x}_U) = (0 \times 26 \cdot 4) + (0 \cdot 6021 \times 74 \cdot 1) - (1 \cdot 8063 \times 100 \cdot 5)/5$
                $= 8 \cdot 30898$.

Now these results can be used to find the components of the sum of squares between doses, as described in § 13.7.

(1) Linear regression, from (13.7.1),

$$\text{SSD} = \frac{(26 \cdot 89672 + 8 \cdot 30898)^2}{1 \cdot 50126 + 0 \cdot 43503} = 640 \cdot 111405.$$

(2) Deviations from parallelism, from (13.7.2),

$$\text{SSD} = \frac{26 \cdot 89672^2}{1 \cdot 50126} + \frac{8 \cdot 30898^2}{0 \cdot 43503} - 640 \cdot 111405 = 0 \cdot 472584.$$

(3) Between standard and unknown, from (13.7.3),

$$\text{SSD} = \frac{187 \cdot 1^2}{10} + \frac{100 \cdot 5^2}{5} - 5514 \cdot 25066 = 6 \cdot 4403.$$

(4) Deviations from linearity, by difference,

$$\text{SSD} = 647\cdot48933 - 640\cdot111405 - 0\cdot472584 - 6\cdot4403$$

$$= 0\cdot465045.$$

These figures can now all be filled into the analysis of variance table (Table 13.13.2), which has the form of Table 13.7.1.

TABLE 13.13.2

Source of variation	d.f.	SSD	MS	F	P
Linear regression	1	640·111	640·11	2388	≪0·001
Deviations from parallelism	1	0·473	0·473	1·76	>0·2
Between S and U	1	6·440	6·440	24·03	<0·001
Deviations from linearity	1	0·465	0·465	1·74	>0·2
Between doses	4	647·489	161·872	604·0	≪0·001
Error (within doses)	10	2·680	0·268		
Total	14	650·169			

The interpretation of the analysis is just as in §§ 13.11 and 13.12. There is no evidence of invalidity, though if the responses to standard and unknown had been more nearly the same it would have increased precision slightly (see § 13.6).

Plotting the results

The average slope of the dose response lines, from (13.4.5), is

$$b = \frac{26\cdot89672 + 8\cdot30898}{1\cdot50126 + 0\cdot43503} = 18\cdot18.$$

The slopes of lines fitted separately to standard and unknown would be $b_S = 26\cdot89672/1\cdot50126 = 17\cdot92$, and $b_U = 8\cdot30898/0\cdot43503 = 19\cdot10$. The lines plotted in Fig. 13.13.1 are therefore, from (13.3.3),

$$Y_S = 18\cdot71 + 18\cdot18(x_S - 0\cdot4908),$$
$$Y_U = 20\cdot10 + 18\cdot18(x_U - 0\cdot3613).$$

This calculation, but not the preceeding ones, has been made rather simpler than in §§ 13.11 and 13.12, because there is no simplifying transformation to bother about.

The potency ratio

From (13.3.7), the potency ratio is estimated to be (because $x = \log_{10}$ dose)

$$R = \text{antilog}_{10}\left[(0\cdot49084 - 0\cdot36126) + \frac{(20\cdot10 - 18\cdot71)}{18\cdot18}\right]$$

$$= \text{antilog}_{10}(0\cdot2060) = 1\cdot607.$$

Confidence limits for the potency ratio

Using the quantities already found,

$s^2[y] = 0\cdot2680$ (the error mean square with 10 d.f. from Table 13.13.2),
$s = \sqrt{(0\cdot2680)} = 0\cdot5177,$
$(\bar{y}_\text{U} - \bar{y}_\text{S})/b = (20\cdot10 - 18\cdot71)/18\cdot18 = 0\cdot076458,$
$\sum_{\text{S,U}}\sum(x - \bar{x})^2 = 1\cdot50126 + 0\cdot43503 = 1\cdot9363,$
$t = 2\cdot228$ for $P = 0\cdot95$ limits and 10 d.f. (from tables of Student's t; see §§ 4.4 and 7.4).

Thus, from (13.6.5),

$$g = \frac{2\cdot228^2 \times 0\cdot2680}{18\cdot18^2 \times 1\cdot9363} = 0\cdot00208$$

so $(1 - g) = 0\cdot9979$.
Logs to the base 10 have been used, so the conversion factor $\log_{10} 10 = 1$. The 95 per cent confidence limits for the population value of R are therefore, from the general formula (13.6.6),

$$\text{antilog}_{10}\left[(0\cdot49084 - 0\cdot36126) + \frac{0\cdot076458}{0\cdot9979} \pm \right.$$

$$\left. \frac{0\cdot5177 \times 2\cdot228}{18\cdot18 \times 0\cdot9979}\sqrt{\left\{0\cdot9979\left(\frac{1}{5} + \frac{1}{10}\right) + \frac{0\cdot076458^2}{1\cdot9363}\right\}}\right]$$

$$= \text{antilog}_{10}(0\cdot2062 \pm 0\cdot03496)$$
$$= 1\cdot484 \text{ and } 1\cdot743.$$

Approximate confidence limits

Because g is small (even smaller than in the last two examples), the approximate formula, its general form, can be used. In this case $M = \log_{10} R$ so, using (13.6.8),

$$\text{var}[\log_{10} R] \simeq \frac{0\cdot268}{18\cdot18^2}\left(\frac{1}{5}+\frac{1}{10}+\frac{0\cdot076458^2}{1\cdot9363}\right)$$

$$= 2\cdot4570 \times 10^{-4}.$$

The confidence limits for $\log_{10} R$ are therefore $\log_{10} R \pm t\sqrt{(\text{var}[\log_{10} R])}$, and $\sqrt{(\text{var}[\log_{10} R])} = \sqrt{(2\cdot457)} \times 10^{-2} = 0\cdot015675$, giving the limits as $0\cdot2060 \pm 2\cdot228 \times 0\cdot015675 = 0\cdot2060 \pm 0\cdot03492 = 0\cdot1711$ and $0\cdot2409$. Taking antilogs gives the approximate confidence limits as $1\cdot483$ and $1\cdot742$.

Summary of the result

The assay is not demonstrably invalid. The potency ratio is estimated to be $1\cdot607$, with 95 per cent Gaussian confidence limits of $1\cdot484$ to $1\cdot743$. The analysis depends on the assumptions discussed in §§ 4.2, 7.2, 11.2, and 12.2 and, as usual, the confidence limits are likely to be too narrow (see § 7.2).

13.14. A numerical example of the standard curve (or calibration curve). Error of a value of *x* read off from the line

In Chapter 12 the method for estimation of the error of a value of Y (the dependent variable) read from the fitted line at a given value of x was described. In § 12.4 it was mentioned that the reverse problem,

TABLE 13.14.1

	x	Observations (y)			Total	n	Mean
	1	2.3	1.7		4.0	2	2.0
	2	5.4	4.7	4.9	15·0	3	5·0
standard	3	7·4	6·6		14·0	2	7·0
	4	9·7	8·9	8·4	27·0	3	9·0
Unknown		8·1	8·5		16·6	2	8·3

estimation of the error of a value of x interpolated from the fitted line for a given (observed or hypothetical)value of y, is more complicated. In fact the method is closely related to that used to estimate confidence limits for the potency ratio, and an example will now be worked out.

The results in Table 13.14.1, which are plotted in Fig. 13.14.1, are results of the sort that are obtained when measurements are made from a standard calibration curve. This method is often used for chemical assays. For example x could be concentration of solute, and y the optical density of the solution measured in a spectrophotometer. In this example x can be any independent variable (see § 12.1), or any transformation of the measured variable, as long as y (the dependent

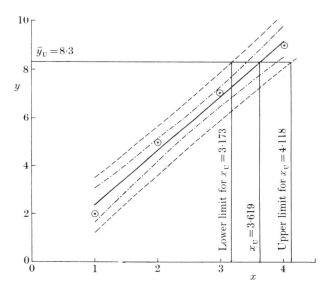

Fig. 13.14.1. The standard calibration curve plotted from the results in Table 13.14.1.

    ⊙ Observed mean responses to standard.
    —— Fitted least squares straight line (see text).
    —·—·— 95 per cent Gaussian confidence limits for the population (true) line, i.e. for the population value of y at any given x value (see text).
    — — — 95 per cent Gaussian confidence limits for the mean of two new observations on y at any given x value (see text).

    The graphical meaning of the confidence limits for the value of x corresponding to the value of y observed for the unknown is illustrated.

variable) is linearly related to x (unlike most of the rest of this chapter, in which the discussion has been confined to parallel line assays in which $x = \log$ dose). It is quite possible to deal with non-linear calibration curves using polynomials (see § 12.7 and Goulden (1952)) but the straight line case only will be dealt with here.

Frequently the standard curve is determined first and it is assumed, as in this section, that it has stayed constant during the subsequent period in which measurements are made on the unknowns. This requires separate verification, and it would obviously be better if standards and unknowns were given in random order or in random blocks. If this is done the unknowns can be incorporated in the analysis of variance as described in § 13.15, the effect of this being to reduce the risk of bias and to improve slightly the estimate of error by taking into account the scatter of replicate observations on the unknown. It will of course be an assumption that the scatter of responses is the same for all of the standards and for the unknowns, in addition to the other assumptions of the Gaussian analysis of variance which have been described in §§ 11.2 and 12.2.

The straight line and its analysis of variance

First a straight line is fitted to the results for the standard. The method has already been described in § 12.6, so only the bare bones of the calculations will be given here. The basic design is a one way classification with $k_S = 4$ independent groups (see § 11.4).

(1) *Correction factor*

$$n_S = 10$$
$$\Sigma y_S = 4 \cdot 0 + 15 \cdot 0 + 14 \cdot 0 + 27 \cdot 0 = 60 \cdot 0,$$
$$\text{correction factor} = \frac{60 \cdot 0^2}{10} = 360 \cdot 0.$$

(2) *Total sum of squares*, from (2.6.5) (cf.(11.4.3))

$$= 2 \cdot 3^2 + 1 \cdot 7^2 + \ldots + 8 \cdot 4^2 - 360 \cdot 0 = 65 \cdot 6200.$$

(3) *Sum of squares between groups*, from (11.4.5),

$$= \frac{4^2}{2} + \frac{15^2}{3} + \frac{14^2}{2} + \frac{27^2}{3} - 360 \cdot 0 = 64 \cdot 0000.$$

Although the x values are equally spaced, the simplifying transformation described at the end of § 12.6 cannot be used, because the number of observations is not the same at each x value.

(a) *Sum of squares due to linear regression.* First calculate

$$\Sigma x_S = (1 \times 2) + (2 \times 3) + (3 \times 2) + (4 \times 3) = 26 \cdot 0$$
$$\text{and } \bar{x}_S = 26 \cdot 0 / 10 = 2 \cdot 60.$$

The sum of products, from (2.6.7) (see § 12.6), is

$$\Sigma y_s(x_s - \bar{x}_s) = (1 \times 4 \cdot 0) + (2 \times 15 \cdot 0) + (3 \times 14 \cdot 0) + (4 \times 27 \cdot 0) -$$

$$\frac{26 \cdot 0 \times 60 \cdot 0}{10} = 28 \cdot 00.$$

The sum of squares for x is, from (2.6.5),

$$\Sigma(x_s - \bar{x}_s)^2 = (1^2 \times 2) + (2^2 \times 3) + (3^2 \times 2) + (4^2 \times 3) - \frac{26 \cdot 0^2}{10}$$

$$= 12 \cdot 40.$$

Thus, sum of squares due to linear regression, from (12.3.4),

$$= \frac{28 \cdot 0^2}{12 \cdot 4} = 63 \cdot 2258.$$

(b) *Sum of squares for deviations from linearity.* By difference

$$\text{SSD} = 64 \cdot 0000 - 63 \cdot 2258 = 0 \cdot 7742.$$

(4) *Sum of squares for error* (within groups sum of squares). By difference

$$\text{SSD} = 65 \cdot 6200 - 64 \cdot 0000 = 1 \cdot 6200.$$

These results can now be entered in the analysis of variance table, Table 13.14.2.

TABLE 13.14.2

Source	d.f.	SSD	MS	F	P
Lin. regression	1	63·2258	63·2258	234	$\ll 0 \cdot 001$
Dev. from linearity	2	0·7742	0·3871	1·43	$> 0 \cdot 2$
Between groups (x values)	3	64·0000	21·3333	79·0	
Error (within groups)	6	1·6200	0·2700		
Total	9	65·6200			

The interpretation is as in § 12.6. There is strong evidence that y increases with x. If the true line were straight then an F value for deviations from linearity equal to or greater than 1·43 would be expected in more than 20 per cent ($P > 0 \cdot 2$) of repeated experiments (given the assumptions—see § 11.2 and 12.2), so there is no reason to believe the true line is not straight. However this analysis does not distinguish between systematic and unsystematic deviations from linearity. Looking at Fig. 13.14.1 suggests the deviations in this case, though no larger than would be expected on the basis of experimental error, are of a systematic sort. The line appears to be flattening out. Now physical considerations, and past experience suggest that this is just the sort of nonlinearity that would be expected in a plot of, say, optical density against concentration. In a case like this it would be rather rash to fit a straight line, in spite of the fact that there are no grounds for rejecting the null hypothesis that the true (population) line is straight. *This is a good example of the practical importance of the logical fact explained in § 6.1, that if there are no grounds for rejecting a hypothesis this does not mean that there are good grounds for accepting it.* In a small experiment, such as this, with substantial experimental errors, it is more than likely that deviations from linearity that are real, and large enough to be of practical importance, would not be detected with any certainty. The verdict is not proven (see § 6.1). *For purposes of illustration, a straight line will now be fitted, though the foregoing remarks suggest that a polynomial (see above) would be safer.* The least squares estimates of the parameters (see § 12.2) are thus, from (12.2.6),

$$a_S = \bar{y}_S = \frac{\Sigma y_S}{n_S} = \frac{60 \cdot 0}{10} = 6 \cdot 00$$

and, from (12.2.8),

$$b_S = \frac{\Sigma y_S(x_S - \bar{x}_S)}{\Sigma(x_S - \bar{x}_S)^2} = \frac{28 \cdot 00}{12 \cdot 40} = 2 \cdot 2581,$$

so the fitted line is

$$Y_S = a_S + b_S(x_S - \bar{x}_S) = 6 \cdot 00 + 2 \cdot 2581 \ (x_S - 2 \cdot 60) \qquad (13.14.1)$$
$$= 0 \cdot 1289 + 2 \cdot 2581 \ x_S$$

and this is the straight line plotted in Fig. 13.14.1.

Interpolation of the unknown

The mean of the two observations ($n_U = 2$) on the unknown, from Table 13.14.1, is $\bar{y}_U = 8 \cdot 30$. The equation for the standard line, (13.14.1), is $Y = a + b(x - \bar{x})$, and rearranging this to find x gives

$$x = \bar{x}_S + \frac{(Y - a_S)}{b_S}. \qquad (13.14.2)$$

The estimate of x_U (e.g. concentration) corresponding to the mean observation $\bar{y}_U$ (e.g. optical density) on the unknown is therefore, from (13.14.1) and (13.14.2),

$$\begin{aligned}
x_U &= \bar{x}_S + (y_U - \bar{y}_S)/b_S \\
&= 2{\cdot}60 + (8{\cdot}30 - 6{\cdot}00)/2{\cdot}2581 \\
&= 3{\cdot}619,
\end{aligned} \tag{13.14.3}$$

as shown graphically in Fig. 13.14.1.

Gaussian confidence limits for the interpolated x value

The approach is exactly like that in § 13.6. In (13.14.3) $\bar{x}_S$ is an accurately measured constant. If the observations are normally distributed (see § 4.2), then $\bar{y}_U - a = \bar{y}_U - \bar{y}_S$ will be a normally distributed variable, and so will the slope of the standard line, b_S (see § 12.4, especially (12.4.1)). Therefore $(\bar{y}_U - \bar{y}_S)/b_S = m$, say, will be the ratio of two normally distributed variables and Fieller's theorem (see § 13.5) can be used, as in § 13.6, to find confidence limits for its true value. If the error mean square from Table 13.14.2 ($s^2 = 0{\cdot}2700$) is taken as the variance of the observations on the unknown as well as the variance of the observations on the standard then, from (2.7.9), $\mathrm{var}(\bar{y}_S) = s^2/n_S$, $\mathrm{var}(\bar{y}_U) = s^2/n_U$. Because the observations on standard and unknown are assumed independent, it follows from (2.7.3) that $\mathrm{var}(\bar{y}_U - \bar{y}_S) = \mathrm{var}(\bar{y}_U) + \mathrm{var}(\bar{y}_S) = s^2(1/n_U + 1/n_S)$, and so from (13.5.3),

$$\begin{aligned}
v_{11} &= (1/n_U + 1/n_S), \\
v_{22} &= 1/\Sigma(x_S - \bar{x}_S)^2 \quad \text{(from (12.4.2))}, \\
v_{12} &= 0 \quad \text{(as in § 13.6)}.
\end{aligned}$$

In the present example

$s^2 = 0{\cdot}2700$ with 6 d.f. (from (13.14.2)),

$s = \sqrt{(0{\cdot}2700)} = 0{\cdot}5196$,

$t = 2{\cdot}447$ for $P = 0{\cdot}95$ and 6 d.f. (from tables of Student's t; see § 4.4),

$$m = \frac{(\bar{y}_U - \bar{y}_S)}{b} = \frac{(8{\cdot}30 - 6{\cdot}00)}{2{\cdot}2581} = 1{\cdot}01856,$$

$$g = \frac{t^2 s^2 v_{22}}{b^2} = \frac{2{\cdot}447^2 \times 0{\cdot}2700}{2{\cdot}2581^2 \times 12{\cdot}40} = 0{\cdot}02557 \quad \text{(from (13.5.8))},$$

$$(1 - g) = 0{\cdot}9744.$$

The 95 per cent confidence limits for the true value of x_U therefore follow from (13.5.9) (by adding $\bar{x}_S$ to the confidence limits for $(\bar{y}_U - \bar{y}_S)/b$; cf. § 13.6) and are

$$\bar{x}_S + \frac{m}{(1-g)} \pm \frac{st}{b(1-g)} \sqrt{[(1-g)v_{11} + m^2 v_{22}]},$$

i.e in the present case

$$2 \cdot 60 + \frac{1 \cdot 01856}{0 \cdot 9744} \pm \frac{0 \cdot 5196 \times 2 \cdot 447}{2 \cdot 2581 \times 0 \cdot 9744} \sqrt{\left[0 \cdot 9744 \left(\frac{1}{2} + \frac{1}{10} \right) + \frac{1 \cdot 01856^2}{12 \cdot 40} \right]}$$

$$= 3 \cdot 173 \text{ and } 4 \cdot 118.$$

Because g is fairly small, similar limits would have been found by using the approximate formula, from (13.5.12), $\mathrm{var}(x_U) \simeq s^2(v_{11} + m^2 v_{22})/b^2$. The limits are not symmetrical about $x_U = 3 \cdot 619$ unless g is negligibly small (or unless $\bar{y}_U = \bar{y}_S$), as discussed in § 13.5. In this case the limits expressed as percentage deviations from $3 \cdot 619$ are $-12 \cdot 3$ per cent to $+13 \cdot 8$ per cent. The graphical meaning of the limits is discussed below.

Summary of the result

The unknown value of x corresponding to $\bar{y}_U = 8 \cdot 3$ is $x_U = 3 \cdot 619$ with 95 per cent Gaussian confidence limits from $3 \cdot 173$ to $4 \cdot 118$. These results depend on the assumptions described in §§ 7.2, 11.2, and 12.2 and, as usual, must be considered to be optimistic (see § 7.2).

Confidence limits for the population calibration line

Assuming the true line to be straight, limits for its position can be calculated as described in § 12.4. Another example was worked out in § 12.5. In this case $\mathrm{var}(y) = 0 \cdot 2700$, the error mean square with 6 d.f. from Tables 13.14.2, $N = 10$ (the number of observations used to fit the line), $t = 2 \cdot 447$ as above, $\bar{x}_S = 2 \cdot 60$ and $\Sigma(x - \bar{x})^2 = 12 \cdot 40$ as above. Using these values $\mathrm{var}(y)$, and hence the confidence limits, can be calculated at enough values of x to plot the limits, which are shown as dot-dashed lines in Fig. 13.14.1. Two representative calculations follow.

(1) At $x = 1 \cdot 0$. At this point the estimated value of y is, from (13.14.1),

$$Y = 0 \cdot 1289 + (2 \cdot 2581 \times 1 \cdot 0) = 2 \cdot 387$$

and, from (12.4.4),

$$\text{var}(Y) = 0.2700\left(\frac{1}{10} + \frac{(1.0-2.60)^2}{12.40}\right) = 0.082742.$$

The 95 per cent Gaussian confidence limits for the population value of Y at $x = 1.0$ are therefore, from (12.4.5), $2.387 \pm 2.447 \times \sqrt{0.082742}$ $= 1.683$ and 3.091.

(2) At $x = 2.0$, $Y = 0.1289 + (2.2581 \times 2.0) = 4.645$,

$$\text{var}(Y) = 0.2700\left(\frac{1}{10} + \frac{(2.0-2.60)^2}{12.40}\right) = 0.034839,$$

giving confidence limits of $4.645 \pm (2.447 \times \sqrt{0.034839}) = 4.188$ and 5.102.

Confidence limits for the mean of two new observations at a given x. The graphical meaning of Fieller's theorem.

In § 12.4 a method was described for finding limits within which new observations on y (rather than the population value of y), at a given x, would be expected to lie. In the present example there are $n_U = 2$ new observations on the unknown. Using eqn (12.4.6) with $m = 2$, $N = 10$, and the other values as above, these limits can be calculated for enough values ot x for them to be plotted. They are shown as dashed lines in Fig. 13.14.1. Two representative calculations, using (12.4.6), follow.

(1) At $x = 1.0$. At this point $Y = 2.387$ as above The 95 per cent confidence limits for the mean of two new observations are, from (12.4.6),

$$2.387 \pm 2.447\sqrt{\left[0.2700\left\{\frac{1}{10} + \frac{1}{2} + \frac{(1.0-2.60)^2}{12.40}\right\}\right]} = 1.245 \text{ and } 3.529.$$

(2) At $x = 2.0$, $Y = 4.645$ as above, and the limits are, from (12.4.6),

$$4.645 \pm 2.447\sqrt{\left[0.2700\left\{\frac{1}{10} + \frac{1}{2} + \frac{(2.0-2.60)^2}{12.40}\right\}\right]} = 3.637 \text{ and } 5.653.$$

These limits are seen to be wider than the limits for the population value of Y as would be expected when the uncertainty in the new observations is taken into account. They are also less strongly curved.

The mean of the two observations on the unknown in Table 13.14.1, was $\bar{y}_U = 8\cdot3$, and the corresponding value of x_U read off from the line was $3\cdot619$ as calculated above, and as shown in Fig. 13.14.1. The 95 per cent confidence limits for x_U at $y = 8\cdot3$ were found above to be $3\cdot173$ to $4\cdot118$. It can be seen in Fig. 13.14.1 that these are the points where the line for $y = 8\cdot3$ intersects the confidence limits just calculated (the limits for the mean of two new observations at a given x). The limits found from Fieller's theorem (13.5.9) are, in general, the same as those found graphically via (12.4.6).

13.15. The (*k* +1) dose assay and rapid routine assays

In this section the k_S+1 dose parallel line assay will be illustrated using the same results that were used to illustrate the calibration curve analysis in § 13.14.

Routine assays

The $(2+2)$ or $(3+3)$ dose assays should be preferred for accurate assays. The (k_S+1) dose assay probably occurs must frequently in the form of the $(2+1)$ dose assay in which the unknown is interpolated between 2 standards. This is the fastest method and is often used when large number of unknowns have to be assayed. It is rare in practice for the doses to be arranged randomly, or in random blocks of 3 doses $(k_S+1$ doses in general). Even worse, standard and unknowns are often given alternately, so each standard is used to interpolate both the unknown immediately before it and the unknown immediately after it. This introduces correlation between duplicate estimates, making the estimation of error difficult. Quite often the samples to be assayed will come from an experiment in which replicate samples were obtained, and several assays will be done on each of the replicate samples. In this case a reasonable compromise between speed and statistical purity is to do $(2+1)$ dose assays with alternate standard and unknown, and to interpolate each unknown response between the standard responses (one high and one low) on each side of it. The replicate assays on each sample are then simply averaged. An estimate of error can then be obtained from the scatter of the average assay figures for replicate samples rather than doing the calculations described below. The treatments should have been applied in random order (see § 2.3) in the original experiment and the samples should be assayed in random order. If the ratio between the high and low standard doses is small (say less than 2) it will usually be sufficiently accurate to interpolate linearly (rather than logarithmically) between the standards. See Colquhoun and Tattersall (1969) for further discussion.

A numerical example of a (4+1) *dose parallel line assay*

In a parallel line assay $x = \log$ (dose) by definition (see § 13.1), unlike § 13.14 in which x could have been any independent variable. In biological assays it is usual to specify the dose of unknown (e.g. in ml or g of impure solid) and to compute a potency ratio R (see § 13.3),

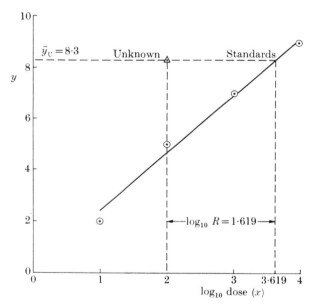

FIG. 13.15.1.   If x in § 13.14 (Table 13.14.1) were log dose, then the results in Table 13.14.1 could be treated as a 4+1 dose parallel line assay, as illustrated, as an alternative to the treatment as a standard curve problem which was worked out in § 13.14. The observations and fitted line are as in Fig. 13.14.1 with the addition that the dose of unknown required to produce the unknown responses has been specified.

rather than to interpolate the unknown response on the standard curve as in § 13.14. (Equation (13.14.3) together with (13.3.7) is seen to imply $\log R = 0$, i.e. $R = 1$, which simply means that a given dose, in terms of the active substance, gives the same response whether it is labelled standard or unknown.) Suppose, for example, that x in Table 13.14.1 represents the $\log_{10}$ of the standard dose (measured in ml) in a (4+1) dose parallel line assay. Suppose further that a $\log_{10}$ (dose in ml) of unknown, $x = 2 \cdot 0$, is administered twice and produces responses $y_U = 8\cdot1$ and $8\cdot5$, so $\bar{y}_U = 8\cdot3$ as in Table 13.14.1. This assay is plotted in Fig. 13.15.1. Using the general formula for the log potency ratio, (13.3.7), gives, using (13.14.3),

$$\log_{10}R = \bar{x}_S - \bar{x}_U + \frac{\bar{y}_U - \bar{y}_S}{b}$$

$$= 3\cdot619 - 2\cdot00 = 1\cdot619.$$

Taking antilogs gives $R = 41\cdot59$ which means that $41\cdot59$ ml of standard must be given to produce the same effect as 1 ml of unknown.

It was mentioned in § 13.14 that it is dangerous to determine the standard curve first and then to measure the unknowns later unless there is very good reason to believe that the standard curve does not change with time. It is preferable to do the standards and unknowns (all 12 measurements in Table 13.14.1) in random order (or in random block of 5 measurements, cf. §§ 13.11 and 13.12). If this had been done the analysis of variance would follow the lines described in § 13.7, except that there can obviously be no test of parallelism with only one dose of standard (a 2+1 dose assay would have no test of deviations from linearity either). There are now 5 groups and 12 observations. The total, between group and error sum of squares are found in the usual way (see §§ 11.4, 13.13, or 13.14) from the 5 groups of observations in Table 13.14.1. The results are shown in Table 13.15.1. The between groups sum of squares can be split up into components using the general formulae (13.7.1)–(13.7.3). In a (k_S+1) dose assay there is only one unknown dose so $\bar{x}_U = x_U$, i.e. $(x_U - \bar{x}_U) = 0$ so the expressions for the slope (13.4.5) and the sum of squares for linear regression, (13.7.1), reduce to those used already in § 13.14, which are entered in Table 13.12.1 (it is only common sense that the observations on the unknown can give no information on the slope of the log dose-response line). The sum of squares for differences in responses to standard and unknown, from (13.7.3), is

$$\frac{60^2}{10} + \frac{16\cdot6^2}{2} - \frac{76\cdot6^2}{12} = 8\cdot8167.$$

When this is entered in Table 13.15.1 the sum of squares for deviations from linearity can be found by difference. It is seen to be identical with that in Table 13.14.2, as expected.

The error variance in Table 13.15.1 is $0\cdot2429$, less than the figure of $0\cdot2700$ from Table 13.14.2. Inclusion of the unknown responses has slightly reduced the estimate of error because they are in relatively good agreement. The interpretation is the same as in § 13.14.

The confidence limits for the log potency ratio be found from the general parallel line assay formula, (13.6.6). The calculation is, with

TABLE 13.15.1

Source	d.f.	SSD	MS	F	P
Linear regression	1	63·2258	63·2258	260	$\ll 0·001$
Bet. std. and unknown	1	8·8167	8·8167	36·3	$<0·001$
Deviations from linearity	2	0·7742	0·3871	1·59	$>0·2$
Between doses	4	72·8167	18·2042	74·9	$\ll 0·001$
Error (within doses)	7	1·7000	0·2429		
Total	11	75·5167			

any luck, seen to be exactly the same as in § 13.14 except $x_U = 2·00$ is subtracted from the result. The limits are therefore $3·173 - 2·00 = 1·173$ and $4·118 - 2·00 = 2·118$. Taking antilogs gives the 95 per cent Gaussian confidence limits for the true value of R (estimated as 41·59) as 14·89 to 131·2—not a very good assay.

14. The individual effective dose, direct assays, all-or-nothing responses and the probit transformation

14.1. The individual effective dose and direct assays

THE quantity of, for example, a drug needed to *just* produce any specified response (e.g. convulsions or heart failure) in an animal is referred to as the individual effective dose (IED) for that animal and will be denoted z. More generally, the amount or concentration of any treatment needed to *just* produce any specified effect on a test object can be treated as an IED. A standard preparation of a drug, and a preparation of the same drug of unknown concentration, can be used to estimate the unknown concentration. This sort of biological assay is usually referred to as a *direct assay*.

A group of animals is divided randomly (see § 2.3) into two sub-groups. On each animal (test object, in general) of one group the IED of a standard solution of the substance to be assayed is measured. The IED of the unknown solution is measured on each animal of the other group.

It is important to notice that in this case the *dose is the variable* not the response as was the case in Chapter 13.

If the doses of both solutions are measured in the same units (see § 13.11) then the dose (z ml, say) needed for a given amount (in mg, say) of substance is inversely proportional to the concentration of the solution. The object of the assay is to find the potency ratio (R) of the solutions, i.e. the ratio of their concentrations. Thus

$$R = \frac{\text{concentration of unknown}}{\text{concentration of standard}} =$$
$$= \frac{\text{population mean IED of standard}}{\text{population mean IED of unknown}}. \qquad (14.1.1)$$

In practice the population mean† IEDs must, of course, be replaced

† See Appendix 1.

by sample estimates, the average, $\bar{z}$, of the observed IEDs. The question immediately arises as to what sort of average should be used.

If the IEDs were normally distributed there are theoretical reasons (see §§ 2.5, 4.5, and 7.1) for preferring to calculate the arithmetic mean IED for each preparation (standard and unknown). In this case the estimated potency ratio would be $R = \bar{z}_S/\bar{z}_U$. Because the IED has been supposed to be a normally distributed variable, this is the ratio of two normally distributed variables. A pooled estimate of the variance $s^2[z]$ could be found from the scatter within groups (as in § 9.4). The confidence limits for R could then be found from Fieller's theorem, eqn. (13.5.9), with $v_{11} = 1/n_S$ and $v_{22} = 1/n_U$, where n_S and n_U are the numbers of observations in each group. (Because each IED is supposed to be independent of the others, $v_{12} = 0$.)

However, if the IEDs are lognormally distributed (see § 4.5) then the problem is simpler. Tests of normality are discussed in § 4.6.

Use of the logarithmic dose scale for direct assays

In those cases in which it has been investigated it has often been found that the logarithm of the IED ($x = \log z$, say) is normally distributed (i.e. z is lognormally distributed, see § 4.5). It therefore tends to be assumed that this will always be so, though, as usual, there is no evidence one way or the other in most cases. If it were so then it would be appropriate to take the logarithm of each observation and carry out the calculations on the $x = \log z$ values, because they will be normally distributed. (In parallel line assays a logarithmic scale is used for the dose, which is the independent variable and has no distribution, for a completely different reason; to make the dose–response curve straight. See §§ 11.2, 12.2, and 13.1, p. 283.)

Taking logarithms of both sides of (14.1.1) gives the log of the potency ratio (M, say) as

$$M \equiv \log R = \log\left(\frac{\text{IED of standard}}{\text{IED of unknown}}\right) =$$

$$= \log (\text{IED of S}) - \log (\text{IED of U}).$$

If the log IED is denoted $x = \log z$ then it follows that the estimated log potency ratio will be

$$M \equiv \log R = \bar{x}_S - \bar{x}_U. \tag{14.1.2}$$

The variance of this will, because the estimates of IED have been assumed to be independent, be

$$\mathrm{var}(M) = \mathrm{var}(\bar{x}_\mathrm{S}) + \mathrm{var}(\bar{x}_\mathrm{U}) = \frac{\mathrm{var}(x_\mathrm{S})}{n_\mathrm{S}} + \frac{\mathrm{var}(x_\mathrm{U})}{n_\mathrm{U}}, \qquad (14.1.3)$$

from (2.7.3) and (2.7.8). It is necessary, as in § 9.4, to assume that the scatter of the measurements (x values) is the same in both groups so a pooled estimate of var(x) is calculated from the scatter of the logs of the observations within groups as in § 9.4, and used as the best estimate of both var(x_S) and var(x_U). The confidence limits for the log potency ratio are then $M \pm t\sqrt{\{\mathrm{var}(M)\}}$ as in § 7.4. Taking antilogarithms of these, and of (14.1.2), gives the estimates of R and its confidence limits. A numerical example is given by Burn, Finney, and Goodwin (1950, pp. 44–8).

14.2. The relation between the individual effective dose and all-or-nothing (quantal) responses

In the sort of experiment described in § 14.1 the individual effective dose (IED) just sufficient to produce a given effect is measured directly on each individual. For example, the amount of digitalis solution needed to produce cardiac arrest can be measured on each of a group of animals by giving it as a slow intravenous infusion and observing the volume administered at the point when the heart stops. The results given in Table 14.2.1 are an idealized version of experimental measurements of 100 individual lethal doses (z) of cocaine cited by J. W. Trevan (1927). The results have been grouped so that a histogram can be plotted from them and the percentage of individual effective doses falling in each dose interval is denoted f. The logarithms (x) of the doses are also given (1 has been added to each of the values to make them all positive).

From the results in Table 14.2.1 the mean individual effective dose is the total of the z values divided by the total number of observations†

$$\bar{z} = \frac{\Sigma fz}{\Sigma f} = \frac{51 \cdot 475}{100} \simeq 0 \cdot 515 \text{ mg.}$$

The median effective dose (dose for $p = 50$ per cent) (interpolated from Fig. 14.2.2) $\simeq 0 \cdot 49$ mg.

The modal effective dose (interpolated from Fig. 14.2.1) $\simeq 0 \cdot 44$ mg.

$\left. \vphantom{\begin{array}{c} 1 \\ 1 \\ 1 \\ 1 \\ 1 \\ 1 \end{array}} \right\}$ (14.2.1)

† This mean is calculated from the grouped results, each IED being assumed to have the central value of the group in which it falls. If the original ungrouped observations were available, the mean of these would be preferred. If it is accepted that z is lognormal (see below) then the mean can also be estimated using the equation on p. 78 with $\mu = \overline{1} \cdot 707$ and $\sigma = 0 \cdot 104$ from Fig. 14.2.6. This gives antilog$_{10}$ $(\overline{1} \cdot 707 + 1 \cdot 1513 \times 0 \cdot 104^2) = 0 \cdot 524$ mg.

A histogram of the distribution of the individual effective doses is plotted in Fig. 14.2.1 and the estimated mean, median, and modal IEDs (see § 2.5) plotted on it. The distribution looks positively skewed and therefore, as expected, mean > median > mode (see § 4.5).

Table 14.2.1

Frequency f = percentage of animals responding in each dose interval. Cumulative frequency p = total percentage of animals responding to dose equal to or less than the upper limit of each dose interval. Probits (see § 14.3) were obtained from Fisher and Yates tables (1963, Table IX, p. 69). The p values are found as the cumulative sum of the observed f values. For example 54 = 38+15+1

Dose interval (mg of cocaine)	Mid-point (z)	log dose interval +1 (x)	f	p	fz	Probit of p
0–0·2	0·10	$-\infty$–0·301	0	0	0	$-\infty$
0·2–0·3	0·25	0·301–0·477	1	1	0·25	2·674
0·3–0·4	0·35	0·477–0·602	15	16	5·25	4·005
0·4–0·5	0·45	0·602–0·699	38	54	17·11	5·100
0·5–0·6	0·55	0·699–0·778	25	79	13·76	5·806
0·6–0·7	0·65	0·778–0·845	11	90	7·15	6·282
0·7–0·8	0·75	0·845–0·903	6·5	96·5	4·88	6·812
0·8–0·9	0·85	0·903–0·954	2·5	99	2·125	7·326
0·9–1·0	0·95	0·954–1·000	1	100	0·95	$+\infty$
			Σf = 100		Σfz = 51·475	

The individual effective dose is of course a *continuous* variable and the distribution of IEDs is a continuous distribution (see § 4.1). However, in order to get an idea of the shape of the distribution it has been necessary to group the observed IEDs so that the histogram in Fig. 14.2.1 can be plotted. A continuous line has been drawn by eye through the histogram as an estimate of what the actual continuous distribution should look like.

In Fig. 14.2.2 the histogram of cumulative frequency (p) is plotted against dose. When a continuous line is drawn through the top right-hand corner (see below) of each block an unsymmetrical sigmoid curve is obtained. This is the cumulative distribution, or distribution function, $F(x)$, (defined in (4.1.4)) corresponding to the distribution of IED shown in Fig. 14.2.1. That is to say the ordinate of the curve in Fig. 14.2.2 for any specified value of the dose, z, is equal to the *area* under the curve in Fig. 14.2.1 below z (cf. Fig. 5.1.2 and its cumulative form, Fig. 5.1.1, and Figs. 4.1.3 and 4.1.4).

The relation between the IED and quantal responses can now be illustrated. When a quantal response is obtained the IED itself is not measured. A fixed dose, z, of drug is given to a group of n subjects and the number, r, of subjects showing the chosen response is observed. The proportion of subjects responding in the group is r/n. This is, of course, a *discontinuous* variable; and if the same dose were given

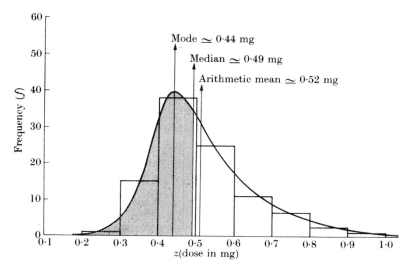

FIG. 14.2.1. Histogram of the individual effective dose measurement in Table 14.2.1. The frequency, f, is plotted against dose (z). The continuous line has been drawn by eye through the histogram as an estimate of the true (continuous) distribution of individual effective doses. The distribution is skew so median effective dose (shaded area = 50 per cent of total area under curve) is less than the mean but greater than the modal effective dose (see § 4.5).

repeatedly to many groups of n subjects then the number showing a response, r, would be expected to vary from trial to trial according to the discontinuous *binomial* distribution (see §§ 3.2–3.4). The subjects that respond will be those in the group whose IED is equal to *or less than* the dose given, z. Therefore if the doses chosen were the upper limits of each interval in Table 14.2.1 (i.e. doses of 0.2, 0.3, 0.4, . . ., 1.0 mg) then the values of r/n observed in each of the 9 groups of animals would be the same (apart from experimental error) as the values of p tin he table (which is why p and probit $[p]$ are plotted against the upper limits of each dose interval in Fig. 14.2.2, 14.2.4, 14.2.5, and 14.2.6).

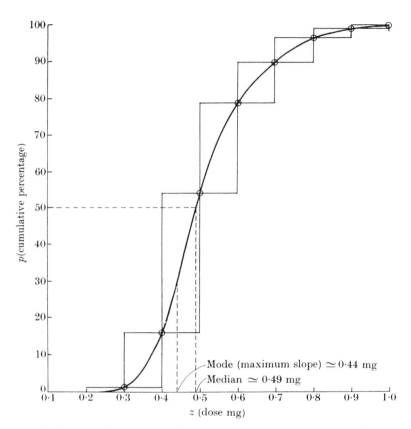

F IG . 14.2.2.   Results from Table 14.2.1. The histogram is plotted using the cumulative frequency p, against dose z. The blocks, each of height f, from Fig. 14.2.1, have been put above each other so that the total height is p. The sigmoid curve has been drawn by eye through the top right-hand corner of each block (see text) as an estimate of the true (continuous) cumulative distribution (i.e. the distribution function, see § 4.1) of individual effective doses, i.e. the ordinate is the percentage of animals with an individual effective dose equal to or less than z.

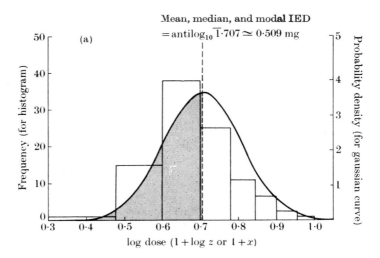

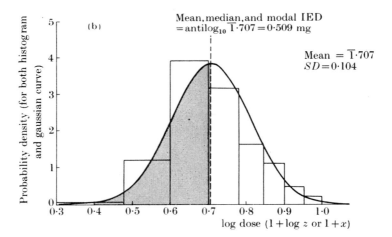

FIG. 14.2.3. Results from Table 14.2.1. Histogram of individual effective dose measurements with dose on a logarithmic scale, rather than on an arithmetic scale as in Fig. 14.2.1 (1 has been added to the logs to avoid negative values). Now that the blocks of the histogram are not of equal width, their area is no longer proportional to their height, so a convention must be adopted as to whether area or height shall represent frequency.

(a) In this figure height represents frequency i.e. frequency (left-hand scale) is plotted against log dose. The heights of the blocks are as in Fig. 14.2.1. The continuous curve is a Gaussian (normal) distribution, calculated using the mean of the log individual effective doses ($\overline{1}\cdot707$), and the standard deviation of the log IED ($0\cdot104$), estimated from Fig. 14.2.6 as described in the text. The probability density (right-hand) scale has been chosen to make the areas under

So if the quantal responses (values of r/n) were plotted against the dose an unsymmetrical sigmoid dose-response curve like the continuous line in Fig. 14.2.2 would be expected.

Thus when quantal responses are measured the dose is fixed by the experimenter and the number (or proportion) of subjects responding is the variable measured. On the other hand, in direct assays the dose is not fixed but is the variable quantity measured by the experimenter.

The subjects responding in the quantal experiment are the subjects in the group with an IED equal to or less than the fixed dose given. No information is obtained about IEDs of a single animals so Fig. 14.2.1 cannot be plotted directly (though it can of course be obtained by plotting the slope of the quantal dose-response curve, Fig. 14.2.2, against dose, i.e. by differentiation of Fig. 14.2.2 (this was shown in (4.1.5)).

The cumulative curve in Fig. 14.2.2 is analogous to an ordinary dose–response curve, for example the tension (a continuous variable) developed by a smooth muscle preparation in response to various doses of histamine. Because it is easier to handle a straight line than a curve, it is usual to look for ways of converting dose–response curves to straight lines. A method of doing this that often works in the case of

---

histogram and the Gaussian curve equal, but the two are still not comparable because area represents frequency for the continuous curve (see § 4.1), but not for the histogram.

(b) The histogram in this figure has been constructed so that the area of each block represents frequency, f, or, more precisely, the proportion $f/\Sigma f = f/100$ in this example. The area is the height (h say) times the width of the log dose interval (Δx say). For example, the first and last blocks represent a frequency of $f = 1$ per cent (see Table 13.2.1) so the first and last blocks are of equal height in Figs. 14.2.1 and 14.2.3(a). However, in Fig. 14.2.3(b) they have equal areas (each have 1 per cent of the total area), and therefore unequal heights. By definition, proportion $= f/100 = h\Delta x = $ area. For example, for the first block $\Delta x = 0.477 -0.301 = 0.176$, so the height (probability density) is $h = f/100\Delta x = 1/17.6 = 0.05682$, as plotted. For the last block $\Delta x = 1.0 - 0.954 = 0.046$, so $h = f/100\Delta x = 1/4.6 = 0.2174$ as plotted.

The area convention shown in Fig. 14.2.3(b) is the preferable one, because it shows the shape of the distribution correctly when the widths of the groups are not equal (though only at the expense of making it not obvious when frequencies are equal, because it is more difficult to judge relative areas than relative heights by eye). The continuous curve is a Gaussian curve with the same mean and standard deviation as in Fig. 14.2.3(a), and it can now be compared directly with the histogram because both have been plotted using the same (area) convention (see § 4.1), and both have a total area of 1.0. The Gaussian curve is seen to fit the observations reasonably well.

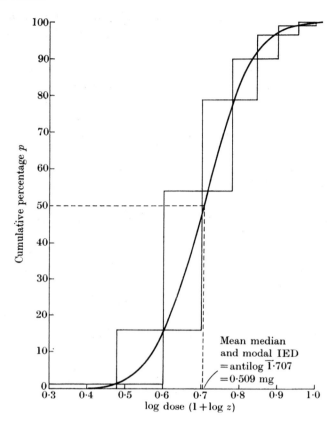

FIG. 14.2.4. Results from Table 14.2.1. Cumulative frequency, p, plotted against dose ($x = \log z$). This figure is related to Fig. 14.2.3 in the same way as Fig. 14.2.2 is related to Fig. 14.2.1.

The blocks (observations) are each the height of the f values in Table 14.2.1, and they are put above each other so the total height gives the p value from Table 14.2.1. The blocks are the same as those in Fig. 14.2.3(a), and the height of each block is proportional to the area of each block (i.e. the frequency, f) in Fig. 14.2.3(b).

The continuous curve is an estimate of the true (continuous) cumulative distribution (i.e. the distribution function, see § 4.1) of log IED values. In other words, the ordinate is the percentage of animals with a log IED equal to or less than x. The continuous curve in this figure is related to that in Fig. 14.2.3(b) in exactly the same way as the blocks are related; it is a calculated Gaussian distribution function (see § 4.1 and text)—the ordinate is the area to the left of x. under the calculated Gaussian distribution in Fig. 14.2.3(b), just as the ordinate for the blocks is the total area of the blocks below x under the histogram in Fig. 14.2.3(b). The calculated Gaussian function fits quite well (the continuous curve in Fig. 14.2.2 fits exactly only because it was drawn through the observations by eye).

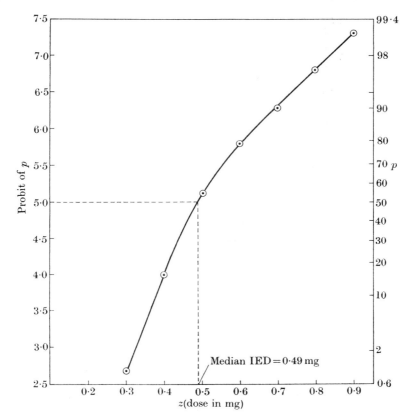

FIG. 14.2.5. Results from Table 14.2.1. Plot of the probit of p against the dose (z). The corresponding percentage scale is shown on the right for comparison. The non-linearity indicates that IED values are not normally distributed. A smooth curve has been drawn through the points by eye and the median IED ($p = 50$ per cent, probit [p] = 5) is estimated to be 0·49 mg, as was also found by interpolation in Fig. 14.2.2 (cf. Fig. 14.2.6, which gives a slightly different estimate).

quantal responses is discussed in § 14.3, and illustrated in Fig. 14.2.3–14.2.6, which show various manipulations of the original results.

14.3. The probit transformation. Linearization of the quantal dose response curve

When dealing with continuously variable responses it is common practice to plot the response against the logarithm of the dose (x = log z, say) in the hope that this will produce a reasonably straight dose response line. If p (from Table 14.2.1) is plotted against the log

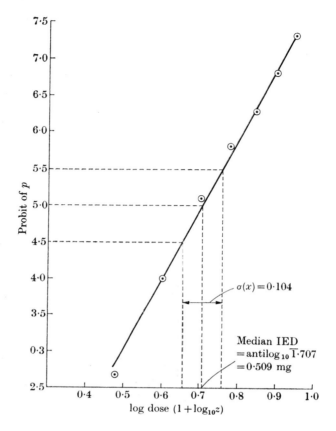

FIG. 14.2.6.   Results from Table 14.2.1. Plot of the probit of p against log
dose ($x = \log z$). The graph is reasonably straight, indicating that log IED
values are approximately normally distributed (i.e. IED values are approximately
lognormal, see § 4.5). The reciprocal of the slope ($1/9\cdot60 = 0\cdot104$) estimates the
standard deviation of the normal distribution of log IED values, and the dose
corresponding to $p = 50$ per cent (probit[p] = 5), i.e. antilog $\bar{1}\cdot707 = 0\cdot509$ mg,
estimates the median (= mean = mode) of the distribution of log IED values.
The distribution plotted with this mean and standard deviation is shown in
Fig. 14.2.3. The estimate of the median effective dose from this plot, $0\cdot509$ mg,
is different from that obtained from Fig. 14.2.2 and 14.2.5 ($0\cdot49$ mg). This is
because a straight line has been drawn in this figure, using all the points; and
the dose corresponding to probit[p] = 5 has been interpolated from the straight
line even though it does not go exactly through the points. This would be the
best procedure if the true line were in fact straight (i.e. if the population of log
IED values were in fact Gaussian). In Figs. 14.2.2 and 14.2.5, curves were
drawn by eye to go exactly through all the points, so effectively on the observations on each side of probit[p] = 5 were being used for interpolation of the
median, whereas when a straight line (or other *specified* function) is fitted, all the
observations are taken into account. In a real quantal experiment the straight

dose, the result, shown in Fig. 14.2.4, is not a straight line but a symmetrical sigmoid curve. (In fact similar results are often observed with continuous responses also.)

A way of converting the results to a straight line is suggested by Fig. 14.2.3, in which f rather than p is plotted against log dose. The histogram has become roughly symmetrical compared with the skewed distribution of IEDs seen in Fig. 14.2.1. The continuous line in Fig. 14.2.3 is a calculated normal (Gaussian) distribution with a mean and standard deviation estimated as described below and illustrated in Fig. 14.2.6. The calculated normal distribution is seen to fit the observed histogram quite well suggesting that the logarithms of the IEDs (values of $x = \log z$) are normally distributed, i.e. that the IEDs (values of z) are lognormally distributed (see § 4.5). *Any* curve can be linearized *if* the mathematical formula describing it is known. The sigmoid curve in Fig. 14.2.4, the cumulative form of the distribution in Fig. 14.2.3, is a cumulative normal distribution. This was illustrated in Fig. 4.1.4, which shows the cumulative form, $p = F(x)$, of the normal distribution in Fig. 4.1.3. If the abscissa in Fig. 4.1.4 is some measure of the effective dose then the ordinate of the cumulative normal distribution is

$$p = F(x) = \text{area under normal curve below } x$$
$$= \text{proportion of animals for which IED} \leqslant x, \qquad (14.3.1)$$

i.e. exactly what is plotted as the ordinate in Fig. 14.2.2 and 14.2.4.

The formula for the integral normal curve shown in Fig. 14.1.4, from (4.1.4) and (4.2.1), is

$$p = F(x) = \int_{-\infty}^{x} \frac{1}{\sigma \sqrt{(2\pi)}} \exp\left[-\frac{(x-\mu)^2}{2\sigma^2} \right] .\mathrm{d}x. \qquad (14.3.2)$$

This curve can be transformed to a straight line if, instead of plotting p against x, the abscissa corresponding to p is read off from a standard normal curve (see § 4.3) and *this* is plotted against x. For example, if a dose of $x = 3$ produces an effect in 16 per cent of a group of animals, the abscissa (viz. $u = -1$, see § 4.3) of the standard normal curve corresponding to an area in the lower tail of the curve of 16 per cent

---

line would be fitted to the points in this figure using the iterative method discussed in § 14.4. In this example, the quantal data has been generated, for illustrative purposes, using actual IED measurements rather than by giving fixed doses to groups of animals, so the best that can be done is to fit an unweighted straight line (shown) as described in § 12.5.

would be read off as shown in Fig. 14.3.1. This value of the abscissa would then be plotted against the dose (or some transformation of it, such as the logarithm of the dose), as shown in Fig. 14.3.2.

The abscissa of the standard normal curve, is, as described in § 4.3, $u = (x-\mu)/\sigma$, where σ is the standard deviation of x (i.e. of the log IED in the present case). So in effect, instead of plotting p against x, *the value of u corresponding to p (which is called the normal equivalent deviation or NED) is plotted against x.* But because the relation between u and x,

$$u = \frac{x-\mu}{\sigma} = \left(\frac{1}{\sigma}\right)x - \left(\frac{\mu}{\sigma}\right), \tag{14.3.3}$$

has the form of the general equation for a straight line $u = bx+a$, the plot of NED against x will be a straight line with slope $1/\sigma$ and intercept $(-\mu/\sigma)$ *if*, and only if, the values of x are normally distributed. This is because the NED corresponding to be observed p were read from a *normal* distribution curve.

The values of u are negative for $p < 50$ per cent response and so, to avoid the inconvenience of handling negative values, 5·0 is added to all values of the NED and the result is called the *probit* corresponding to p or probit $[p]$. Tables of the probit transformation are given, for example, by Fisher and Yates (1963, Table IX, p. 68). From Fig. 14.3.1, it is seen that $p = 50$ per cent response corresponds to $u \equiv$ NED $= 0$, i.e. probit [50 per cent] $= 5$. Thus

$$\text{probit } [p] \equiv u+5 \equiv \text{NED}+5 = 5+\frac{x-\mu}{\sigma}$$

$$= \left(5-\frac{\mu}{\sigma}\right)+\left(\frac{1}{\sigma}\right)x \tag{14.3.4}$$

so the plot of probit $[p]$ against x will be a straight line (if x is Gaussian) with slope $1/\sigma$ (as above) and intercept $(5-\mu/\sigma)$. Here, as above, σ is the standard deviation of the distribution of x, i.e. of the log IED in the present case. It is therefore a measure of the heterogeneity of the subjects (see § 14.4 also).

From Fig. 14.3.1, it can be seen that the NED of a 16 per cent response (i.e. 16 per cent of individuals affected) is -1 or, in other words, the probit of a 16 per cent response is $+4$. This follows from the fact (see § 4.3) that about 68 per cent of the area under a normal distribution curve is within $\pm\sigma$ (i.e. within ± 1 on a standard normal

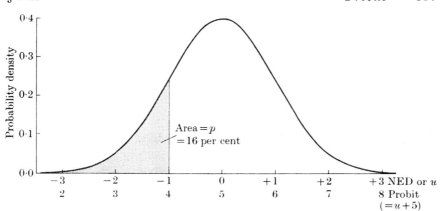

FIG. 14.3.1. Standard Gaussian (normal) distribution (see Chapter 4). Sixteen per cent of individuals responding corresponds to a value of u of -1 (the NED), i.e. to a probit of 4.

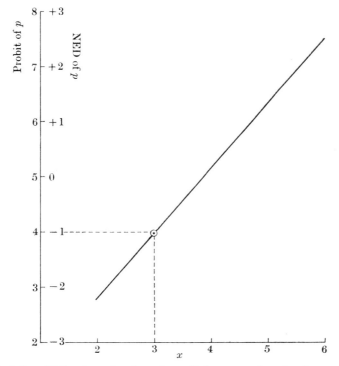

FIG. 14.3.2. If the dose (or transformed dose, e.g. log dose) $x = 3$ caused 16 per cent of individuals to respond, the probit of 16 per cent, i.e. 4·0, from Fig. 14.3.1, would be plotted against $x = 3$. See complete plots in Figs. 14.2.5 and 14.2.6.

curve), and of the remaining 32 per cent of the area, 16 per cent is below $u = -1$ and 16 per cent is above $+1$.

In Fig. 14.2.5 the probit of the percentage response is plotted against the dose z. The curve is not straight, implying that individual effective doses do not follow the Gaussian distribution in the animal population. This has already been inferred by inspection of the distribution shown in Fig. 14.2.1 which is clearly skew. However, in the usual quantal response experiment the distribution in Fig. 14.2.1 is not itself observed. The directly observed results are of the form shown in Fig. 14.2.2; and it is not immediately obvious from Fig. 14.2.2 that individual effective doses are not normally distributed. When the probit of the percentage response is plotted against log dose, in Fig. 14.2.6, the line is seen to be approximately straight, showing that (in this particular instance) the logarithms of the individual effective doses are approximately normally distributed (cf. Fig. 14.2.3).

The use of this line is discussed in the next section.

14.4. Probit curves. Estimation of the median effective dose and quantal assays

The probit transformation described in § 14.3 can be used to estimate the *median effective dose* or concentration; that is, the dose (or concentration) estimated to produce the effect in 50 per cent of the population of individuals (see § 2.5). This dose is referred to as the ED50 (or, if the effect happens to be death, as the LD50, the *median lethal dose*). If the individual effective doses (IEDs) are measured on a scale on which they are normally distributed the ED50 will be the same as the *mean* effective dose (e.g. in § 14.3 the median effective log dose is the same as the mean effective log dose, see § 4.5 and Figs. 14.2.1 and 14.2.3).

The procedure is to measure the proportion ($p = r/n$) of individuals showing the effect in response to each of a range of doses. The proportions are converted to probits and plotted against the dose or some function (usually the logarithm) of the dose. If the curve does not deviate from a straight line by more than is reasonable on the basis of experimental error (see below) the ED50 can be read from the fitted straight line. From Fig. 14.2.6 it can be seen that the graphical estimate of the ED50 is 0·509 mg, the antilog of the log dose corresponding to a probit of 5 as explained in § 14.3.

Furthermore, the slope of Fig. 14.2.6 is an estimate of $1/\sigma$, according to (14.3.4); so the reciprocal of this slope (i.e. $1/9\cdot60 = 0\cdot104 \log_{10}$

units, from Fig. 14.2.6) is an estimate of σ, the standard deviation of the distribution of the log IED, and this value of σ was used to plot the distribution in Fig. 14.2.3. This standard deviation is a measure of the variability of the individuals, i.e. of the extent to which they do not all have the same individual effective dose.

Assays of the sort described in Chapter 13 can also be done using quantal responses, and if a log dose scale is used they will be parallel line assays (see § 13.1).

In all the applications discussed the problem arises of how to fit the 'best' line to the observed points. Methods for doing this have been described in Chapters 12 and 13 but they all assume that the scatter of the observations is the same at every value of x, i.e. that the results are homoscedastic (see §§ 12.2 and 13.1). This is not the case for probit plots (see § 14.5 for an exception) and this complicates the process of curve fitting. Numerical examples of the methods are given by Burn, Finney, and Goodwin (1950, p. 114), and Finney (1964, Chapters 17–21).

The reason for the heteroscedasticity is not difficult to see. The number of individuals (r) responding, out of a *randomly selected* (notice that random selection is, as usual, essential for the analysis) group of n, should follow the binomial distribution (§§ 3.2–3.4), and the variance of the proportion responding, $p = r/n$, would be estimated from (3.4.5) to be $\text{var}[p] = p(1-p)/n$. Because the line is to be fitted to the plot of $\text{probit}[p](= y$, say) against dose metameter, it is the variance of $y = \text{probit}[p]$ that is of interest. From (2.7.13) it is seen that $\text{var}[y] \simeq \text{var}[p].(\mathrm{d}y/\mathrm{d}p)^2 = (\mathrm{d}y/\mathrm{d}p)^2.p(1-p)/n$. Now the standard normal curve, in Fig. 14.3.1 can be written (by (4.1.1)) as $\mathrm{d}p = f\,\mathrm{d}y$, and thus $\mathrm{d}y/\mathrm{d}p = 1/f$, where f is the ordinate of the standard normal curve (the probability density, see § 4.1 and (4.2.1); f was used with a different meaning in §§ 14.2 and 14.3). This result follows, slightly more rigorously, from (14.3.1) and (4.1.5). Therefore $\text{var}[y] \simeq p(1-p)/nf^2$, and this is *not* a constant but varies with p. The probit plot is therefore heteroscedastic and each probit (y value) must be given a weight $1/\text{var}[y] \simeq nf^2/p(1-p)$ when fitting the dose response lines (cf. §§ 2.5 and 13.4); it is this that gives rise to the complications. When a line is fitted it will lead to a better estimate of the y corresponding to each x, and hence to better estimates of the weights and hence to a better fitting line. The calculation is therefore iterative.

It is because of the existence of this theoretical estimate (cf. § 3.7) of $\text{var}[y]$ that the deviations from linearity of Fig. 14.2.6 can be tested even though there is only one observation (y value) at each x value (cf. §§ 12.5 and 12.6).

If the weight is plotted against p (Fisher and Yates (1963, p. 71) give a table of $f^2/p(1-p)$), it is found to have a maximum value when $p = 0.5$, i.e. 50 per cent response rate. This is the reason why the ED50 is calculated as a measure of effectiveness. It is the quantity that can be determined most precisely.

The minimum effective dose

This term is fairly obviously meaningless† as it stands (unless the IED is the same for all individuals). The larger the sample the larger the chance that it will contain a very susceptible individual (from the lower tail of Fig. 14.2.3) so the lower the estimate of the minimum effective dose will be. Clearly it is necessary to specify the *proportion* of individuals affected in the population. It was 50 per cent in the discussion above.

Unfortunately, it is often not of interest to known the ED50. If one were interested in the proportion of individuals suffering harmful radiation effects from the fall-out nuclear explosions it is (or should be) only of secondary interest to known what dose of radiation will harm 50 per cent of the population. What *is* required is an estimate of the dose of radiation that will not harm anyone. No answer other than zero dose is consistent with the lognormal distribution of individual effective radiation doses usually assumed because the normal distribution of the log IED is asymptotic to the dose axis (see § 4.2), zero effect being produced only by log dose $= -\infty$, i.e. zero dose. The question is not compatible with a normal distribution of doses either, as this would imply the existence of negative doses. This is a very real problem because when dealing with a very large population a very small *proportion* harmed means a very large *number* of people harmed. Suppose that it is decided that the ED0·01 shall be estimated, i.e. the dose affecting 0·1 per cent of the population (about 0.0001×3500 million $= 350\ 000$ people on a world scale !). The weight, $f^2/p(1-p)$, corresponding to $p = 0.0001$ (i.e. probit $[p] \simeq 1.3$) is seen from the tables to be 0·00167, compared with 0·6366 (about 380 times larger) for $p = 0.5$. Thus to estimate the ED0·01 with the same precision as the ED50, the

---

† This necessitates the abandonment of the conventional definition of the unit of beauty (Marlowe 1604), viz. one milliHelen = that quantity of beauty just sufficient to launch a single ship. An alternative definition could follow the lines of the purity in heart index discussed in § 7.8.

sample size (n) would have to be much bigger. And this is not the only problem. Working with small proportions means working with the tails of the distribution where assumptions about its form are least reliable. For example, the straight line in Fig. 14.2.6 might be extra-polated—a very hazardous process, as was shown in § 12.5, Fig. 12.5.1.

14.5. Use of the probit transformation to linearize other sorts of sigmoid curve

The probit transformation can be tried quite empirically in attempt-ing to linearize any sigmoid curve if the ordinate can be expressed as a proportion (see § 14.6). If this is done the variability will not usually be binomial in origin so the method discussed in § 14.4 cannot be used, and curves should not be fitted by the methods described in books on quantal bioassay. It would be necessary to have several observations at each x value to test for deviations from linearity, and the assumptions discussed in § 12.2 must be tested empirically.

An example is provided by the osmotic lysis of red blood cells by dilute salt solutions. It is often found that the plot of the probit of the proportion (p) of cells not haemolysed against salt concentration (not log concentration) is straight over a large part of its length. This implies (see § 14.3) that the concentration of salt just sufficient to prevent lysis of individual cells (the IED) is approximately normally distributed in the population of cells with a standard deviation esti-mated, by (14.3.4), as the reciprocal of the slope of the plot. In this sort of experiment each test would usually be done on a very large number (n) of cells so the variability expected from the binomial distribution, $p(1-p)/n$, would be very small. However, in this case most of the variability (random scatter of observations about the probit—concentration line) would not be binomial in origin but would be the result of factors that do not enter into the sort of animal experi-ment described in § 14.3, such as variability of n from sample to sample, and errors in counting the number of cells showing the specified response (not lysed).

14.6. Logits and other transformations. Relationship with the Michaelis–Menten hyperbola

The use of the probit transformation for linearizing quantal dose response curves is, in real life, completely empirical. The probit of the proportion responding is plotted against some function (x, say) of the dose, and the plot tested for deviations from linearity. However, there are many other curves that closely resemble the sigmoid cumulative normal curve in Figs. 4.1.4 and 14.2.4. One example is the logistic curve defined† by

$$p = \frac{1}{1+e^{-(a+bx)}}. \tag{14.6.1}$$

This plotted in Fig. 14.6.1, curve (b), and is seen to be very like the cumulative normal curve. If the relation between p and x was represented by (14.6.1) then it could be linearized by plotting logit[p]

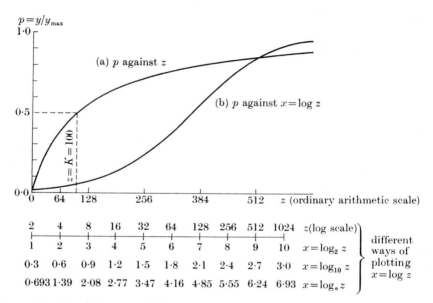

FIG. 14.6.1.   Curve (a) Plot of p against z from eqn. (14.6.3). When $b = 1$ this curve is part of a hyperbola.

Curve (b) Plot of p against x from eqn. (14.6.1). This curve is the same as curve (a) with z plotted on a logarithmic scale (three equivalent ways of plotting $x = \log z$ are shown). It is a logistic curve and can be linearized by plotting logit[p] against x.

The particular values used to plot the graphs were $K = 100$, $b = 1$.

† An equivalent definition of the logistic curve that may be encountered is

$$p = \frac{1}{2}\left[1+\tanh\left(\frac{a+bx}{2}\right)\right]$$

where the hyperbolic tangent is defined by $\tanh\phi = (e^{2\phi}-1)/(e^{2\phi}+1)$. In terms of this definition logit [p] $\equiv 2\tanh^{-1}(2p-1) = a+bx$.

(instead of probit) against x, where logit$[p]$ is defined as $\log_e\{p/(1-p)\}$. This follows from (14.6.1) which implies

$$\text{logit}[p] = \log_e\left(\frac{p}{1-p}\right) = \log_e\left(\frac{1}{e^{-(a+bx)}}\right) = a+bx \qquad (14.6.2)$$

which is a straight line with slope b and intercept a. (Remember that, in general, $\log_e e^x = x$ because the log is defined as the power to which base must be raised to give argument. This implies, also, that e^x can be written, in general as antilog$_e x$, which is used below in deriving (14.6.3).) The use of this and other transformations for analysing quantal response experiments is described by Finney (1964). The probit and logit transformations are too similar for it to be possible to detect which fits the results better, with quantal experiments of the usual size.

The logit transformation is also a linearizing transformation for the hyperbola discussed in § 12.8, and plotted in Fig. 14.6.1, curve (a). In this application the response, y, is a continuous variable, not a quantal variable. The linearity follows by taking $x = \log_e z$ (using z to represent dose, or concentration, rather than x which was used for this purpose in § 12.8), and $p = y/y_{\max}$, i.e. y expressed as a proportion of its maximum possible value, the value approached as z becomes very large (in § 12.8 $y_{\max}$ was called V). If the constant, a, is redefined as $-\log_e K$ then putting these quantities into (14.6.1) gives

$$p = \frac{y}{y_{\max}} = \frac{1}{1+e^{\log K - b\log z}} = \frac{1}{1+e^{\log K - \log z^b}}$$

$$= \frac{1}{1+e^{\log K/z^b}} = \frac{1}{1+K/z^b}$$

$$= \frac{z^b}{K+z^b} \qquad (14.6.3)$$

which is the hyperbola (12.8.1), in the special case when $b = 1$. As mentioned in § 12.8, this is the Michaelis–Menten equation of biochemistry (when $b = 1$). The more general form, (14.6.3), has been used, for example, in biochemistry and pharmacology (the Hill equation). The plot of logit $[p]$ against log z is known as the Hill plot. The use and physical interpretation of the Hill plot are discussed by Rang and Colquhoun (1973).

Summarizing these arguments, if the response y, plotted against dose or concentration z, follows (14.6.3) (which in the special case $b = 1$ is the hyperbola plotted in Fig. 14.6.1, curve (a), and in Fig. 12.8.1), then the response plotted against log concentration, $x = \log z$, will be a sigmoid logistic curve defined by (14.6.1) and plotted in Fig. 14.6.1, curve (b). And logit $[y/y_{max}]$ plotted against x will be a straight line with intercept $a = -\log K$, and slope b. Quite empirically, equations like (14.6.3) are often found to represent dose–response curves in pharmacology reasonably well (the extent to which this justifies physical models is discussed by Rang and Colquhoun (1973)), so plots of response against log dose are sigmoid like Fig. 14.6.1, curve (b). The central portion of this sigmoid curve is sufficiently nearly straight to be not grossly incompatible with the assumption, made in most of Chapter 13, that response is linearly related to log dose.

It is worth noticing that the sigmoid plot of y against x in Figs. 14.2.4 or 4.1.4 (the cumulative normal curve, linearized by plotting probit$[p]$ against x) looks very like the sigmoid plot of y against x in Fig. 14.6.1, curve (b) (the logistic curve, linearized by plotting logit $[p]$ against x). However, if x is log z, then the corresponding plots of y against z (e.g. response against dose, rather than log dose) are quite distinct. The corresponding plots are, respectively, that in Fig. 14.2.2 (the cumulative lognormal distribution, see § 4.5), which has an obvious 'foot', it flattens off at low z values; and the hyperbola in Fig. 14.6.1, curve (a), which rises straight from the origin with no trace of a 'foot' or threshold'. This distinction is effectively concealed when a logarithmic scale is used for the abscissa (e.g. dose).

In order to use the logit transformation for continuously variable responses it is necessary to have an estimate of the maximum response, y_{max}. This introduces statistical complications (see, for example, Finney (1964, pp. 69–70)). A simple solution is not to bother with linearizing transformations except as a convenient method for preliminary assessment and display of results, but to estimate the parameters y_{max}, K, and b directly by the method of least squares as described in § 12.8.

Appendix 1

Expectation, variance, and non-experimental bias

THE object of this appendix is to provide a brief account of some rather more mathematical ideas which, although they are not necessary for following the main body of the book, will be useful to anyone wanting to go further. Also some of the following results will be useful in Appendix 2. Further explanation will be found, for example, in Brownlee (1965, pp. 51, 57, and 87), Mood and Graybill (1963, p. 103), or Kendall and Stuart (1963, Chapter 2). All the ideas discussed in this section require that the distribution of the variable be specified.

A1.1. Expectation—the population mean

The population mean value of a variable is called its *expectation* and is defined as†

$$E(x) = \sum_{\text{all } x} x \, P(x) \text{ for discontinuous distributions,} \qquad (A1.1.1)$$

$$E(x) = \int_{-\infty}^{+\infty} x \, f(x) \, \mathrm{d}x \text{ for continuous distributions.} \qquad (A1.1.2)$$

This can be regarded as the arithmetic mean of an indefinitely large number of observations on the variable x, the distribution of which is specified by the probability $P(x)$ (discontinuous), or the probability density $f(x)$ (continuous), as explained in §§ 3.1 and 4.1. The reasonableness of the definition (A1.1.1) is obvious if a large but finite number of observations, N, is considered. On the average proportion $P(x)$ of the observations will have the value x, so the *number* of observations with the value x will be $f = NP(x)$ and the total of the f observations will be fx. The total of all N observations will be Σfx, and their mean will therefore be $\Sigma fx/N$, which is exactly eqn. (A1.1.1) if f/N is substituted for P. The form for continuous variables, (A1.1.2), is just the same as (A1.1.1) except that the P is replaced by $\mathrm{d}P = f(x) \, \mathrm{d}x$ (from (4.1.1)), and consequently summation is replaced by integration.

As a numerical example take the binomial distribution with $n = 3$ and $\mathscr{P}(B) = 0 \cdot 9$ from Table 3.2.2 and Fig. 3.2.4. From (A1.1.1)

$$E(r) = \sum_{r=0}^{r=3} r P(r) = (0 \times 0 \cdot 001) + (1 \times 0 \cdot 027) + (2 \times 0 \cdot 243) + (3 \times 0 \cdot 729) = 2 \cdot 7,$$

---

† If x, considered as a random variable, is denoted $\tilde{x}$ to distinguish it from x considered as an algebraic symbol (as in (4.1.4) and § A2.6), the definition of expectation can be written in the preferable form $E(\tilde{x}) = \Sigma x P(x)$, etc.

which is $n\mathscr{P}$ $(= 3\times0\cdot9 = 2\cdot7)$, the population mean value of r (number of successes in three trials), as mentioned in § 3.4. Notice that in this case the mean value is never actually observed. All observations must be integers.

Several properties follow directly from the definition of expectation. For example, for a linear function, where a and b are constants,

$$\mathrm{E}[a+bx] = \mathrm{E}[a]+\mathrm{E}[bx] = a+b\mathrm{E}[x] \tag{A1.1.3}$$

and also, more generally,

$$\mathrm{E}\left[\sum_{i=1}^{i=N} x_i\right] = \sum_{i=1}^{i=N} (\mathrm{E}[x_i]) \text{ in general} \tag{A1.1.4}$$

$$= N\mathrm{E}[x] \text{ if all } x_i, \text{ have the same mean.} \tag{A1.1.5}$$

But for a nonlinear function, $g(x)$ say,†

$$\mathrm{E}[g(x)] \neq g(\mathrm{E}[x]), \tag{A1.1.6}$$

so averaging a function of x will not give the same answer as averaging x first and then finding the function of the average (cf. (2.5.4); the arithmetic mean of $\log x$ is not the log of the arithmetic mean of x, but the log geometric mean). See also (A1.2.2).

Mean of the Poisson distribution

It was stated in §§ 3·5 and 5·1 that m in (3.5.1) was the mean of the Poisson distribution. This follows, using (A1.1.1) and (3.5.1), giving

$$\mathrm{E}(r) = \sum_{r=0}^{\infty} r\, P(r) = \sum_{r=0}^{\infty} \left(r.\frac{m^r}{r!}\mathrm{e}^{-m}\right)$$

$$= \mathrm{e}^{-m}\left[0.\frac{m^0}{0!}+1.\frac{m^1}{1!}+2.\frac{m^2}{2!}+...\right]$$

$$= \mathrm{e}^{-m}\left[0+m\left(1+m+\frac{m^2}{2!}+...\right)\right] = \mathrm{e}^{-m}.m.\mathrm{e}^{m} \tag{A1.1.7}$$

$$= m$$

Mean of the normal distribution

Using (A1.1.2) the statement that the parameter μ in (4.2.1) can be interpreted as the mean of the normal distribution can be justified. From (A1.1.2),

$$\mathrm{E}(x) = \int_{-\infty}^{\infty} xf(x)\mathrm{d}x = \int_{-\infty}^{\infty} [\mu+(x-\mu)]f(x)\ \mathrm{d}x$$

$$= \mu\int_{-\infty}^{\infty} f(x)\mathrm{d}x+\int_{-\infty}^{\infty} (x-\mu)\,f(x)\ \mathrm{d}x$$

$$= \mu+0 = \mu \tag{A1.1.8}$$

† The expectation of a function of x is defined in (A1.1.16) and (A1.1.17).

because the first integral is the area under the whole distribution curve, i.e. 1. The second integral is zero because, using (4.2.1) for the density, $f(x)$, of the normal distribution and putting $y = (x-\mu)^2$ so that $\mathrm{d}y = 2(x-\mu)\,\mathrm{d}x$, it becomes

$$\frac{1}{\sigma\sqrt{(2\pi)}}\int (x-\mu)e^{-(x-\mu)^2/2\sigma^2}\mathrm{d}x = \frac{1}{\sigma\sqrt{(2\pi)}}\int \frac{1}{2}e^{-y/2\sigma^2}\mathrm{d}y$$

$$= -\frac{\sigma}{\sqrt{(2\pi)}}\left[e^{-(x-\mu)^2/2\sigma^2}\right]_{-\infty}^{\infty} = 0. \tag{A1.1.9}$$

Mean and median of the exponential distribution

The exponential distribution of intervals between random events was introduced in Chapter 5 and is discussed in more detail in Appendix 2. It was defined in (5.1.3) by the probability density

$$\begin{aligned} f(x) &= \lambda e^{-\lambda x} \text{ for } x \geqslant 0, \\ f(x) &= 0 \qquad \text{for } x < 0, \end{aligned} \tag{A1.1.10}$$

which is plotted in Fig. 5.1.2. It was argued in § 5.1 that the population mean interval between events must be λ^{-1}. This follows from (A1.12) which gives

$$\mathrm{E}(x) = \int_0^\infty x.\lambda e^{-\lambda x}\mathrm{d}x.$$

The lower limit can be taken as 0 rather than $-\infty$ because, from (A1.1.10), $f(x)$, and hence the integral, is 0 for $x < 0$. This can be evaluated using integration by parts. See, for example, Thompson (1965, p. 188), or Massey and Kestelman (1964, pp. 332 and 402).) Putting $u = x$, so $\mathrm{d}u = \mathrm{d}x$, and $\mathrm{d}v = \lambda^{-\lambda x}\mathrm{d}x$ so $v = \int \lambda e^{-\lambda x}\mathrm{d}x = [-e^{-\lambda x}]$, gives

$$\mathrm{E}(x) = \int u\mathrm{d}v = [uv] - \int v\mathrm{d}u$$

$$= \left[-xe^{-\lambda x}\right]_0^\infty - \int_0^\infty (-e^{-\lambda x})\mathrm{d}x$$

$$= 0 - \left[\frac{e^{-\lambda x}}{\lambda}\right]_0^\infty = \frac{1}{\lambda} = \lambda^{-1}. \tag{A1.1.11}$$

To evaluate this notice that $xe^{-\lambda x} \to 0$ as $x \to \infty$; see, for example, Massey and Kestelman (1964, p. 122). The area under the distribution curve up to any value x, i.e. the probability that an interval is equal to or less than x is, from (5.1.4),

$$F(x) = 1 - e^{-\lambda x} \tag{A1.1.12}$$

so the proportion of all intervals that are shorter than the mean interval, putting $x = \lambda^{-1}$ in (A1.1.12), is

$$F(\lambda^{-1}) = 1 - e^{-1} = 0.6321, \tag{A1.1.13}$$

i.e. 63·21 per cent of the area under the distribution in Fig. 5.1.2 lies below the mean (1.0 in Fig. 5.1.2).

The median (see § 2.5, p. 26) length of the intervals between random events is the length such that 50 per cent of intervals are longer, and 50 per cent shorter than it, i.e. it is the value of x bisecting the area under the distribution curve. If the population median value of x is denoted x_m then, from (A1.1.12),

$$1 - e^{-\lambda x_m} = 0 \cdot 5,$$
$$\text{i.e. } x_m = \lambda^{-1} \log 2 = 0 \cdot 69315 \lambda^{-1}. \tag{A1.1.14}$$

This is shown on Fig. 5.1.2. As expected for a positively skewed distribution (see § 4.5), the population median is less than (in fact 69·315 per cent of) the population mean, λ^{-1}. The mode of the distribution is even lower at $x = 0$, as seen from Fig. 5.1.2.

The variance of an exponentially distributed variable, from (A1.2.2), is

$$var(x) = \lambda^{-2}. \tag{A1.1.15}$$

For details see, e.g. Brownlee (1965, p. 59).

The expectation of a function of x

The expectation, or long run mean, of the value of any function of x, say $g(x)$, can be found without first finding the probability density of $g(x)$. The derivation is given, for example, by Brownlee (1965, p. 55). The results, analogous to (A1.1.1) and (A1.1.2), are

$$E[g(x)] = \Sigma g(x) \, P(x) \text{ for discontinuous distributions}, \tag{A1.1.16}$$

$$E[g(x)] = \int_{-\infty}^{+\infty} g(x) \, f(x) \, \mathrm{d}x \text{ for continuous distributions}. \tag{A1.1.17}$$

The expectation of a function of *two* random variables is discussed in § A1.4.

A1.2. Variance

For any variable x, with expectation μ, the population variance of x is defined as the expected value (long run mean) of the square of the deviation of x from $\mu = E(x)$, i.e.

$$var(x) = E[(x - \mu)^2] \tag{A1.2.1}$$

$$= E[x^2] - (E[x])^2. \tag{A1.2.2}$$

The second form of the definition follows from the first by expanding the square. It shows that $E[x^2]$ is not the same, in general, as $(E[x])^2$ (this is an example of the relation (A1.1.6)).

Use of this definition in conjunction with (A1.1.1) or (A1.1.2) gives, for example, the variance of the binomial distribution as $n\mathscr{P}(1 - \mathscr{P})$, of the Poisson as m, of the exponential as λ^{-2}, and of the normal as σ^2, as asserted in (3.4.4), (3.5.3), (A1.1.15), and (4.2.1). For details of the derivations see the references at the beginning of this section. The mean and variance of a function of two random variables is discussed in § A1.4.

The *standardized form* of any random variable, x, can be defined as X say, where

$$X = \frac{x - E[x]}{\sqrt{var(x)}} \tag{A1.2.3}$$

(see for example, the standard normal distribution, § 4.3). X must always have a population mean of zero, and population variance of one because

$$E[X] = E\left[\frac{x - E[x]}{\sqrt{var(x)}}\right] = \frac{E[x] - E[x]}{\sqrt{var(x)}} = 0 \tag{A1.2.4}$$

and, from (A1.2.2), (A1.2.4), and (A1.2.1),

$$var(X) = E[X^2] - (E[X])^2 = E[X^2]$$

$$= E\left[\frac{(x - E[x])^2}{var(x)}\right] = \frac{var(x)}{var(x)} = 1. \tag{A1.2.5}$$

A1.3. Non-experimental bias

It has been mentioned in § 2.6 in connection with the standard deviation, and in § 12.8, that estimates of quantities calculated from observations may be biased even when the observations themselves have no bias at all. In this case the estimation method (i.e. the formula used to calculate the sample estimate, say $\hat{\theta}$, of a parameter θ from the observations) is said to be biased. An estimation method is said to have a bias $= E[\hat{\theta}] - \theta$, and it is said to be unbiased if

$$E[\hat{\theta}] = \theta. \tag{A1.3.1}$$

For example, the sample arithmetic mean is an unbiased estimate of the parameter $E[x]$ (whatever the distribution of x) because $E[\bar{x}] = E[x]$. Using (A1.1.4),

$$E[\bar{x}] = E\left[\sum_{i=1}^{i=N} x/N\right] = E[\Sigma x]/N = \Sigma(E[x])/N = NE[x]/N = E[x] = \mu. \tag{A1.3.2}$$

Furthermore, (2.6.3) gives an unbiased estimate of the population variance, $var(x)$ or $\sigma^2(x)$, for any distribution because, from (A1.1.3), (A1.1.4), and (A1.2.1),

$$E\left[\frac{\Sigma(x - \mu)^2}{N}\right] = \frac{E[\Sigma(x - \mu)^2]}{N} = \frac{\Sigma E[(x - \mu)^2]}{N} = \frac{N var(x)}{N} = var(x). \tag{A1.3.3}$$

However, if μ is replaced by its (unbiased) sample estimate, $\bar{x}$, an unbiased estimate of $var(x)$ is no longer obtained (as discussed in § 2.6). If $\hat{\sigma}^2 = \Sigma(x - \bar{x})^2/N$ then

$$N\hat{\sigma}^2 = \sum_{i=1}^{i=N} (x - \bar{x})^2 = \Sigma[(x - \mu) - (\bar{x} - \mu)]^2$$

$$= \Sigma(x - \mu)^2 - N(\bar{x} - \mu)^2$$

because $2(\bar{x}-\mu)$. $\Sigma(x-\mu) = 2(\bar{x}-\mu)\ (\Sigma x-N\mu) = 2(\bar{x}-\mu)\ (N\bar{x}-N\mu) = 2N(\bar{x}-\mu)^2$. Thus, using (A1.1.3), (A1.1.4), (A1.2.1), and (2.7.8),

$$N\mathrm{E}[\hat{\sigma}^2] = \mathrm{E}[\Sigma(x-\mu)^2 - N(\bar{x}-\mu)^2]$$
$$= \Sigma\mathrm{E}[(x-\mu)^2] - N\mathrm{E}[(\bar{x}-\mu)^2]$$
$$= N\mathrm{E}[(x-\mu)^2] - N\mathrm{E}[(\bar{x}-\mu)^2]$$
$$= N\ var(x) - N\ var(\bar{x}) = N\ var(x) - N\ var(x)/N$$

and so $\qquad \mathrm{E}[\hat{\sigma}^2] = var(x)\left(\dfrac{N-1}{N}\right) \text{or } \sigma^2\left(\dfrac{N-1}{N}\right).$ \qquad (A1.3.4)

Because $\hat{\sigma}^2$ is a biased estimate of $var(x)$, its expectation being less than $var(x)$, it is not used. Instead it is multiplied by $N/(N-1)$ to correct the bias, giving the usual estimate, (2.6.2), with $N-1$ rather than N in the denominator.

A1.4. Expectation and variance with two random variables. The sum of a variable number of random variables†

In dealing with a function of two random variables, a proper procedure is to average over one of them holding the other fixed, and then to average over the other. This is rather like averaging the rows in a square table, and then averaging the row averages to find the grand average. The proof will be outlined for the case of the sum of a randomly variable number of random variables, but the result (§§ A1.4.11 and A1.4.12) is general. The result is used and illustrated in § 3.6. Relevant information will be found, for example, in Mood and Graybill (1963, p. 117) and Bailey (1964).

It will be necessary, as on pp. 68 and 388, to distinguish between random variables denoted $\tilde{z}$, $\tilde{m}$, etc., and particular values that these variables may take, denoted z, m, etc.

Suppose we are interested, as in § 3.6, in the sum, itself a random variable denoted $\tilde{S}_{\tilde{m}}$, of m values of z, where $\tilde{m}$ and $\tilde{z}$ are random variables, i.e.

$$\tilde{S}_{\tilde{m}} = z_1 + z_2 + ... + z_m. \qquad (A1.4.1)$$

The population means and variances of the variables will be denoted, for brevity

$$\mathrm{E}(\tilde{m}) \equiv \mu_m \quad var(\tilde{m}) \equiv \sigma_m^2 \qquad (A1.4.2)$$

$$\mathrm{E}(\tilde{z}) \equiv \mu_z \quad var(\tilde{z}) \equiv \sigma_z^2 \qquad (A1.4.3)$$

We shall deal only with the case where the z values are independent, and each S value is made up of a random sample (of variable size) from the population of z values. It is assumed in (A1.4.3) that the z values all have the same mean and variance, for example, that they are from a single population.

The probability that $\tilde{S}_{\tilde{m}}$ is equal or less than a specified value S, i.e. the distribution function of the sum (see § 4.1), can be written, looking at each possible value of m separately, as

$$P[\tilde{S}_{\tilde{m}} \leqslant S] = P[(\tilde{m} = 1 \text{ and } \tilde{S}_1 \leqslant S) \text{ or } (\tilde{m} = 2 \text{ and } \tilde{S}_2 \leqslant S) \text{ or}...].$$
$$(A1.4.4)$$

† I am very grateful to R. Galbraith, Department of Statistics University College, London for showing me how to obtain the results in this section.

The events in parentheses are mutually exclusive so, using the addition rule (2.4.2), this becomes a sum (over all possible m values), viz.

$$\sum_m P[\tilde{m} = m \text{ and } \tilde{S}_m \leqslant S]. \tag{A1.4.5}$$

Now, using the multiplication rule in its general form (2.4.4) shows that $P[\tilde{m} = m \text{ and } \tilde{S}_m \leqslant S]$ can be written in terms of the conditional probabilities as $P[\tilde{S}_m \leqslant S | \tilde{m} = m] . P[\tilde{m} = m]$ and so (A1.4.5) becomes

$$P[\tilde{S}_{\tilde{m}} \leqslant S] = \sum_m P[\tilde{S}_{\tilde{m}} \leqslant S | \tilde{m} = m] . P[\tilde{m} = m]. \tag{A1.4.6}$$

This can be written in terms of distribution functions (see § 4.1) as

$$F(S) = \sum_m F(S|m) . P[\tilde{m} = m] \tag{A1.4.7}$$

and differentiating this with respect to S gives, as in (4.1.5), the probability density function as

$$f(S) = \sum_m f(S|m) . P[\tilde{m} = m]. \tag{A1.4.8}$$

To find the expectation of any function of the sum, $g(\tilde{S}_{\tilde{m}})$, we now simply use this result in the definition of expectation, (A1.1.17), giving

$$\begin{aligned}
\mathrm{E}[g(\tilde{S}_{\tilde{m}})] &= \int_S g(S) \, f(S) \mathrm{d}S \\[2mm]
&= \int_S g(S) . \sum_m f(S|m) . P[\tilde{m} = m] . \mathrm{d}S \\[2mm]
&= \sum_m \left\{ \int_S g(S) \, f(S|m) \, \mathrm{d}S \right\} . P[\tilde{m} = m] \tag{A1.4.9} \\[2mm]
&= \sum_m \{\mathrm{E}_S[g(\tilde{S}_{\tilde{m}}) | \tilde{m} = m]\} . P[\tilde{m} = m]. \tag{A1.4.10}
\end{aligned}$$

The last step follows because the term in curly brackets in (A1.4.9) is simply the expectation of the function g, when $\tilde{m}$ has a fixed value, m. The value of this will of course depend on (be a function of) the value, m, chosen, so (A1.4.10) has the form $\sum_m$ (function of m. $P[\tilde{m} = m]$), just like (A1.1.16), so it means that the term in curly brackets is being averaged over all m values and (A1.4.10) can therefore be written

$$\mathrm{E}[g(\tilde{S}_{\tilde{m}})] = \mathrm{E}_m\{\mathrm{E}_S[g(\tilde{S}_{\tilde{m}}) | \tilde{m} = m]\}, \tag{A1.4.11}$$

which describes in symbols the two-stage averaging process mentioned at the beginning of the section. The result is much more general than it appears from this derivation. If $\tilde{x}$ and $\tilde{y}$ are any two random variables, continuous or discontinuous, then, analagously to (A1.4.11) we have

$$\mathrm{E}[g(\tilde{x},\tilde{y})] = \mathrm{E}_y\{\mathrm{E}_x[g(\tilde{x},\tilde{y}) | \tilde{y} = y]\}. \tag{A1.4.12}$$

The mean value of the sum follows directly from (A1.4.11) if the function $g(\tilde{S}_{\tilde{m}})$ is simply identified with $\tilde{S}_{\tilde{m}}$. Averaging the sum for a fixed value of m, using the definitions in (A1.4.1)–(A1.4.3), gives the term in curly brackets in (A1.4.11) as

$$\mathrm{E}[\tilde{S}_{\tilde{m}} | \tilde{m} = m] = m\mu_z, \tag{A1.4.13}$$

i.e. the average value of the total of a fixed number, m, of values of z is m times the average value of z, fairly obviously. Actually, this step is not quite as obvious as it looks. Written out in full we have

$$\mathrm{E}[\tilde{S}_{\tilde{m}}|\tilde{m} = m] = \mathrm{E}[(z_1 + z_2 + \ldots + z_m)|\tilde{m} = m]$$
$$= \mathrm{E}[z_1|\tilde{m} = m] + \mathrm{E}[z_2|\tilde{m} = m] + \ldots + \mathrm{E}[z_m|\tilde{m} = m]$$

(A1.4.14)

and only if m is independent of the z values, i.e. if the size of z values does not depend on whether m is large or small, can this be written

$$= \mathrm{E}[z_1] + \mathrm{E}[z_2] + \ldots + \mathrm{E}[z_m]$$

(A1.4.15)

and if all the z values have the same mean, μ_z, as assumed, this is simply $m\mu_z$ as stated in (A1.4.13).

Having found this for a fixed value of m, we now do the second stage of averaging, over m values, treating m as a random variable though μ_z is, of course, a constant. This gives, using (A1.4.11) with (A1.4.13) and (A1.4.2),

$$\mathrm{E}[\tilde{S}_{\tilde{m}}] = \mathrm{E}_m[\tilde{m}\mu_z]$$
$$= \mu_z \mathrm{E}_m[\tilde{m}] = \mu_z \mu_m,$$

(A1.4.16)

which is just what would be expected for the average value of the sum of m values of z.

To find the variance of $\tilde{S}$ we use the definition (A1.2.2) which is

$$var(\tilde{S}_{\tilde{m}}) = \mathrm{E}[\tilde{S}_{\tilde{m}}^2] - (\mathrm{E}[\tilde{S}_{\tilde{m}}])^2.$$

(A1.4.17)

The only thing needed now is to find the expectation of $\tilde{S}_{\tilde{m}}^2$. To do this we use (A1.4.11) again, but this time $g(\tilde{S}_{\tilde{m}})$ is identified with $\tilde{S}_{\tilde{m}}^2$. So first we want to find the term in curly brackets in (A1.4.11), the expectation of $\tilde{S}_{\tilde{m}}^2$ when $\tilde{m}$ has a fixed value, m. This we find by rearranging the definition of variance (A1.2.2) to give the general relation

$$\mathrm{E}[\tilde{x}^2] = var(\tilde{x}) + (\mathrm{E}[\tilde{x}])^2 \equiv \sigma_x^2 + \mu_x^2.$$

(A1.4.18)

The term in curly brackets is therefore

$$\mathrm{E}[\tilde{S}_{\tilde{m}}^2|\tilde{m} = m] = var(\tilde{S}_{\tilde{m}}|\tilde{m} = m) + (\mathrm{E}[\tilde{S}_{\tilde{m}}|\tilde{m} = m])^2$$
$$= m\sigma_z^2 + m^2\mu_z^2,$$

(A1.4.19)

the first term being the variance of the sum of m independent variables, from (2.7.4), and the second term following from (A1.4.13) above. (This step again assumes that the z_i are independent of $\tilde{m}$, as in (A1.4.13).) Now we average over m values, i.e. we now treat m as a random variable though σ_z^2 and μ_z are constants of course. Thus (A1.4.11) gives, using (A1.4.19) and (A1.4.2),

$$\mathrm{E}[\tilde{S}^2] = \mathrm{E}_m[\tilde{m}\sigma_z^2 + \tilde{m}^2\mu_z^2]$$
$$= \sigma_z^2 \mathrm{E}_m[\tilde{m}] + \mu_z^2 \mathrm{E}_m[\tilde{m}^2]$$
$$= \sigma_z^2 \mu_m + \mu_z^2(\sigma_m^2 + \mu_m^2),$$

(A1.4.20)

the last line following by the use of (A1.4.18) to find $\mathrm{E}[\tilde{m}^2]$.

The variance of $\tilde{S}_{\tilde{m}}$ can now be found by substituting (A1.4.16) and (A1.4.20) into (A1.4.17) giving the required result

$$var(\tilde{S}) = \sigma_z^2 \mu_m + \mu_z^2(\sigma_m^2 + \mu_m^2) - \mu_z^2 \mu_m^2$$
$$= \sigma_z^2 \mu_m + \sigma_m^2 \mu_z^2$$

(A1.4.21)

Using the coefficients of variation defined in (2.6.4), i.e.

$$\mathscr{C}(\tilde{z}) \equiv \sigma_z/\mu_z, \ \mathscr{C}(\tilde{m}) \equiv \sigma_m/\mu_m \text{ and } \mathscr{C}(\tilde{S}_{\tilde{m}}) \equiv \sqrt{[var(\tilde{S}_{\tilde{m}})]}/\mathrm{E}[\tilde{S}_{\tilde{m}}],$$

we get, using (A1.4.21) and (A1.4.16),

$$\mathscr{C}^2(\tilde{S}_{\tilde{m}}) = \frac{\mathscr{C}^2(\tilde{z})}{\mu_m} + \mathscr{C}^2(\tilde{m}). \qquad (A1.4.22)$$

An illustration of the use of this result is given in § 3.6 (p. **59**).

Appendix 2

Stochastic (or random) processes

Some basic results and an attempt to explain the unexpected properties of random processes

---

The Science of the age, in short, is physical, chemical, physiological; in all shapes mechanical. Our favourite Mathematics, the highly prized exponent of all these other sciences, has also become more and more mechanical. Excellence in what is called its higher departments depends less on natural genius than on acquired expertness in wielding its machinery. Without under-valuing the wonderful results which a Lagrange or Laplace educes by means of it, we may remark, that their calculus, differential and integral, is little else than a more cunningly-constructed arithmetical mill; where the factors being put in, are, as it were, ground into the true product, under cover, and without other effort on our part than steady turning of the handle.

THOMAS CARLYLE 1829
(Signs of the Times, *Edinburgh Review*, No. 98).

THE following discussions require more calculus than is needed to follow the main body of the book so they have been confined to an appendix to avoid scaring the faint-hearted. However, the *principles* involved are the important thing, so do not worry if you cannot see, for example, how an integral is evaluated. That is merely a techincal matter that can always be cleared up if it it becomes necessary.

A2.1. Scope of the stochastic approach

In many cases the probabilistic approach is necessary for, or at least is enlightening in, the description of processes that are variable by their nature rather than because of experimental error. This approach might, for example, involve consideration of (1) the probability of birth and death in the study of populations, (2) the probability of becoming ill in the study of epidemics, (3) the probability that a queue (e.g. for hospital appointments) will have a particular length and that the waiting time before a queue member is served has a particular value, (4) the random movement of a molecule undergoing Brownian motion in the study of diffusion processes, and (5) the probability that a molecule will undergo a chemical reaction within a specified period of time (see examples in §§ A2.3 and A2.4).

The appendix will deal with aspects of only one particular stochastic process, the Poisson process which has already been discussed in Chapters 3 and 5. It is a characteristic of this process that events occurring in non-overlapping intervals of time are quite independent of each other. The same idea can also be expressed by saying, at the risk of being anthropomorphic, that the process has 'no memory' and therefore is unaffected by what has happened in the past, or that the process 'does not age' (see also Cox (1962, pp. 3–5 and 29)).

Examples of Poisson processes discussed in Chapters 3 and 5 were the disintegration of radioactive atoms at random intervals and the random occurrence of miniature end plate potentials (MEPP). Other examples are (1) the random length of time that a molecule remains adsorbed on a membrane before being desorbed (e.g. an atropine molecule on its cellular receptor site, see § A2.4), and (2) the random length of time that elapses before a drug molecule is broken down in the experiment described in § 12.6.

The lifetime of a molecule on its adsorption site (or of a drug molecule in solution, or of a radioactive atom) is a random variable with the same properties as the random intervals between MEPP (see § 5.1). In the case of the adsorbed molecule, this implies that the complex formed between molecule and adsorption site does not age, and the probability of the complex breaking up in the next 5 seconds, say, is a constant and does *not* depend on how long the molecule has already been adsorbed, just as the probability of a penny coming down heads was supposed to be constant at each throw, regardless of how many heads have already occurred, when discussing the binomial distribution in Chapter 3. Consequently the Poisson distribution can be derived from the binomial as explained in §3.5. Another derivation is given in § A2.2 below.

The arrival of buses would not, in general, be a Poisson Process, although it often seems pretty haphazard. The waiting time problem for *randomly arriving* buses, discussed in § 5.2, is typical of the sort of result that is usually surprising and puzzling to people who have not got used to the properties of random processes. I certainly found it surprising and puzzling until recently, and so I hope the reader will find the results presented below as enlightening as I did.

Fur further reading on the subject see, for example, Cox (1962), Feller (1957, 1966), Bailey (1964, 1967), Cox and Lewis (1966), and Brownlee (1965, p. 190).

A2.2. A derivation of the Poisson distribution

As mentioned in § 3.5, the distribution follows directly from the condition that events in non-overlapping intervals of time or space are independent each, using the definition of independence discussed in § 2.4.

The probability of one event occurring in the time interval between t and $t+\Delta t$ can be defined as $\lambda \Delta t$, if Δt is small enough. From the discussion of the nature of the Poisson process in §§ 3.5 and A2.1, it follows that λ must be a *constant* (i.e. it does not vary with time, and does not depend on the past or present state of the system) that characterizes the rate of occurrence of events. More properly, it should be said that the probability of one occurrence in the infinitesimal time interval, dt, between t and $t+dt$ is constant and can be written λdt. This definition, plus the condition of independence, is sufficient to define the Poisson distribution. If finite time intervals, Δt, are considered then the probability of *one* event in the interval between t and $t+\Delta t$ should be written $\lambda\Delta t+o(\Delta t)$ (see (A2.2.9)). Furthermore, the probability of *more than one* event occurring in the interval Δt becomes negligible when the interval is very short, and so it is also written $o(\Delta t)$, as shown in (A2.2.11).

The symbol $o(\Delta t)$, which occurs often when discussing stochastic processes, is used to stand for *any* quantity that becomes negligible relative to Δt when the interval length Δt becomes very small (it does not always stand for the same quantity, and may be used twice in the same expression standing for a different quantity each time). More precisely, any quantity is written $o(\Delta t)$ if it obeys the definition

$$\lim_{\Delta t \to 0}\left(\frac{o(\Delta t)}{\Delta t}\right) = 0 \tag{A2.2.1}$$

so no approximation will be involved in the limit in ignoring $o(\Delta t)$ terms.

The probability that there will be *no events* between t and $t+\Delta t$ is thus, from the addition rule ((2.4.2) and (2.4.3)), $1-$probability of one or more events $= 1-\lambda\Delta t-o(\Delta t)$.

The probability that r events occur between 0 (the time when measurement is started) and t will be symbolized $P(r, t)$, an extension of the notation used in § 3.5 and Chapter 5. Using this notation, $P(0, t+\Delta t)$ stands for the probability that zero events occur between 0 and $t+\Delta t$. For this to happen there must be both

(zero events between 0 and t) **and** (zero events between t and $t+\Delta t$).

The probability of the first of these contingencies is $P(0, t)$, and the probability of the second is, as above, $1-\lambda\Delta t-o(\Delta t)$. If the events in the non-overlapping time intervals from 0 to t from t to $t+\Delta t$ are *independent* (this is the crucial, and very strong, assumption), the probability that both contingencies will happen follows from the multiplication rule (2.4.6), and is the product of separate possibilities, i.e.

$$P(0, t+\Delta t) = P(0, t).\,[1-\lambda\Delta t-o(\Delta t)]. \tag{A2.2.2}$$

Rearranging this gives

$$\frac{P(0, t+\Delta t)-P(0, t)}{\Delta t} = -\lambda P(0, t)-P(0, t).\frac{o(\Delta t)}{\Delta t}.$$

In the limit, letting $\Delta t \to 0$, the left-hand side becomes, by definition of differentiation, d $P(0, t)/dt$ (see, for example, Massey and Kestelman (1964, p. 59)), and the second term on the right-hand side becomes zero (from (A2.2.1)) so

$$\frac{\mathrm{d}\,P(0, t)}{\mathrm{d}t} = -\lambda P(0, t) \tag{A2.2.3}$$

and the solution of this differential equation is

$$P(0, t) = \mathrm{e}^{-\lambda t}. \tag{A2.2.4}$$

This is found using the condition that $P(0, 0) = e^0 = 1$ (i.e. it is certain that zero events will occur in zero time). The solution is easily checked by differentiating (A2.2.4), giving (A2.2.3) back again, thus $dP(0, t)/dt = de^{-\lambda t}/dt = -\lambda P(0, t)$. Equation (A2.2.4) is just the probability of zero events occurring in time t given by the Poisson distribution (3.5.1), if λ is interpreted as the average number of events in unit time (see §§ 3.5, and 5.1 and eqn (A1.1.7)), so $m = \lambda t$ is the mean number of events in time t.

To find the Poisson distribution when $r > 0$ notice that r events will occur between 0 and $t + \Delta t$ if either

[(r events occur between 0 and t) **and** (zero events occur between t and $t + \Delta t$)]

or

  [($r-1$ events occur between 0 and t)

                  and (one event occurs between t and $t + \Delta t$)].

The probabilities of the four events in brackets have been defined as $P(r, t)$, $(1 - \lambda \Delta t - o(\Delta t))$, $P(r-1, t)$, and $\lambda \Delta t + o(\Delta t)$ respectively. Therefore, using the addition rule (2.4.2) and the multiplication rule for independent events (2.4.6), the probability of r events occurring between 0 and $t + t\Delta t$ becomes

$$P(r, t + \Delta t) = P(r, t) . [1 - \lambda \Delta t - o(\Delta t)] + P(r-1, t).[\lambda \Delta t + o(\Delta t)].$$

$$(A2.2.5)$$

Rearranging this gives

$$\frac{P(r, t + \Delta t) - P(r, t)}{\Delta t} = -\lambda P(r, t) + \lambda P(r-1, t) + \frac{o(\Delta t).}{\Delta t}[P(r-1, t) - P(r, t)].$$

$$(A2.2.6)$$

Again letting $\Delta t \to 0$ gives, using (A2.2.1) as above,

$$\frac{d\ P(r, t)}{dt} = -\lambda P(r, t) + \lambda P(r-1, t). \qquad (A2.2.7)$$

This holds for any r greater than 0, so putting $r = 1$ gives an equation for $P(1, t)$, the probability of $r = 1$ event occurring in a time interval of length t. Inserting the value of $P(r-1, t) = P(0, t) = e^{-\lambda t}$ from (A2.2.4) into (A2.2.7) results in an equation that can be solved giving $P(1, t) = (\lambda t)e^{-\lambda t}$, which is the Poisson probability for $r = 1$ defined in eqn (3.5.1) and § 5.2. This can be inserted into (A2.2.7) with $r = 2$ to find $P(2, t)$, the next term of the Poisson series. Alternatively, simply notice that the probability of r events in a time interval of length t, the solution of (A2.2.7) for *any* value of r, (greater than 0) is

$$P(r, t) = \frac{(\lambda t)^r}{r!}e^{-\lambda t}, \qquad (A2.2.8)$$

which is the Poisson distribution defined in (3.5.1) (see also § 5.1), because it
has been shown in (A2.2.4) that (A2.2.8) does actually hold for $r = 0$ as well.
This solution is easily checked by differentiating (A2.2.8) giving

$$\frac{\mathrm{d}P(r,t)}{\mathrm{d}t} = \frac{\mathrm{d}}{\mathrm{d}t}\left(\frac{(\lambda t)^r}{r!}.e^{-\lambda t}\right) = \frac{\lambda^r}{r!}.rt^{r-1}.e^{-\lambda t} + \frac{(\lambda t)^r}{r!}.(-\lambda e^{-\lambda t})$$

$$= \lambda.\frac{(\lambda t)^{r-1}}{(r-1)!}.e^{-\lambda t} - \lambda.\frac{(\lambda t)^r}{r!}.e^{-\lambda t}$$

$$= \lambda P(r-1,t) - \lambda P(r,t).$$

Thus (A2.2.8) is a solution of (A2.2.7).

Why the remainder terms can be neglected

Having derived the Poisson distribution, the remainder terms, which were
written $o(\Delta t)$ above, can be written explicitly, so it can be seen that they do in
fact become negligible relative to Δt when $\Delta t \to 0$, as stated in (A2.2.1).

The probability of $r = 1$ event occurring in the interval Δt is found by putting
$r = 1$ in (A2.2.8). The exponential is then expanded in series (as in (3.5.2)) giving

$$\lambda\Delta t e^{-\lambda\Delta t} = \lambda\Delta t\left(1 - \lambda\Delta t + \frac{(\lambda\Delta t)^2}{2!} - \frac{(\lambda\Delta t)^3}{3!} + ...\right)$$

$$= \lambda\Delta t - (\lambda\Delta t)^2 + \frac{(\lambda\Delta t)^3}{2!} - ...$$

$$= \lambda\Delta t + o(\Delta t) \tag{A2.2.9}$$

as stated at the beginning of this section. All the terms but the first on the
penultimate can be written as $o(\Delta t)$ because they obey the definition (A2.2.1),
thus

$$\lim_{\Delta t\to 0}\left(\frac{o(\Delta t)}{\Delta t}\right) = \lim_{\Delta t\to 0}\left(\frac{-(\lambda\Delta t)^2 + (\lambda\Delta t)^3/2! - ...}{\Delta t}\right)$$

$$= \lim_{\Delta t\to 0}\left(-\lambda\Delta t + \frac{(\lambda\Delta t)^2}{2!} - ...\right)$$

$$= 0 \tag{A2.2.10}$$

because every term is zero when Δt becomes zero.

The probability that more than one event $(r > 1)$ occurs in Δt is, from (A2.2.8),

$$\frac{(\lambda\Delta t)^r}{r!}e^{-\lambda\Delta t}$$

and for all $r > 1$ this can *also* be written $o(\Delta t)$. For example, for $r = 2$ we have,
using the definition (A2.2.1),

$$\lim_{\Delta t\to 0}\left(\frac{o(\Delta t)}{\Delta t}\right) = \lim_{\Delta t\to 0}\left(\frac{(\lambda\Delta t)^2/2!.e^{-\lambda\Delta t}}{\Delta t}\right)$$

$$= \lim_{\Delta t\to 0}\left(\frac{\lambda\Delta t}{2!}.e^{-\lambda\Delta t}\right)$$

$$= 0 \tag{A2.2.11}$$

as stated at the beginning of this section.

A2.3. The connection between the lifetimes of individual adrenaline molecules and the observed breakdown rate and half-life of adrenaline

In the experiment analysed in § 12.6 it is found that when adrenaline was incubated with liver slices *in vitro* the concentration of adrenaline fell exponentially or, to be more precise, there was no evidence that the relationship was not exponential. The estimated rate constant was $k = 0\cdot07219$ min^{-1} (from (12.6.14)), i.e. the estimated time constant was $1/k = 13\cdot85$ min (from (12.6.16)), and the estimated half-life, from (12.6.6), was $0\cdot69315/k = 9\cdot602$ min. The arguments in this section apply equally to the disintegration of radioisotopes since the number of radioactive nuclei is observed to fall with an exponential time course. One then considers the lifetimes of individual unstable nuclei.

Focus attention on single adrenaline molecules. Suppose that they are perfectly stable until, at zero time, the adrenaline solution is added to the liver preparation that contains enzymes catalysing its catabolism. Suppose that after the addition of enzymes at $t = 0$, there is a constant probability,† $\lambda\Delta t + o(\Delta t)$ say, that any individual adrenaline molecule will be catabolized in any short interval of time Δt. As before, λ is a constant (it does not vary with time) that characterizes the rate of catabolism. The probability that the molecule will *not* be catabolized, from (2.4.3), is therefore $1 - \lambda\Delta t - o(\Delta t)$. The argument is now exactly like that in § A2.2. Denote as $P(t)$ the probability that the molecule is still intact at time t. The molecule will still be intact at time $t + \Delta t$ if

(it is still intact at time t) **and** (it is not catabolized between t and $t + \Delta t$).

If these events are independent, then the multiplication rule of probability, (2.4.6), implies

$$P(t + \Delta t) = P(t).[1 - \lambda\Delta t - o(\Delta t)]. \tag{A2.3.1}$$

This is like eqn. (A2.2.2). Rearranging gives

$$\frac{P(t + \Delta t) - P(t)}{\Delta t} = -\lambda P(t) - P(t).o\frac{(\Delta t)}{\Delta t}$$

and, using (A2.2.1) just as in § A2.2, when $\Delta t \to 0$ this becomes $\mathrm{d}P(t)/\mathrm{d}t = -\lambda P(t)$ (see, for example, Massey and Kestelman (1964, p. 59)). The solution (using the condition that $P(0) = 1$, i.e. it is certain that the molecule is still intact at zero time) is, as in § A2.2,

$$P(t) = \mathrm{e}^{-\lambda t}. \tag{A2.3.2}$$

Now in a large population of molecules the *probability* that a molecule will be still intact at time t can be identified with the *proportion* of molecules

---

† See § A2.2. A fuller explanation of the nature of the term $o(\Delta t)$, which becomes negligible for short enough time intervals, is given in § A2.5.

that are still intact at time t, i.e. y/y_0 where y is the concentration of adrenaline at time t, and y_0 is the initial concentration. Equation (A2.3.2) is now seen to be identical with the observed exponential decline of concentration (eqn. (12.6.4)) if the rate constant, k, is identified with λ.

Furthermore, the probability that a molecule is still intact at time t, given by (A2.3.2), can be identified, just as in § 5.1, with the probability that a molecule has a lifetime greater than t (if it did not it would not still be intact). The probability that the lifetime is equal to or *less* than t is therefore, from the addition rule (2.4.3), and (A2.3.2),

$$1 - P(t) = 1 - \mathrm{e}^{-\lambda t} \equiv F(t), \qquad (A2.3.3)$$

which is exactly like (5.1.4) (the distribution function, F, was defined in (4.1.4)). This is consistent with (see § 5.1) the hypothesis that lifetimes of individual adrenaline molecules are random variables following the exponential distribution (see Fig. 5.1.1 and 5.1.2), with probability density (from (4.1.5) and (A2.3.3))

$$f(t) = \frac{\mathrm{d}F(t)}{\mathrm{d}t} = \lambda \mathrm{e}^{-\lambda t} \; (t \geqslant 0) \qquad (A2.3.4)$$

as previously defined ((5.1.3) and (A1.1.10)). In other words, the mean lifetime of molecules is λ^{-1} (as explained in § 5.1 and proved in (A1.1.11)).

Referring again to the example in § 12.6, it can now be seen that the time constant for the observed exponential fall in adrenaline concentration, $k^{-1} \equiv \lambda^{-1} = 13 \cdot 85$ min (from (12.6.16)), can be interpreted as the *mean* value of the lifetimes of individual adrenaline molecules (measured from the time of addition of enzyme at $t = 0$, or, as shown in the following sections of this appendix, from any other arbitrary time). It follows from the arguments in § A2.6 that if adrenaline molecules were being synthesized in the system, their mean lifetime measured from the moment of synthesis to the moment of catabolism would also be $\lambda^{-1} = 13 \cdot 85$ min.

Furthermore, the half-time for the observed decay of concentration $0 \cdot 69315/k = 9 \cdot 602$ min (from (12.6.6) and (12.6.17)), can be interpreted as the *median* value of the lifetimes of individual adrenaline molecules, because it was shown in (A1.1.4) that the population median of the exponential distribution is $0 \cdot 69315/\lambda$. Fifty per cent of molecules survive longer than $9 \cdot 602$ min.

A2.4. A stochastic view of the adsorption of molecules from solution

Suppose that a surface (e.g. cell membrane) containing many identical and independent adsorption sites is immersed in a solution and is continually bombarded with solute molecules. Some of these will become adsorbed on to adsorption sites, remain on the sites for a time, and then desorb back into the solution. Macroscopic observations of the amount of material adsorbed can be related to what happens to individual molecules using the same sort of approach as in §§ A2.2 and A2.3. This is, for example, the simplest model for the interaction of drug molecules with cell receptor sites and, as such, it is discussed by Rang and Colquhoun (1973).

Consider a single site. The probability that a site is occupied by an adsorbed molecule at time t will be denoted $P_1(t)$, and the probability that the site is empty at time t will be denoted $P_0(t)$. Thus, from (2.4.3),

$$P_0(t) = 1 - P_1(t). \tag{A2.4.1}$$

The probability that an empty site will become occupied would be expected to be proportional to the rate at which solute molecules are bombarding the surface, i.e. to the concentration, c say, of the solute (assumed constant). The probability that an empty site will become occupied during the short interval of time Δt, between t and $t + \Delta t$, will therefore be written $\lambda c \Delta t$ where λc, as in §§ A2.2 and A2.3, is a constant (i.e. does not change with time). The probability that an occupied site becomes empty during the interval Δt will not depend on the concentration of solute, and so will be written $\mu \Delta t$, where μ is another constant. The probability that an occupied site does *not* become empty during Δt is therefore, from (2.4.3), $1 - \mu \Delta t$.† Now a site will be occupied at time $t + \Delta t$ if either [(site was empty at time t) **and** (site is occupied during interval between t and $t + \Delta t$)] **or** [(site was occupied at time t) **and** (site does *not* become empty between t and $t + \Delta t$)].

Now the probabilities of the four events in parentheses have been defined as $P_0(t)$, $\lambda c \Delta t$, $P_1(t)$, and $(1 - \mu \Delta t)$ respectively. So, by application of the addition rule (2.4.2), and the multiplication rule (2.4.6) (assuming, as in §§ A2.2 and A2.3, that the events happening in the non-overlapping intervals of time, from 0 to t and from t to $t + \Delta t$, are independent), it follows that the probability that a site will be occupied at time $t + \Delta t$ will be

$$P_1(t + \Delta t) = P_0(t).\lambda c \Delta t + P_1(t)(1 - \mu \Delta t) + o(\Delta t), \tag{A2.4.2}$$

where $o(\Delta t)$ is a remainder term that includes the probability of several transitions between occupied and empty states during Δt. As in §§ A2.2 and A2.3, $o(\Delta t)$ becomes negligible when Δt is made very small. Rearranging (A2.4.2) gives

$$\frac{P_1(t + \Delta t) - P_1(t)}{\Delta t} = P_0(t).\lambda c - P_1(t).\mu + \frac{o(\Delta t)}{\Delta t}.$$

Now let $\Delta t \to 0$. As before the left-hand side becomes, by definition of differentiation (e.g. Massey and Kestelman (1964, p. 59)), $\mathrm{d}P_1/\mathrm{d}t$, so, using (A2.4.1) and (A2.2.1),

$$\frac{\mathrm{d}P_1(t)}{\mathrm{d}t} = \lambda c(1 - P_1(t)) - \mu P_1(t). \tag{A2.4.3}$$

---

† The probabilities should really be written $\lambda c \Delta t + o(\Delta t)$, $\mu \Delta t + o(\Delta t)$ and $1 - \mu \Delta t - o(\Delta t)$, as in §§ A2.2 and A2.3, if the time interval, Δt, is finite. Alternatively, it could be said, as in § A2.2, that the probability that an occupied site becomes empty during the infinitesimal interval between t and $t + \mathrm{d}t$ can be written $\mu \mathrm{d}t$, etc. A fuller discussion of the nature of the $o(\Delta t)$ terms is given in § A2.5. All these terms have been gathered together and written as $o(\Delta t)$ in (A2.4.2), which holds for finite time intervals.

If $P_1(t)$, the *probability* that an individual site is occupied at time t, is interpreted as the *proportion* of a large population of sites that is occupied at time t, then (A2.4.3) is exactly the same as the equation arrived at by a conventional deterministic approach through the law of mass action, if λ and μ are identified with the mass action adsorption and desorption rate constants. Thus λ/μ is the law of mass action affinity constant for the solute molecule-site reaction. The derivations and solution of (A2.4.3), and its experimental verification in pharmacology is discussed by Rang and Colquhoun (1973).

The length of time for which an adsorption site is occupied; its distribution and mean

In order to investigate the length of time for which a molecule remains adsorbed consider the special case of (A2.4.3) with $\lambda c = 0$. The probability of an adsorbed molecule desorbing does not depend on the probability, λ, that an empty site will be filled, or on the concentration of solute, so this does not spoil the generality of the argument. For example, at $t = 0$ the surface, with a certain number of adsorbed molecules, might be transferred to a solute-free medium (i.e. $c = 0$) so that adsorbed molecules are gradually desorbed, but no further molecules can be adsorbed, so that a site that becomes empty remains empty. When $c = 0$, (A2.4.3) becomes

$$\frac{\mathrm{d}P_1(t)}{\mathrm{d}t} = -\mu P_1(t). \tag{A2.4.4}$$

This equation has already been encountered in §§ A2.2 and A2.3. Integration gives the probability that a site will be occupied, at time t after transfer to solute free medium, as

$$P_1(t) = P_1(0)\mathrm{e}^{-\mu t} \tag{A2.4.5}$$

where $P_1(0)$ is the probability that a site will be occupied at the moment of transfer ($t = 0$). In other words, the *proportion* of sites occupied, and therefore the amount of solute adsorbed, would be expected to fall exponentially with rate constant μ. Such exponential desorption has, in some cases, been observed experimentally.

Now if the total number of adsorption sites is N_{tot}, then the number of sites occupied at time t will be $N(t) = N_{tot}P_1(t)$, and the number occupied at $t = 0$ will be $N(0) = N_{tot}P_1(0)$. The *proportion* of initially occupied sites, that are still occupied after time t will, from (A2.4.5), be

$$\frac{N(t)}{N(0)} = \frac{P_1(t)}{P_1(0)} = \mathrm{e}^{-\mu t}, \tag{A2.4.6}$$

and this will also be the *probability* that an individual site, that was occupied at $t = 0$, will still be occupied after time t†.

A site will only be occupied after time t if the length for which the molecule remains adsorbed (its lifetime) is greater than t, so (A2.4.6) is the probability

† i.e. has been continuously occupied between 0 and t.

that the lifetime of an adsorbed molecule is *longer* than t. Analogous situations were met in §§ 5.1 and A2.3. The probability that the lifetime of an adsorbed molecule is t or *less* is therefore, from (2.4.3),

$$P(0 \leq \text{lifetime} \leq t) \equiv F(t) = 1 - e^{-\mu t}. \tag{A2.4.7}$$

This is exactly like (5.1.4) and (A2.3.3), and is consistent with (see § 5.1) the hypothesis implied by the physical model of identical and *independent* adsorption sites, that the lifetime of individual adsorbed molecules is an exponentially distributed variable, with probability density as before, from (4.1.5),

$$f(t) = \frac{\mathrm{d} F(t)}{\mathrm{d}t} = \mu e^{-\mu t}. \tag{A2.4.8}$$

The *mean* lifetime of a molecule on an adsorption site is therefore μ^{-1} (from (A1.1.11)), the observed time constant (see (12.6.4)) for desorption of adsorbed molecules into a solute-free medium; and, just as in § A2.3, the observed half-time for desorption, $0 \cdot 69315/\mu$ (from (12.6.6)) can be interpreted, using (A1.1.14), as the *median* lifetime of a molecule on an adsorption site. Fifty per cent of molecules stick for a longer time than $0 \cdot 69315/\mu$.

What is meant by lifetime? In the discussion above, the lifetime of an adsorbed molecule was measured from the arbitrary instant ($t = 0$) when the surface was transferred to solute-free medium until the instant when the molecule desorbed. The average length of this *residual lifetime* (see § A2.7) was μ^{-1}. It is of more fundamental interest to known the average length of time a molecule remains adsorbed, i.e. the lifetime measured from the instant of adsorption to the instant of desorption. The mean length of this lifetime is also μ^{-1}, as implied in § 5.1. It might be expected that, because the adsorbed molecules have already been adsorbed for some time at the time that the surface is transferred to the solute-free medium, and the lifetime measured from moment of adsorption to moment of desorption would be longer than μ^{-1} (see Fig. A2.7.3). This cannot be because of the 'lack of memory' or 'lack of ageing' of the Poisson process. It is nevertheless surprising to most people, in just the same way as the analogous bus-waiting time 'paradox' described in § 5.2 is, at first sight, surprising.

If the mean interval between bus arrivals (supposed random) is 10 min then the waiting time from an arbitrary moment until the next bus was stated in § 5.2 to be 10 min also, just as the waiting time from an arbitrary moment until desorption (residual lifetime) is the same (μ^{-1}) as the mean time between adsorption and desorption (lifetime). In words, the reason for this is that if one looks at the surface at an arbitrary moment of time† it is more likely that it will contain long-lived molecule-adsorption site complexes than short-lived ones which, because they exist for only a short time do not stand such a good chance of being in existence at any specified arbitrary moment. Similarly in § 5.2, it is more probable that a person will arrive at

---

† An arbitrary moment of time means a time chosen by any method at all as long as it is independent of the occurrence of events, i.e. independent of the times when molecules move on and off adsorption sites in this case.

the bus stop during a long interval than a short one. In fact, the mean life-time (from the moment of adsorption to the moment of desorption) *of molecules present at an arbitrary moment* (such as the moment when the surface with its adsorbed molecules is transferred to solute-free medium) is exactly twice the mean lifetime of all molecules, i.e. it is $2\mu^{-1}$, so the average residual waiting time until desorption is μ^{-1} as stated, and the mean length of time that a molecule has already been adsorbed at the arbitrary moment is also μ^{-1}, cf. § 5.2). These statements are further discussed and proved in §§ A2.6 and A2.7.

The length of time for which an adsorption site is empty

The argument follows exactly the same lines as that presented above for the average length of time for which a site is occupied. As above it is convenient to consider the special case when the combination of molecule with adsorption site is irreversible so that once occupied a site remains occupied, i.e. $\mu = 0$ (the probability of an empty site becoming occupied does not depend on μ so this does not spoil the generality of the arguments). In this case, because it follows from (A2.4.1) that $\mathrm{d}P_0/\mathrm{d}t = -\mathrm{d}P_1/\mathrm{d}t$, equation (A2.4.3) becomes

$$\frac{\mathrm{d}P_0}{\mathrm{d}t} = -\lambda c P_0(t), \qquad (A2.4.8)$$

which has exactly the same form as (A2.4.4). Using the same arguments as above, it follows that the length of time for which a site remains empty is an exponentially distributed random variable with a mean length of $(\lambda c)^{-1}$. The mean length is inversely proportional to the concentration of solute (c). As above, this is the lifetime measured either from an arbitrary moment, or from the time when the site was last vacated by a desorbing molecule.

Adsorption at equilibrium

After a long time ($t \to \infty$) equilibrium will be reached, i.e. the rate at which molecules are desorbed will be the same as the rate at which they are adsorbed. Therefore, the proportion of sites occupied, P_1, will be constant, i.e. $\mathrm{d}P_1/\mathrm{d}t = 0$. Equation (A2.4.3) gives

$$\lambda c(1-P_1)-\mu P_1 = 0$$

from which it follows that, at equilibrium,

$$P_1 = \frac{\lambda c}{\lambda c + \mu} = \frac{Kc}{Kc+1} \qquad (A2.4.10)$$

if $K = \lambda/\mu$, the law of mass action affinity constant. This equation is the hyperbola in §§ 12.8 and in 14.6. Now it has been shown that the mean length of time for which an individual site is occupied is μ^{-1}, and the mean length of time for which it is empty is $(\lambda c)^{-1}$. These values hold whether or not equilibrium has been reached.† After transferring a membrane with empty sites, to a solution containing a constant concentration, c, of solute, the empty sites will have to wait, on average, $(\lambda c)^{-1}$ seconds before they become occupied so equilibration will take time; see Rang and Colquhoun (1973). Using these values, it follows that

$$Kc = (\text{occupied time/empty time}) \qquad (A2.4.11)$$

and therefore (A2.4.10) can be written

$$P_1 = \frac{1}{1 + (\text{empty time/occupied time})}. \qquad (A2.4.12)$$

For example, if the probability that a site is occupied is $P_1 = 0\cdot5$ (this will be independent of time at equilibrium), i.e. 50 per cent of sites, on the average, are occupied at any moment of time, it follows from (A2.4.12) that empty time = occupied time, i.e any given site is occupied for 50 per cent of the time. This state is attained, at equilibrium, when $(\lambda c)^{-1} = \mu^{-1}$, i.e. when the concentration of solute is $c = \mu/\lambda = 1/K$ (as inferred directly from (A2.4.10)).

A2.5. The relation between the lifetime of individual radioisotope molecules and the interval between disintegrations

The examples of random intervals between miniature and plate potentials (MEPP) discussed in § 5.1) and between bus arrivals (in § 5.2) were straightforward in that there was in each case a single continuous stream of events. In the case of radioisotope disintegration (§§ 3.5–3.7), catabolism of adrenaline (§ A2.3), or adsorption of solute molecules (§ A2.4) the situation is not quite the same. For each isotopic atom there is only one event, disintegration. Nevertheless the random intervals between MEPP or buses have the same properties as the random intervals defined as the lifetimes of isotope atoms (or adrenaline molecules, or solute molecule-adsorption site complexes).

The mean lifetime of isotope molecules, measured from any arbitrary time (see §§ A2.3, A2.4, A2.6, and A2.7) may be thousands of years. For example the half-life (i.e. median lifetime, see § A2.3) of carbon-14 molecules is 5760 years, so the mean lifetime of a molecule is $\lambda^{-1} = 5760/0\cdot69135 = 8310$ years (from (A1.1.11) and (A1.1.14)), i.e. multiplying by the number of seconds in a Gregorian year, $8310 \times 3\cdot155695 \times 10^7 = 2\cdot6224 \times 10^{11}$ s. This is obviously independent of the amount of ^{14}C present.

However, in § 3.7 the Poisson distribution considered was that of the number of disintegrations per second (it will be assumed, for the sake of example, that the isotope involved was ^{14}C). Because this variable is Poisson-distributed with mean, for the example in § 3.7, of $\lambda' t = \lambda' = 2089\cdot5$ disintegra-

---

† If the movements of molecules could be observed, the mean length of time for which a site was occupied could be measured, but the average would obviously have to be taken over a long period of time, relative to μ^{-1}, and $(\lambda c)^{-1}$, even if many sites, rather than just one, were observed. It can be shown that the time constant for equilibration of the sites is $(\lambda c + \mu)^{-1}$ (see, for example, Rang and Colquhoun (1973)), so in fact the average can only be given a frequency interpretation over a period that is long relative to the time taken to reach equilibrium.

tions per second, the mean number of events in $t = 1$ second (assuming that the counter detects all disintegrations), it follows from the arguments in § 5.1 that the intervals between disintegrations are exponentially distributed with mean interval $(\lambda')^{-1} = 1/2089 \cdot 5 = 0 \cdot 000478583$ second (this obviously depends on the amount of ^{14}C present). Compare this with the lifetimes of individual molecules that are also exponentially distributed with mean lifetime $\lambda^{-1} = 8310$ years. These two exponential distributions are, as expected, closely related. This will now be shown.

The probability that any individual ^{14}C atom disintegrates in an interval of time of length Δt, from the arguments in §§ 5.1 and A2.3, must be $\lambda\Delta t$.†
Suppose that at time t a sample of ^{14}C contains $N(t)$ undisintegrated ^{14}C atoms. Define as an 'event' the disintegration of any of these atoms, i.e. if the atoms could be numbered, the disintegration of either atom number 1 **or** atom number 2 **or** . . . **or** atom number $N(t)$. The probability of this event occurring in an interval of time of length Δt, is, from the addition rule (2.4.1),

$$\lambda\Delta t + \lambda\Delta t + ... + \lambda\Delta t - o(\Delta t) = N(t)\lambda\Delta t - o(\Delta t), \qquad (A2.5.1)$$

where $o(\Delta t)$ is a remainder term (see (A2.2.1)) that includes all the probabilities of more than one disintegration occurring during Δt, which will be negligible when Δt is made very small. The argument now follows exactly the same lines as in § A2.3. Define the probability that no event occurs up to time t as $P(t)$. No event will occur up to time $t + \Delta t$ if (no event occurs up to t) **and** (no event occurs between t and $t + \Delta t$), and the probability of this is, from the multiplication rule (2.4.6),

$$P(t + \Delta t) = P(t)[1 - N(t)\lambda\Delta t + o(\Delta t)]. \qquad (A2.5.2)$$

Rearranging this and allowing $\Delta t \to 0$ gives, as in (A2.2.2) and (A2.3.1), $dP(t)/dt = -N(t)\lambda P(t)$. Now *if* the length of time considered is short enough for the decay of the radioisotope to be negligible (as assumed in § 3.7) then $N(t)$ can be treated as a constant. It follows that the solution for $P(t)$, using the condition that $P(0) = 1$ (i.e. it is certain that no events will occur in zero time), will be, as before,

$$P(t) = e^{-N(t)\lambda t}, \qquad (A2.5.3)$$

just as (A2.3.2). This probability, that no disintegration will occur up to time t, can be identified with the probability that the interval between disintegrations is longer than t. Using the same arguments as in § A2.3 it follows that the interval between disintegrations is an exponentially distributed variable with a mean length, defined above as $(\lambda')^{-1}$, of $(N(t)\lambda)^{-1}$, and the mean number of disintegrations per second is therefore

$$\lambda' = N(t)\lambda \qquad (A2.5.4)$$

† This probability should really be written $\lambda\Delta t + o(\Delta t)$, if Δt is finite, as in §§ A2.3 and A2.4. The nature of the $o(\Delta t)$ terms, and a more rigorous derivation of (A2.5.1), are discussed at the end of this section.

which decreases, as expected, as the total number of isotope molecules, $N(t)$, decreases. The intervals, will of course, only be exponentially distributed, and the disintegration rate will only be Poisson distributed, over time intervals short enough for $N(t)$ to be substantially constant. Using (A2.5.4) and the figures given above for the example in § 3.7 shows that the number of ^{14}C atoms present at the time the sample was counted must have been

$$N(t) = \lambda'/\lambda = 2089\cdot5 \text{ (atoms s}^{-1}) \times 2\cdot6224 \times 10^{11} \text{ (s)}.$$
$$= 5\cdot4795 \times 10^{14} \text{ atoms}.$$

Therefore the weight of ^{14}C was

$$5\cdot4795 \times 10^{14}/6\cdot023 \times 10^{23} = 9\cdot098 \times 10^{-10} \text{ gramme molecules, or}$$
$$9\cdot098 \times 10^{-10} \times 14 = 12\cdot74 \times 10^{-9} \text{ g}.$$

A more careful look at the nature of the $o(\Delta t)$ terms in processes like the catabolism of adrenaline, the decay of radio-isotopes and the adsorption of molecules

The basic Poisson process consists of a continuous stream of events, such as the occurrence of miniature end plate potentials (see § 5.1) or the random arrivals of buses at a bus stop. It was shown in § A2.2 that in this sort of process the probability of one event occurring in a finite time interval Δt can be written as $\lambda\Delta t + o(\Delta t)$. Obviously this probability cannot be written simply as $\lambda\Delta t$ because this would become indefinitely large, if long enough time intervals were considered, whereas all probabilities must be less than 1.

In processes like the catabolism of adrenaline, the decay of radioisotopes, or the adsorption of molecules, the situation is not quite the same. Each adrenaline molecule can only be destroyed once, so one cannot consider the probability of it being "destroyed r times during Δt" as in § A2.2. Nevertheless it clearly will not do to say that the probability of catabolism (decay, adsorption, etc.) during Δt is $\lambda\Delta t$, because, as above, this can be greater than 1. Suppose that this probability can be written $\lambda\Delta t + o(\Delta t)$. The argument in the first part of this section can now be made more rigorous.

The catabolism (decay, etc.) of different atoms during a finite time Δt are not mutually exclusive events, so the simple addition rule cannot be used. Instead, the binomial theorem should be used. In the language of §§ 3.2–3.4, let a 'trial' be the observing of a molecule during the time Δt, and let a 'success' be the occurrence of catabolism (decay, etc.) during this period. If, as above, there are N molecules present altogether, then the probability that one of them will be catabolized (decay, etc.) during Δt can be identified with the probability of $r = 1$ success occurring in N trials, and this is given by the binomial distribution, (3.4.3), as $N\mathscr{P}(1-\mathscr{P})^{N-1}$ where it has been supposed that the probability of success at each trial can be written $\mathscr{P} = \lambda\Delta t + o(\Delta t)$. This probability is the same at every trial as discussed in § 3.5. Substituting it for $\mathscr{P}$ in $N\mathscr{P}(1-\mathscr{P})^{N-1}$, and expanding the resulting expression (use the binomial expansion on $(1-\mathscr{P})^{N-1}$), it is found that the required probability of one of the N molecules being catabolized during Δt can indeed be written

$$N\mathscr{P}(1-\mathscr{P})^{N-1} = N\lambda\Delta t + o(\Delta t) \tag{A2.5.5}$$

as asserted, on the basis of a simplified argument, in (A2.5.1).

This argument can now be turned upside down starting from the experimental observations and working backwards. The decay of radioisotopes, and, in some circumstances at least, the catabolism of molecules, and the desorbtion of adsorbed molecules, are *observed* to follow an exponential time course. In each case the implication is that the probability that a molecule is still intact at time t, $1 - F(t)$, is $e^{-\lambda t}$. This is consistent, as described in earlier sections, with the physical model that specifies that the lifetime of individual molecules is an exponentially distributed variable with mean λ^{-1}. In the case of radioisotope decay, this can be confirmed experimentally by the observation that the number of disintegrations in unit time is Poisson distributed (over times during which N is substantially constant). Now, if the number of molecules catabolized, etc. during Δt is Poisson distributed, and the mean number of events during Δt is $\lambda' \Delta t$ as above, then the probability that one molecule of the N present will be catabolized, etc. during Δt is given by the Poisson distribution, (3.5.1), with $r = 1$ and $m = \lambda' \Delta t$, i.e. it is $\lambda' \Delta t e^{-\lambda' \Delta t}$. Substituting $\lambda' = N\lambda$ from (A2.5.4), and expanding the exponential term exactly as in (A2.2.9), gives the probability of one of the N molecules being catabolized, etc. during Δt as

$$N\lambda \Delta t . e^{-N\lambda \Delta t} = N\lambda \Delta t + o(\Delta t). \qquad (A2.5.6)$$

just as in (A2.5.5) and (A2.5.1). Now, according to the argument above, this can be equated with $N\mathscr{P}(1 - \mathscr{P})^{N-1}$, where $\mathscr{P}$ is the probability of any individual molecule being catabolized during Δt. The only two solutions of this equation for $\mathscr{P}$ are $\mathscr{P} = \lambda \Delta t$ or $\mathscr{P} = \lambda \Delta t + o(\Delta t)$. The former will not do, as explained above, so the probability must be written $\lambda \Delta t + o(\Delta t)$ as asserted in §§ A2.3–A2.5.

A2.6. Why the waiting time until the next event does not depend on when the timing is started, for a Poisson process

The assertion that waiting time does not depend on when timing was started has been made repeatedly in Chapter 5 and this appendix. For example, the mean waiting time until a molecule is desorbed does not depend on the arbitrary time when the timing is started, and will be the same, λ^{-1}, as if the timing were started from the moment the molecule was adsorbed.

Suppose that the interval from one event to the next is exponentially distributed with mean λ^{-1}. It will be convenient, as at the end of § 4.1, to use l to stand for time measured from the last event considered as a random variable, and t, t_0, etc., to stand for particular values of l. Suppose that a time t_0 is known to have elapsed from the last event. Given this fact, what is the probability that the time from t_0 until the next event (the *residual lifetime*) is less than any specified time t, i.e. what is the probability that $l < t_0 + t$ (event E_1 say) given that $l > t_0$ (event E_2 say)? In symbols this is $P(E_1 | E_2)$, i.e. from the definition of conditional probability (2.4.4),

$$P(l < t_0 + t \,|\, l > t_0) = \frac{P(l < t_0 + t \text{ and } l > t_0)}{P(l > t_0)}. \qquad (A2.6.1)$$

Now the event that $(l < t_0 + t \text{ and } l > t_0)$ is the same as the event $t_0 < l < t_0 + t$ and, because the intervals between events are being supposed to follow the exponential distribution (5.1.3), $f(t) = \lambda e^{-\lambda t}$ with mean interval between events $= \lambda^{-1}$, the probability of this is, as in (4.1.2),

$$P(t_o < \tilde{t} < t_o + t) = \int_{t_o}^{t_o + t} \lambda e^{-\lambda t} dt$$

$$= \left[-e^{-\lambda t} \right]_{t_o}^{t_o + t} = e^{-\lambda t_o} - e^{-\lambda(t_o + t)}$$

$$= e^{-\lambda t_o} - e^{-\lambda t_o} e^{-\lambda t}. \tag{A2.6.2}$$

The denominator of (A2.6.1) is (cf. (5.1.4)),

$$P(\tilde{t} > t_o) = \int_{t_o}^{\infty} \lambda e^{-\lambda t} \, dt$$

$$= \left[-e^{-\lambda t} \right]_{t_o}^{\infty} = e^{-\lambda t_o}. \tag{A2.6.3}$$

Substituting (A2.6.2) and (A2.6.3) into (A2.6.1) gives the required conditional distribution function (cf. (4.1.4)) for the residual life-time, t, (*measured from t_o to the next event*) as

$$P(\tilde{t} < t_o + t | \tilde{t} > t_o) = \frac{e^{-\lambda t_o} - e^{-\lambda t_o} e^{-\lambda t}}{e^{-\lambda t_o}}$$

$$= 1 - e^{-\lambda t}, \tag{A2.6.3}$$

which is identical with the distribution function ((5.1.4) or (A2.3.3)) for the intervals between events (*measured from the last event to the next event*). Differentiating, as in (A2.3.4), gives the probability density for the residual lifetime, t, as $f(t) = \lambda e^{-\lambda t}$, the exponential distribution with mean λ^{-1} (from (A1.1.11)), exactly the same as the distribution of intervals between events. The common-sense reason for this curious result has been discussed in words in §§ 5.2 and A2.4, and is proved in § A2.7.

A2.7. Length-biased sampling. Why the average length of the interval in which an arbitrary moment of time falls is twice the average length of all intervals for a Poisson process

In § 5.2 it was stated that if buses arrive randomly with an *average* interval of 10 min then, if a person arrives at the bus stop at an arbitrary time, the mean length of the interval in which he arrives is 20 min. Similarly, in § A2.4 it was asserted that the mean lifetime of adsorbed molecule-adsorption site complexes in existence at a specified arbitrary moment of time was twice the average lifetime. In each case this was explained by saying that a long interval has a better chance than a short one of including the arbitrary moment, i.e. the interval lengths are *not* randomly sampled by choosing one that includes an arbitrary time, just as rods of different length would, doubtless, not be randomly sampled by picking a rod out of a bag containing well mixed rods. The long rods would stand a better chance of being picked. Sampling of this sort is described as *length-biased* (see, for example, Cox 1962, p. 65).

The specifying of the arbitrary moment of time constitutes the choice of an interval (the interval in which the time falls) from the population of intervals between events. Imagine that intervals are repeatedly chosen in this way. What will their average length be? First, the distribution of their length must be found.

The distribution of intervals chosen by length-biased sampling

One difficulty in deriving the required result arises because it is necessary to consider an infinite population of intervals. It will be much easier to start off with a finite population. Imagine a finite set of N intervals, and call the length of an interval (the ith interval) t_i. The total length of time occupied by the intervals is thus $\sum_{\text{all } t_i} t_i$. The fraction of this total time occupied by the ith interval will be

$$\frac{t_i}{\sum_{\text{all } t_i} t_i}. \tag{A2.7.1}$$

If these fractions are added up for all intervals that are longer than some specified length t, the result is the proportion of time occupied by intervals longer than t:

$$\frac{\sum_{t_i > t} t_i}{\sum_{\text{all } t} t_i} = \frac{\text{time occupied by intervals longer than } t}{\text{total time}} \tag{A2.7.2}$$

$$= \text{probability that a point chosen at random}\dagger$$
$$\text{falls in an interval longer than } t \tag{A2.7.3}$$

$$\equiv 1 - F_1(t) \tag{A2.7.4}$$

if $F_1(t)$ stands for the distribution function of intervals chosen by length-biased sampling (defined as the proportion of intervals thus chosen with length *less* than the specified value, t, so $1 - F_1(t)$ is the proportion with length greater than t; see (4.1.4) and (5.1.4)).

The crucial step, the equating of (A2.7.2) and (A2.7.3), certainly looks reasonable. Another way of looking at it is to suppose that the probability of choosing any particular interval is directly proportional to its length, t_i, so longer intervals are more likely to be chosen. The proportionality constant must be chosen so that all the probabilities add up to 1, because it is certain that one interval or another will be chosen. The proportionality constant is therefore $1/\sum_{\text{all } t_i} t_i$ giving

---

$\dagger$ i.e. a point chosen at random with the uniform (or rectangular) distribution over the interval $0, \sum_{\text{all } t_i} t_i$.

probability of choosing an interval of length t_i is

$$\text{constant} \times t_i = \frac{t_i}{\sum\limits_{\text{all } t_i} t_i}. \qquad (A2.7.5)$$

It follows, using the addition rule, (2.4.2), that the probability of choosing an interval longer than t is found by adding these probabilities for all intervals longer than t giving

$$\text{probability of choosing an interval longer than } t = \frac{\sum\limits_{t_i > t} t_i}{\sum\limits_{\text{all } t} t_i}, \qquad (A2.7.6)$$

which is exactly the same as found above, eqn. (A2.7.3).

Now suppose that in the finite population of N intervals, some of the intervals are of identical lengths. There are f_i intervals of length t_i, say (so $\Sigma f_i = N$). The time occupied by the f_i intervals of length t_i must be $f_i t_i$, and the total time occupied by all N intervals must be $\sum\limits_{\text{all } t_i} f_i t_i$. The proportion of the total time occupied by intervals longer than a specified value, t, by modificating of (A2.7.2) (or A2.7.6)) must now be written

$$\frac{\text{time occupied by intervals longer than } t}{\text{total time}}$$

$$= \text{probability that an interval chosen by} \\ \text{length-biased sampling is longer than } t \qquad (A2.7.7)$$

$$\equiv 1 - F_1(t)$$

$$= \frac{\sum\limits_{t_i > t} f_i t_i}{\sum\limits_{\text{all } t_i} f_i t_i} = \frac{\sum\limits_{t_i > t} P_i t_i}{\sum\limits_{\text{all } t_i} P_i t_i} \qquad (A2.7.8)$$

if P_i is defined as f_i/N, the *proportion* of intervals of length t_i in the population. The values of P_i define the (discontinuous) distribution of interval lengths t_i, in the finite population under consideration.

It is now possible, at last, to revert to the real problem, in which there is an infinite population of intervals and the intervals can potentially be of any length, i.e. they have a continuous distribution (see Chapters 4 and 5). All that is necessary is to replace P_i by $dP = f(t) \, dt$ (from (4.1.1)). As described in Chapter 4, dP is the probability that the length of an interval will lie within the very narrow range between t and $t + dt$. When this is substituted in (A2.7.8) the summations must, of course, be replaced by integrations.

The result is

proportion of time occupied by intervals longer than t

= probability that an interval chosen by length biased sampling is longer than t

$\equiv 1 - F_1(t)$

$$= \frac{\displaystyle\int_t^\infty tf(t)\mathrm{d}t}{\displaystyle\int_0^\infty tf(t)\mathrm{d}t}.$$ (A2.7.9)

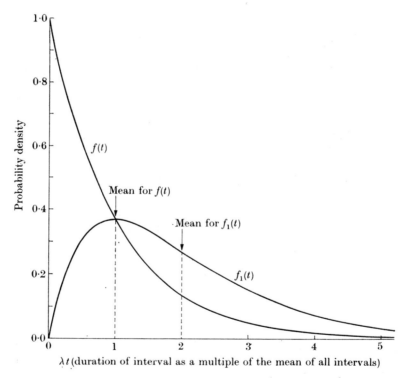

Fɪɢ. A2.7.1.   Distributions of the length of random intervals. The abscissa is plotted as in Figs. 5.1.1 and A2.7.2. The distribution of durations in the population, $f(t)$, is the exponential distribution, exactly as in Fig. 5.1.2. The distribution of the lengths of intervals chosen by length-biased sampling, $f_1(t)$, shows that relatively few short intervals will be chosen, and the mean interval is twice as long as the mean of the whole population. If the abscissa is multiplied by λ^{-1} to convert it into time units, the probability density would be divided by λ^{-1}, so the area under the curves remained 1·0.

For the exponential distribution of intervals in the population, which is what we are interested in, substitute the definition of this distribution, $f(t) = \lambda e^{-\lambda t}$, in (A2.7.9). The integral in the denominator of (A2.7.9) has already been shown in (A1.1.11) to be λ^{-1}. The numerator of (A2.7.9), integrating by parts exactly as in (A1.1.11), is

$$\left[-te^{-\lambda t}\right]_t^\infty - \left[\frac{e^{-\lambda t}}{\lambda}\right]_t^\infty = -0 + te^{-\lambda t} - 0 + \lambda^{-1}e^{-\lambda t}$$

$$= (\lambda^{-1} + t)e^{-\lambda t}. \tag{A2.7.10}$$

Substituting these results in (A2.7.9) gives

$$1 - F_1(t) = \frac{(\lambda^{-1} + t)e^{-\lambda t}}{\lambda^{-1}} = (1 + \lambda t)e^{-\lambda t} \tag{A2.7.11}$$

as the proportion of intervals longer than t, when the intervals are chosen by length-biased sampling. Compare this with the proportion of intervals longer than t in the whole population which, from (5.1.4) or (A1.1.12), is $1 - F(t) = e^{-\lambda t}$. The cumulative distributions are plotted in Fig. A2.7.2.

The proportion of intervals longer than the mean interval

The mean length of all intervals in an exponentially distributed population is λ^{-1}, as proved in (A1.1.11). It was shown in (A1.1.13) that 63·21 per cent of all intervals are shorter than the mean, λ^{-1}. Therefore $100 - 63·21 = 36·79$ per cent of all intervals are longer than the mean length. The proportion of time occupied by intervals that are longer than the mean follows directly from (A2.7.11) and (A2.7.9.), putting $t = \lambda^{-1}$, and is thus

$$(1 + \lambda\lambda^{-1})e^{-\lambda\lambda^{-1}} = 2e^{-1} = 0·7358, \text{ i.e. } 73·58 \text{ per cent.}$$
$$\tag{A2.7.12}$$

Thus, although only 36·79 per cent of intervals in the population are longer than the mean length, this 36·79 per cent occupy 73·58 per cent of the time, and this is one way of looking at the reason for there being a greater chance of an arbitrary time falling in a long interval than a short interval.

The mean length of an interval chosen by length-biased sampling

The question posed at the beginning of this section can now be answered. The probability density function (see § 4.1) defining the distribution of lengths of intervals chosen by length-biased sampling follows from (A2.7.11), using (4.1.5), and is

$$f_1(t) = \frac{d}{dt}F_1(t) = \frac{d}{dt}[1 - (1 + \lambda t)e^{-\lambda t}]$$

$$= \lambda^2 te^{-\lambda t} \text{ (for } t \geqslant 0). \tag{A2.7.13}$$

This distribution curve is drawn in Fig. A2.7.1, and compared with the distribution curve, $f(t) = \lambda e^{-\lambda t}$, for all intervals in the population.

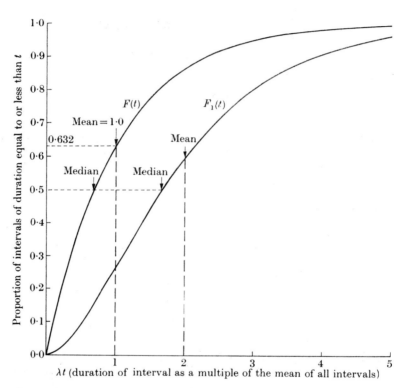

FIG. A2.7.2. Cumulative distributions of the lengths of random intervals. The distribution function, $F(t)$, for the lengths of all intervals is exactly as in Fig. 5.1.1. The abscissa is the interval length as a multiple of the mean length of all intervals, i.e. it is λt as in Fig. 5.1.1. If the mean length of all intervals were $\lambda^{-1} = 10$ s, the figures on the abscissa would be multiplied by 10 to convert them to seconds. The cumulative distribution, $F_1(t)$, for intervals chosen by length-biased sampling, is seen to have more long intervals than there are in the whole population, the mean being $2\lambda^{-1}$ (i.e. 20 s in the example above).

The *mean* length of an interval chosen by length-biased sampling now follows from (A1.1.2), and is

$$E(t) = \int_0^\infty tf_1(t)\mathrm{d}t = \int_0^\infty \lambda^2 t^2 \mathrm{e}^{-\lambda t}\mathrm{d}t. \tag{A2.7.14}$$

To solve this, integrate by parts (see, for example, Massey and Kestelman, (1964, pp. 332, 402)), as in (A1.1.11). Put $u = t^2$, so $\mathrm{d}u = 2t\,\mathrm{d}t$, and put $\mathrm{d}v = \lambda^2 \mathrm{e}^{-\lambda t}\mathrm{d}t$, so $v = \int \lambda^2 \mathrm{e}^{-\lambda t}\mathrm{d}t = [-\lambda \mathrm{e}^{-\lambda t}]$. Thus

$$\begin{aligned}
E(t) &= \int_0^\infty u\mathrm{d}v = [uv] - \int v\mathrm{d}u \\
&= \left[-t^2 \lambda \mathrm{e}^{-\lambda t} \right]_0^\infty - \int_0^\infty (-\lambda \mathrm{e}^{-\lambda t})\,(2t\mathrm{d}t) \\
&= 0 + 2\int_0^\infty t\lambda \mathrm{e}^{-\lambda t}\mathrm{d}t \\
&= 2\lambda^{-1}, \tag{A2.7.15}
\end{aligned}$$

i.e twice the mean (λ^{-1}) of all intervals, as stated. In the evaluation of the first term on the second line of (A2.7.15) notice that $t^2 \mathrm{e}^{-\lambda t} \to 0$ as $t \to \infty$; see, for example, Massey and Kestelman (1964, p. 122). The integral on the third line of (A2.7.15) is simply the mean of the exponential distribution, shown in (A1.1.11) to be λ^{-1}.

Tables

TABLE A1

Nonparametric confidence limits for the median

See §§ 7.3 and 10.2. Rank the n observations and take the rth from each end as limits. †With samples smaller than $n = 6$, 95 per cent limits cannot be found, but the P value for the limits formed by the largest and smallest ($r = 1$) observations are given (Nair 1940).

Sample size n	P approx. 95 per cent r	100 P	P approx. 99 per cent r	100 P	Sample size n	P approx. 95 per cent r	100 P	P approx. 99 per cent r	100 P
					31	10	97·06	8	99·66
2	1†	50·0			32	10	98·00	9	99·30
3	1†	75·0			33	11	96·50	9	99·54
4	1†	87·5			34	11	97·56	10	99·10
5	1†	93·75			35	12	95·90	10	99·40
6	1	96·88			36	12	97·12	10	99·60
7	1	98·44			37	13	95·30	11	99·24
8	1	99·22	1	99·22	38	13	96·64	11	99·50
9	2	96·10	1	99·60	39	13	97·62	12	99·06
10	2	97·86	1	99·80	40	14	96·16	12	99·36
11	2	98·82	1	99·90	41	14	97·24	12	99·52
12	3	96·14	2	99·36	42	15	95·64	13	99·20
13	3	97·76	2	99·66	43	15	96·84	13	99·46
14	3	98·70	2	99·82	44	16	95·12	14	99·04
15	4	96·48	3	99·26	45	16	96·44	14	99·34
16	4	97·88	3	99·58	46	16	97·42	14	99·54
17	5	95·10	3	99·76	47	17	96·00	15	99·20
18	5	96·92	4	99·24	48	17	97·06	15	99·44
19	5	98·08	4	99·56	49	18	95·56	16	99·06
20	6	95·86	4	99·74	50	18	96·72	16	99·34
21	6	97·34	5	99·28	51	19	95·12	16	99·54
22	6	98·30	5	99·56	52	19	96·36	17	99·22
23	7	96·54	5	99·74	53	19	97·30	17	99·46
24	7	97·74	6	99·34	54	20	95·98	18	99·10
25	8	95·68	6	99·60	55	20	97·00	18	99·36
26	8	97·10	7	99·06	56	21	95·60	18	99·54
27	8	98·08	7	99·40	57	21	96·68	19	99·24
28	9	96·44	7	99·62	58	22	95·20	19	99·46
29	9	97·58	8	99·18	59	22	96·36	20	99·14
30	10	95·72	8	99·48	60	22	97·26	20	99·38

Sample size n	P approx. 95 per cent r	P approx. 95 per cent $100\,P$	P approx. 99 per cent r	P approx. 99 per cent $100\,P$	Sample size n	P approx. 95 per cent r	P approx. 95 per cent $100\,P$	P approx. 99 per cent r	P approx. 99 per cent $100\,P$
61	23	96·04	21	99·02	71	27	96·80	25	99·14
62	23	97·00	21	99·28	72	28	95·56	25	99·36
63	24	95·70	21	99·48	73	28	96·56	26	99·04
64	24	96·72	22	99·18	74	29	95·26	26	99·30
65	25	95·36	22	99·40	75	29	96·30	26	99·48
66	25	96·44	23	99·08					
67	26	95·02	23	99·32					
68	26	96·16	23	99·50					
69	26	97·06	24	99·24					
70	27	95·86	24	99·44					

TABLE A2

Confidence limits for the parameter of a binomial distribution, i.e. the population proportion of 'successes'

See 7.7, 7.8, 10.2 and 3.2–3.4. If r 'successes' are observed in a sample of n 'trials', the confidence limits ($100\mathscr{P}_L$ and $100\,\mathscr{P}_U$ from eqns. (7.7.1) and (7.7.2)) for $\mathscr{P}$, the proportion of 'successes' in the population (see § 3.2) from which the sample was drawn, can be found from the table. Reproduced from *Documenta Geigy Scientific Tables*, 6th edn by permission of J. R. Geigy S. A., Basle, Switzerland. The Geigy tables give limits for all n from 2 to 1000.

		95% limits	99% limits
r	$100r/n$	$100\mathscr{P}_L$ $100\mathscr{P}_U$	$100\mathscr{P}_L$ $100\mathscr{P}_U$
		$n=2$	
0	0.00	0.00– 84.19	0.00– 92.93
1	50.00	1.26– 98.74	0.25– 99.75
2	100.00	15.81–100.00	7.07–100.00
		$n=3$	
0	0.00	0.00– 70.76	0.00– 82.90
1	33.33	0.84– 90.57	0.17– 95.86
2	66.67	9.43– 99.16	4.14– 99.83
3	100.00	29.24–100.00	17.10–100.00
		$n=4$	
0	0.00	0.00– 60.24	0.00– 73.41
1	25.00	0.63– 80.59	0.13– 88.91
2	50.00	6.76– 93.24	2.94– 97.06
3	75.00	19.41– 99.37	11.09– 99.87
4	100.00	39.76–100.00	26.59–100.00
		$n=5$	
0	0.00	0.00– 52.18	0.00– 65.34
1	20.00	0.51– 71.64	0.10– 81.49
2	40.00	5.27– 85.34	2.29– 91.72
3	60.00	14.66– 94.73	8.28– 97.71
4	80.00	28.36– 99.49	18.51– 99.90
5	100.00	47.82–100.00	34.66–100.00
		$n=6$	
0	0.00	0.00– 45.93	0.00– 58.65
1	16.67	0.42– 64.12	0.08– 74.60
2	33.33	4.33– 77.72	1.87– 85.64
3	50.00	11.81– 88.19	6.63– 93.37
4	66.67	22.28– 95.67	14.36– 98.13
5	83.33	35.88– 99.58	25.40– 99.92
6	100.00	54.07–100.00	41.35–100.00
		$n=7$	
0	0.00	0.00– 40.96	0.00– 53.09
1	14.29	0.36– 57.87	0.07– 68.49
2	28.57	3.67– 70.96	1.58– 79.70
3	42.86	9.90– 81.59	5.53– 88.23
4	57.14	18.41– 90.10	11.77– 94.47
5	71.43	29.04– 96.33	20.30– 98.42
6	85.71	42.13– 99.64	31.51– 99.93
7	100.00	59.04–100.00	46.91–100.00
		$n=8$	
0	0.00	0.00– 36.94	0.00– 48.43
1	12.50	0.32– 52.65	0.06– 63.15
2	25.00	3.19– 65.09	1.37– 74.22
3	37.50	8.52– 75.51	4.75– 83.03
4	50.00	15.70– 84.30	9.99– 90.01
5	62.50	24.49– 91.48	16.97– 95.25
6	75.00	34.91– 96.81	25.78– 98.63
7	87.50	47.35– 99.68	36.85– 99.94
8	100.00	63.06–100.00	51.57–100.00

		95% limits	99% limits
r	$100r/n$	$100\mathscr{P}_L$ $100\mathscr{P}_U$	$100\mathscr{P}_L$ $100\mathscr{P}_U$
		$n=9$	
0	0.00	0.00– 33.63	0.00– 44.50
1	11.11	0.28– 48.25	0.06– 58.50
2	22.22	2.81– 60.01	1.21– 69.26
3	33.33	7.49– 70.07	4.16– 78.09
4	44.44	13.70– 78.80	8.68– 85.39
5	55.56	21.20– 86.30	14.61– 91.32
6	66.67	29.93– 92.51	21.91– 95.84
7	77.78	39.99– 97.19	30.74– 98.79
8	88.89	51.75– 99.72	41.50– 99.94
9	100.00	66.37–100.00	55.50–100.00
		$n=10$	
0	0.00	0.00– 30.85	0.00– 41.13
1	10.00	0.25– 44.50	0.05– 54.43
2	20.00	2.52– 55.61	1.09– 64.82
3	30.00	6.67– 65.25	3.70– 73.51
4	40.00	12.16– 73.76	7.68– 80.91
5	50.00	18.71– 81.29	12.83– 87.17
6	60.00	26.24– 87.84	19.09– 92.32
7	70.00	34.75– 93.33	26.49– 96.30
8	80.00	44.39– 97.48	35.18– 98.91
9	90.00	55.50– 99.75	45.57– 99.95
10	100.00	69.15–100.00	58.87–100.00
		$n=11$	
0	0.00	0.00– 28.49	0.00– 38.22
1	9.09	0.23– 41.28	0.05– 50.86
2	18.18	2.28– 51.78	0.98– 60.85
3	27.27	6.02– 60.97	3.33– 69.33
4	36.36	10.93– 69.21	6.88– 76.68
5	45.45	16.75– 76.62	11.45– 83.07
6	54.55	23.38– 83.25	16.93– 88.55
7	63.64	30.79– 89.07	23.32– 93.12
8	72.73	39.03– 93.98	30.67– 96.67
9	81.82	48.22– 97.72	39.15– 99.02
10	90.91	58.72– 99.77	49.14– 99.95
11	100.00	71.51–100.00	61.78–100.00
		$n=12$	
0	0.00	0.00– 26.46	0.00– 35.69
1	8.33	0.21– 38.48	0.04– 47.70
2	16.67	2.09– 48.41	0.90– 57.29
3	25.00	5.49– 57.19	3.03– 65.52
4	33.33	9.92– 65.11	6.24– 72.75
5	41.67	15.17– 72.33	10.34– 79.15
6	50.00	21.09– 78.91	15.22– 84.78
7	58.33	27.67– 84.83	20.85– 89.66
8	66.67	34.89– 90.08	27.25– 93.76
9	75.00	42.81– 94.51	34.48– 96.97
10	83.33	51.59– 97.91	42.71– 99.10
11	91.67	61.52– 99.79	52.30– 99.96
12	100.00	73.54–100.00	64.31–100.00

r	100r/n	95% limits 100$\mathscr{P}_L$ 100$\mathscr{P}_U$	99% limits 100$\mathscr{P}_L$ 100$\mathscr{P}_U$	r	100r/n	95% limits 100$\mathscr{P}_L$ 100$\mathscr{P}_U$	99% limits 100$\mathscr{P}_L$ 100$\mathscr{P}_U$
		$n = 13$				$n = 17$ (continued)	
0	0.00	0.00- 24.71	0.00- 33.47	5	29.41	10.31- 55.96	6.97- 63.10
1	7.69	0.19- 36.03	0.04- 44.90	6	35.29	14.21- 61.67	10.14- 68.46
2	15.38	1.92- 45.45	0.83- 54.10	7	41.18	18.44- 67.08	13.71- 73.44
3	23.08	5.04- 53.81	2.78- 62.06	8	47.06	22.98- 72.19	17.64- 78.07
4	30.77	9.09- 61.43	5.71- 69.13	9	52.94	27.81- 77.02	21.93- 82.36
5	38.46	13.86- 68.42	9.42- 75.46	10	58.82	32.92- 81.56	26.56- 86.29
6	46.15	19.22- 74.87	13.83- 81.13	11	64.71	38.33- 85.79	31.54- 89.86
7	53.85	25.13- 80.78	18.87- 86.17	12	70.59	44.04- 89.69	36.90- 93.03
8	61.54	31.58- 86.14	24.54- 90.58	13	76.47	50.10- 93.19	42.68- 95.74
9	69.23	38.57- 90.91	30.87- 94.29	14	82.35	56.57- 96.20	48.96- 97.91
10	76.92	46.19- 94.96	37.94- 97.22	15	88.24	63.56- 98.54	55.87- 99.37
11	84.62	54.55- 98.08	45.90- 99.17	16	94.12	71.31- 99.85	63.70- 99.97
12	92.31	63.97- 99.81	55.10- 99.96	17	100.00	80.49-100.00	73.22-100.00
13	100.00	75.29-100.00	66.53-100.00				
		$n = 14$				$n = 18$	
0	0.00	0.00- 23.16	0.00- 31.51	0	0.00	0.00- 18.53	0.00- 25.50
1	7.14	0.18- 33.87	0.04- 42.40	1	5.56	0.14- 27.29	0.03- 34.63
2	14.29	1.78- 42.81	0.76- 51.23	2	11.11	1.38- 34.71	0.59- 42.17
3	21.43	4.66- 50.80	2.57- 58.92	3	16.67	3.58- 41.42	1.97- 48.84
4	28.57	8.39- 58.10	5.26- 65.79	4	22.22	6.41- 47.64	4.00- 54.92
5	35.71	12.76- 64.86	8.66- 72.01	5	27.78	9.69- 53.48	6.54- 60.55
6	42.86	17.66- 71.14	12.67- 77.66	6	33.33	13.34- 59.01	9.51- 65.79
7	50.00	23.04- 76.96	17.24- 82.76	7	38.89	17.30- 64.25	12.84- 70.68
8	57.14	28.86- 82.34	22.34- 87.33	8	44.44	21.53- 69.24	16.49- 75.26
9	64.29	35.14- 87.24	27.99- 91.34	9	50.00	26.02- 73.98	20.47- 79.53
10	71.43	41.90- 91.61	34.21- 94.74	10	55.56	30.76- 78.47	24.74- 83.51
11	78.57	49.20- 95.34	41.08- 97.43	11	61.11	35.75- 82.70	29.32- 87.16
12	85.71	57.19- 98.22	48.77- 99.24	12	66.67	40.99- 86.66	34.21- 90.49
13	92.86	66.13- 99.82	57.60- 99.96	13	72.22	46.52- 90.31	39.45- 93.46
14	100.00	76.84-100.00	68.49-100.00	14	77.78	52.36- 93.59	45.08- 96.00
		$n = 15$		15	83.33	58.58- 96.42	51.16- 98.03
0	0.00	0.00- 21.80	0.00- 29.76	16	88.89	65.29- 98.62	57.83- 99.41
1	6.67	0.17- 31.95	0.03- 40.16	17	94.44	72.71- 99.86	65.37- 99.97
2	13.33	1.66- 40.46	0.71- 48.63	18	100.00	81.47-100.00	74.50-100.00
3	20.00	4.33- 48.09	2.39- 56.05			$n = 19$	
4	26.67	7.79- 55.10	4.88- 62.73	0	0.00	0.00- 17.65	0.00- 24.34
5	33.33	11.82- 61.62	8.01- 68.82	1	5.26	0.13- 26.03	0.03- 33.11
6	40.00	16.34- 67.71	11.70- 74.39	2	10.53	1.30- 33.14	0.56- 40.37
7	46.67	21.27- 73.41	15.87- 79.49	3	15.79	3.38- 39.58	1.86- 46.82
8	53.33	26.59- 78.73	20.51- 84.13	4	21.05	6.05- 45.57	3.78- 52.71
9	60.00	32.29- 83.66	25.61- 88.30	5	26.32	9.15- 51.20	6.17- 58.18
10	66.67	38.38- 88.18	31.18- 91.99	6	31.58	12.58- 56.55	8.95- 63.29
11	73.33	44.90- 92.21	37.27- 95.12	7	36.84	16.29- 61.64	12.07- 68.09
12	80.00	51.91- 95.67	43.95- 97.61	8	42.11	20.25- 66.50	15.49- 72.60
13	86.67	59.54- 98.34	51.37- 99.29	9	47.37	24.45- 71.14	19.19- 76.84
14	93.33	68.05- 99.83	59.84- 99.97	10	52.63	28.86- 75.55	23.16- 80.81
15	100.00	78.20-100.00	70.24-100.00	11	57.89	33.50- 79.75	27.40- 84.51
		$n = 16$		12	63.16	38.36- 83.71	31.91- 87.93
0	0.00	0.00- 20.59	0.00- 28.19	13	68.42	43.45- 87.42	36.71- 91.05
1	6.25	0.16- 30.23	0.03- 38.14	14	73.68	48.80- 90.85	41.82- 93.83
2	12.50	1.55- 38.35	0.67- 46.28	15	78.95	54.43- 93.95	47.29- 96.22
3	18.75	4.05- 45.65	2.23- 53.44	16	84.21	60.42- 96.62	53.18- 98.14
4	25.00	7.27- 52.38	4.55- 59.91	17	89.47	66.86- 98.70	59.63- 99.44
5	31.25	11.02- 58.66	7.45- 65.85	18	94.74	73.97- 99.87	66.89- 99.97
6	37.50	15.20- 64.57	10.86- 71.32	19	100.00	82.35-100.00	75.66-100.00
7	43.75	19.75- 70.12	14.71- 76.38			$n = 20$	
8	50.00	24.65- 75.35	18.97- 81.03	0	0.00	0.00- 16.84	0.00- 23.27
9	56.25	29.88- 80.25	23.62- 85.29	1	5.00	0.13- 24.87	0.03- 31.71
10	62.50	35.43- 84.80	28.68- 89.14	2	10.00	1.23- 31.70	0.53- 38.71
11	68.75	41.34- 88.98	34.15- 92.55	3	15.00	3.21- 37.89	1.76- 44.95
12	75.00	47.62- 92.73	40.09- 95.45	4	20.00	5.73- 43.66	3.58- 50.66
13	81.25	54.35- 95.95	46.56- 97.77	5	25.00	8.66- 49.10	5.83- 55.98
14	87.50	61.65- 98.45	53.72- 99.33	6	30.00	11.89- 54.28	8.46- 60.96
15	93.75	69.77- 99.84	61.86- 99.97	7	35.00	15.39- 59.22	11.39- 65.66
16	100.00	79.41-100.00	71.81-100.00	8	40.00	19.12- 63.95	14.60- 70.09
		$n = 17$		9	45.00	23.06- 68.47	18.06- 74.28
0	0.00	0.00- 19.51	0.00- 26.78	10	50.00	27.20- 72.80	21.77- 78.23
1	5.88	0.15- 28.69	0.03- 36.30	11	55.00	31.53- 76.94	25.72- 81.94
2	11.76	1.46- 36.44	0.63- 44.13	12	60.00	36.05- 80.88	29.91- 85.40
3	17.65	3.80- 43.43	2.09- 51.04	13	65.00	40.78- 84.61	34.34- 88.61
4	23.53	6.81- 49.90	4.26- 57.32	14	70.00	45.72- 88.11	39.04- 91.54
				15	75.00	50.90- 91.34	44.02- 94.17
				16	80.00	56.34- 94.27	49.34- 96.42
				17	85.00	62.11- 96.79	55.05- 98.24

r	100r/n	95% limits 100𝒫_L 100𝒫_U	99% limits 100𝒫_L 100𝒫_U
		n = 20 (continued)	
18	90·00	68·30– 98·77	61·29– 99·47
19	95·00	75·13– 99·87	68·29– 99·97
20	100·00	83·16–100·00	76·73–100·00

n = 21

r	100r/n	100𝒫_L 100𝒫_U	100𝒫_L 100𝒫_U
0	0·00	0·00– 16·11	0·00– 22·30
1	4·76	0·12– 23·82	0·02– 30·43
2	9·52	1·17– 30·38	0·50– 37·18
3	14·29	3·05– 36·34	1·68– 43·22
4	19·05	5·45– 41·91	3·39– 48·76
5	23·81	8·22– 47·17	5·53– 53·92
6	28·57	11·28– 52·18	8·01– 58·78
7	33·33	14·59– 56·97	10·78– 63·37
8	38·10	18·11– 61·56	13·81– 67·72
9	42·86	21·82– 65·98	17·07– 71·85
10	47·62	25·71– 70·22	20·55– 75·76
11	52·38	29·78– 74·29	24·24– 79·45
12	57·14	34·02– 78·18	28·15– 82·93
13	61·90	38·44– 81·89	32·28– 86·19
14	66·67	43·03– 85·41	36·63– 89·22
15	71·43	47·82– 88·72	41·22– 91·99
16	76·19	52·83– 91·78	46·08– 94·47
17	80·95	58·09– 94·55	51·24– 96·61
18	85·71	63·66– 96·95	56·78– 98·32
19	90·48	69·62– 98·83	62·82– 99·50
20	95·24	76·18– 99·88	69·57– 99·98
21	100·00	83·89–100·00	77·70–100·00

n = 22

r	100r/n	100𝒫_L 100𝒫_U	100𝒫_L 100𝒫_U
0	0·00	0·00– 15·44	0·00– 21·40
1	4·55	0·12– 22·84	0·02– 29·24
2	9·09	1·12– 29·16	0·48– 35·77
3	13·64	2·91– 34·91	1·60– 41·61
4	18·18	5·19– 40·28	3·23– 46·99
5	22·73	7·82– 45·37	5·26– 52·01
6	27·27	10·73– 50·22	7·61– 56·74
7	31·82	13·86– 54·87	10·24– 61·23
8	36·36	17·20– 59·34	13·10– 65·49
9	40·91	20·71– 63·65	16·18– 69·54
10	45·45	24·39– 67·79	19·46– 73·40
11	50·00	28·22– 71·78	22·93– 77·07
12	54·55	32·21– 75·61	26·60– 80·54
13	59·09	36·35– 79·29	30·46– 83·82
14	63·64	40·66– 82·80	34·51– 86·90
15	68·18	45·13– 86·14	38·77– 89·76
16	72·73	49·78– 89·27	43·26– 92·39
17	77·27	54·63– 92·18	47·99– 94·74
18	81·82	59·72– 94·81	53·01– 96·77
19	86·36	65·09– 97·09	58·39– 98·40
20	90·91	70·84– 98·88	64·23– 99·52
21	95·45	77·16– 99·88	70·76– 99·88
22	100·00	84·56–100·00	78·60–100·00

n = 23

r	100r/n	100𝒫_L 100𝒫_U	100𝒫_L 100𝒫_U
0	0·00	0·00– 14·82	0·00– 20·58
1	4·35	0·11– 21·95	0·02– 28·14
2	8·70	1·07– 28·04	0·46– 34·46
3	13·04	2·78– 33·59	1·53– 40·12
4	17·39	4·95– 38·78	3·08– 45·34
5	21·74	7·46– 43·70	5·02– 50·22
6	26·09	10·23– 48·41	7·25– 54·83
7	30·43	13·21– 52·92	9·74– 59·21
8	34·78	16·38– 57·27	12·46– 63·38
9	39·13	19·71– 61·46	15·37– 67·36
10	43·48	23·19– 65·51	18·48– 71·16
11	47·83	26·82– 69·41	21·76– 74·79
12	52·17	30·59– 73·18	25·21– 78·24
13	56·52	34·49– 76·81	28·84– 81·52
14	60·87	38·54– 80·29	32·64– 84·63
15	65·22	42·73– 83·62	36·62– 87·54
16	69·57	47·08– 86·79	40·79– 90·26
17	73·91	51·59– 89·77	45·17– 92·75
18	78·26	56·30– 92·54	49·78– 94·98
19	82·61	61·22– 95·05	54·66– 96·92
20	86·96	66·41– 97·22	59·88– 98·47
21	91·30	71·96– 98·93	65·54– 99·54
22	95·65	78·05– 99·89	71·86– 99·98
23	100·00	85·18–100·00	79·42–100·00

n = 24

r	100r/n	95% limits 100𝒫_L 100𝒫_U	99% limits 100𝒫_L 100𝒫_U
0	0·00	0·00– 14·25	0·00– 19·81
1	4·17	0·11– 21·12	0·02– 27·13
2	8·33	1·03– 27·00	0·44– 33·24
3	12·50	2·66– 32·36	1·46– 38·73
4	16·67	4·74– 37·38	2·95– 43·79
5	20·83	7·13– 42·15	4·79– 48·55
6	25·00	9·77– 46·71	6·92– 53·04
7	29·17	12·62– 51·09	9·30– 57·32
8	33·33	15·63– 55·32	11·88– 61·40
9	37·50	18·80– 59·41	14·65– 65·30
10	41·67	22·11– 63·36	17·59– 69·04
11	45·83	25·55– 67·18	20·70– 72·62
12	50·00	29·12– 70·88	23·96– 76·04
13	54·17	32·82– 74·45	27·38– 79·30
14	58·33	36·64– 77·89	30·96– 82·41
15	62·50	40·59– 81·20	34·70– 85·35
16	66·67	44·68– 84·37	38·60– 88·12
17	70·83	48·91– 87·38	42·68– 90·70
18	75·00	53·29– 90·23	46·96– 93·08
19	79·17	57·85– 92·87	51·45– 95·21
20	83·33	62·62– 95·26	56·21– 97·05
21	87·50	67·64– 97·34	61·27– 98·54
22	91·67	73·00– 98·97	66·76– 99·56
23	95·83	78·88– 99·89	72·78– 99·98
24	100·00	85·75–100·00	80·19–100·00

n = 25

r	100r/n	100𝒫_L 100𝒫_U	100𝒫_L 100𝒫_U
0	0·00	0·00– 13·72	0·00– 19·10
1	4·00	0·10– 20·35	0·02– 26·18
2	8·00	0·98– 26·03	0·42– 32·10
3	12·00	2·55– 31·22	1·40– 37·43
4	16·00	4·54– 36·08	2·82– 42·35
5	20·00	6·83– 40·70	4·59– 46·98
6	24·00	9·36– 45·13	6·63– 51·36
7	28·00	12·07– 49·39	8·89– 55·53
8	32·00	14·95– 53·50	11·35– 59·52
9	36·00	17·97– 57·48	13·99– 63·35
10	40·00	21·13– 61·33	16·79– 67·02
11	44·00	24·40– 65·07	19·74– 70·54
12	48·00	27·80– 68·69	22·83– 73·93
13	52·00	31·31– 72·20	26·07– 77·17
14	56·00	34·93– 75·60	29·46– 80·26
15	60·00	38·67– 78·87	32·98– 83·21
16	64·00	42·52– 82·03	36·65– 86·01
17	68·00	46·50– 85·05	40·48– 88·65
18	72·00	50·61– 87·93	44·47– 91·11
19	76·00	54·87– 90·64	48·64– 93·37
20	80·00	59·30– 93·17	53·02– 95·41
21	84·00	63·92– 95·46	57·65– 97·18
22	88·00	68·78– 97·45	62·57– 98·60
23	92·00	73·97– 99·02	67·90– 99·58
24	96·00	79·65– 99·90	73·82– 99·98
25	100·00	86·28–100·00	80·90–100·00

n = 26

r	100r/n	100𝒫_L 100𝒫_U	100𝒫_L 100𝒫_U
0	0·00	0·00– 13·23	0·00– 18·44
1	3·85	0·10– 19·64	0·02– 25·29
2	7·69	0·95– 25·13	0·41– 31·04
3	11·54	2·45– 30·15	1·34– 36·21
4	15·38	4·36– 34·87	2·71– 41·00
5	19·23	6·55– 39·35	4·40– 45·50
6	23·08	8·97– 43·65	6·35– 49·77
7	26·92	11·57– 47·79	8·52– 53·85
8	30·77	14·33– 51·79	10·87– 57·75
9	34·62	17·21– 55·67	13·38– 61·50
10	38·46	20·23– 59·43	16·05– 65·10
11	42·31	23·35– 63·08	18·86– 68·57
12	46·15	26·59– 66·63	21·81– 71·91
13	50·00	29·93– 70·07	24·89– 75·11
14	53·85	33·37– 73·41	28·09– 78·19
15	57·69	36·92– 76·65	31·43– 81·14
16	61·54	40·57– 79·77	34·90– 83·95
17	65·38	44·33– 82·79	38·50– 86·62
18	69·23	48·21– 85·67	42·25– 89·13
19	73·08	52·21– 88·43	46·15– 91·48
20	76·92	56·35– 91·03	50·23– 93·65
21	80·77	60·65– 93·45	54·50– 95·60
22	84·62	65·13– 95·64	59·00– 97·29

r	100r/n	95% limits $100\mathscr{P}_L$ $100\mathscr{P}_U$	99% limits $100\mathscr{P}_L$ $100\mathscr{P}_U$
		n = 26 (continued)	
23	88·46	69·85– 97·55	63·79– 98·66
24	92·31	74·87– 99·05	68·96– 99·59
25	96·15	80·36– 99·90	74·71– 99·98
26	100·00	86·77–100·00	81·56–100·00
		n = 27	
0	0·00	0·00– 12·77	0·00– 17·82
1	3·70	0·09– 18·97	0·02– 24·46
2	7·41	0·91– 24·29	0·39– 30·04
3	11·11	2·35– 29·16	1·29– 35·07
4	14·81	4·19– 33·73	2·60– 39·73
5	18·52	6·30– 38·08	4·23– 44·11
6	22·22	8·62– 42·26	6·10– 48·28
7	25·93	11·11– 46·28	8·17– 52·26
8	29·63	13·75– 50·18	10·42– 56·08
9	33·33	16·52– 53·96	12·83– 59·75
10	37·04	19·40– 57·63	15·38– 63·28
11	40·74	22·39– 61·20	18·07– 66·69
12	44·44	25·48– 64·67	20·88– 69·98
13	48·15	28·67– 68·05	23·81– 73·14
14	51·85	31·95– 71·33	26·86– 76·19
15	55·56	35·33– 74·52	30·02– 79·12
16	59·26	38·80– 77·61	33·31– 81·93
17	62·96	42·37– 80·60	36·72– 84·62
18	66·67	46·04– 83·48	40·25– 87·17
19	70·37	49·82– 86·25	43·92– 89·58
20	74·07	53·72– 88·89	47·74– 91·83
21	77·78	57·74– 91·38	51·72– 93·90
22	81·48	61·92– 93·70	55·89– 95·77
23	85·19	66·27– 95·81	60·27– 97·40
24	88·89	70·84– 97·65	64·93– 98·71
25	92·59	75·71– 99·09	69·96– 99·61
26	96·30	81·03– 99·91	75·54– 99·98
27	100·00	87·23–100·00	82·18–100·00
		n = 28	
0	0·00	0·00– 12·34	0·00– 17·24
1	3·57	0·09– 18·35	0·02– 23·69
2	7·14	0·88– 23·50	0·38– 29·11
3	10·71	2·27– 28·23	1·25– 33·99
4	14·29	4·03– 32·67	2·51– 38·53
5	17·86	6·06– 36·89	4·07– 42·80
6	21·43	8·30– 40·95	5·86– 46·87
7	25·00	10·69– 44·87	7·86– 50·76
8	28·57	13·22– 48·67	10·02– 54·49
9	32·14	15·88– 52·35	12·32– 58·08
10	35·71	18·64– 55·93	14·77– 61·55
11	39·29	21·50– 59·42	17·33– 64·90
12	42·86	24·46– 62·82	20·02– 68·14
13	46·43	27·51– 66·13	22·82– 71·26
14	50·00	30·65– 69·35	25·72– 74·28
15	53·57	33·87– 72·49	28·74– 77·18
16	57·14	37·18– 75·54	31·86– 79·98
17	60·71	40·58– 78·50	35·10– 82·67
18	64·29	44·07– 81·36	38·45– 85·23
19	67·86	47·65– 84·12	41·92– 87·68
20	71·43	51·33– 86·78	45·51– 89·98
21	75·00	55·13– 89·31	49·24– 92·14
22	78·57	59·05– 91·70	53·13– 94·14
23	82·14	63·11– 93·94	57·20– 95·93
24	85·71	67·33– 95·97	61·47– 97·49
25	89·29	71·77– 97·73	66·01– 98·75
26	92·86	76·50– 99·12	70·89– 99·62
27	96·43	81·65– 99·91	76·31– 99·98
28	100·00	87·66–100·00	82·76–100·00
		n = 29	
0	0·00	0·00– 11·94	0·00– 16·70
1	3·45	0·09– 17·76	0·02– 22·96
2	6·90	0·85– 22·77	0·36– 28·23
3	10·34	2·19– 27·35	1·20– 32·98
4	13·79	3·98– 31·66	2·42– 37·40
5	17·24	5·85– 35·77	3·92– 41·57
6	20·69	7·99– 39·72	5·65– 45·54
7	24·14	10·30– 43·54	7·56– 49·33
8	27·59	12·73– 47·24	9·64– 52·99
9	31·03	15·28– 50·83	11·85– 56·51
10	34·48	17·94– 54·33	14·20– 59·91

r	100r/n	95% limits $100\mathscr{P}_U$ $100\mathscr{P}_U$	99% limits $100\mathscr{P}_L$ $100\mathscr{P}_U$
		n = 29 (continued)	
11	37·93	20·69– 57·74	16·66– 63·20
12	41·38	23·52– 61·06	19·23– 66·38
13	44·83	26·45– 64·31	21·91– 69·46
14	48·28	29·45– 67·47	24·69– 72·43
15	51·72	32·53– 70·55	27·57– 75·31
16	55·17	35·69– 73·55	30·54– 78·09
17	58·62	38·94– 76·48	33·62– 80·77
18	62·07	42·26– 79·31	36·80– 83·34
19	65·52	45·67– 82·06	40·09– 85·80
20	68·97	49·17– 84·72	43·49– 88·15
21	72·41	52·76– 87·27	47·01– 90·36
22	75·86	56·46– 89·70	50·67– 92·44
23	79·31	60·28– 92·01	54·46– 94·35
24	82·76	64·23– 94·15	58·43– 96·08
25	86·21	68·34– 96·11	62·60– 97·58
26	89·66	72·65– 97·81	67·02– 98·90
27	93·10	77·23– 99·15	71·77– 99·64
28	96·55	82·24– 99·91	77·04– 99·98
29	100·00	88·06–100·00	83·30–100·00
		n = 30	
0	0·00	0·00– 11·57	0·00– 16·19
1	3·33	0·08– 17·22	0·02– 22·27
2	6·67	0·82– 22·07	0·35– 27·40
3	10·00	2·11– 26·53	1·16– 32·03
4	13·33	3·76– 30·72	2·33– 36·34
5	16·67	5·64– 34·72	3·78– 40·40
6	20·00	7·71– 38·57	5·45– 44·28
7	23·33	9·93– 42·28	7·29– 47·99
8	26·67	12·28– 45·89	9·29– 51·56
9	30·00	14·73– 49·40	11·42– 55·01
10	33·33	17·29– 52·81	13·67– 58·34
11	36·67	19·93– 56·14	16·04– 61·57
12	40·00	22·66– 59·40	18·50– 64·70
13	43·33	25·46– 62·57	21·07– 67·73
14	46·67	28·34– 65·67	23·73– 70·67
15	50·00	31·30– 68·70	26·48– 73·52
16	53·33	34·33– 71·66	29·33– 76·27
17	56·67	37·43– 74·54	32·27– 78·93
18	60·00	40·60– 77·34	35·30– 81·50
19	63·33	43·86– 80·07	38·43– 83·96
20	66·67	47·19– 82·71	41·66– 86·33
21	70·00	50·60– 85·27	44·99– 88·58
22	73·33	54·11– 87·72	48·44– 90·71
23	76·67	57·72– 90·07	52·01– 92·71
24	80·00	61·43– 92·29	55·72– 94·55
25	83·33	65·28– 94·36	59·60– 96·22
26	86·67	69·28– 96·24	63·66– 97·67
27	90·00	73·47– 97·89	67·97– 98·84
28	93·33	77·93– 99·18	72·60– 99·65
29	96·67	82·78– 99·92	77·73– 99·98
30	100·00	88·43–100·00	83·81–100·00
		n = 1000 (extract)	
0	*0·00*	*0·00– 0·37*	*0·00– 0·53*
1	*0·10*	*0·00– 0·56*	*0·00– 0·74*
2	*0·20*	*0·02– 0·72*	*0·01– 0·92*
3	*0·30*	*0·06– 0·87*	*0·03– 1·09*
4	*0·40*	*0·11– 1·02*	*0·07– 1·25*
5	*0·50*	*0·16– 1·17*	*0·09– 1·39*
6	*0·60*	*0·22– 1·31*	*0·14– 1·54*
7	*0·70*	*0·28– 1·44*	*0·18– 1·69*
8	*0·80*	*0·34– 1·58*	*0·24– 1·83*
9	*0·90*	*0·41– 1·71*	*0·29– 1·97*
10	*1·00*	*0·48– 1·84*	*0·35– 2·11*
11	*1·10*	*0·55– 1·97*	*0·41– 2·25*
12	*1·20*	*0·62– 2·09*	*0·48– 2·38*
13	*1·30*	*0·69– 2·22*	*0·54– 2·51*
14	*1·40*	*0·77– 2·34*	*0·60– 2·65*
15	*1·50*	*0·84– 2·47*	*0·67– 2·78*
16	*1·60*	*0·92– 2·59*	*0·74– 2·91*
17	*1·70*	*0·99– 2·71*	*0·81– 3·04*
18	*1·80*	*1·07– 2·84*	*0·88– 3·17*
19	*1·90*	*1·15– 2·96*	*0·95– 3·30*
20	*2·00*	*1·22– 3·08*	*1·02– 3·42*
.	.	.	.
.	.	.	.
.	.	.	.

TABLE A3

The Wilcoxon test for two independent samples

See § 9.3. The sample sizes are n_1 and n_2. If the sample sizes are not equal n_1 is taken as the smaller. If the rank sum for sample 1 (that with n_1 observations) is equal to or less than the smaller tabulated value, or equal to or greater than the larger tabulated value, then P (two tail) is equal to or less than the figure at the head of the column. If the null hypothesis were true P would be the probability of observing a rank sum equal to or greater than the larger figure, or equal to or less than the smaller. If one or both samples contain more than 20 observations, use the method described at the end of § 9.3. M. I. Sutcliffe's table reproduced from Mainland (1963) by permission of the author and publisher.

n_1	n_2	P (approx.) 0·10	0·05	0·10
2	4	—	—	—
	5	3; 13	—	—
	6	3; 15	—	—
	7	3; 17	—	—
	8	4; 18	3; 19	—
	9	4; 20	3; 21	—
	10	4; 22	3; 23	—
	11	4; 24	3; 25	—
	12	5; 25	4; 26	—
	13	5; 27	4; 28	—
	14	6; 28	4; 30	—
	15	6; 30	4; 32	—
	16	6; 32	4; 34	—
	17	6; 34	5; 35	—
	18	7; 35	5; 37	—
	19	7; 37	5; 39	3; 41
	20	7; 39	5; 41	3; 43
3	3	6; 15	—	—
	4	6; 18	—	—
	5	7; 20	6; 21	—
	6	8; 22	7; 23	—
	7	8; 25	7; 26	—
	8	9; 27	8; 28	—
	9	10; 29	8; 31	6; 33
	10	10; 32	9; 33	6; 36
	11	11; 34	9; 36	6; 39
	12	11; 37	10; 38	7; 41
	13	12; 39	10; 41	7; 44
	14	13; 41	11; 43	7; 47
	15	13; 44	11; 46	8; 49
	16	14; 46	12; 48	8; 52
	17	15; 48	12; 51	8; 55

n_1	n_2	P (approx.) 0·10	0·05	0·01
3	18	15; 51	13; 53	8; 58
	19	16; 53	13; 56	9; 60
	20	17; 55	14; 58	9; 63
4	4	11; 25	10; 26	—
	5	12; 28	11; 29	—
	6	13; 31	12; 32	10; 34
	7	14; 34	13; 35	10; 38
	8	15; 37	14; 38	11; 41
	9	16; 40	14; 42	11; 45
	10	17; 43	15; 45	12; 48
	11	18; 46	16; 48	12; 52
	12	19; 49	17; 51	13; 55
	13	20; 52	18; 54	13; 59
	14	21; 55	19; 57	14; 62
	15	22; 58	20; 60	15; 65
	16	24; 60	21; 63	15; 69
	17	25; 63	21; 67	16; 72
	18	26; 66	22; 70	16; 76
	19	27; 69	23; 73	17; 79
	20	28; 72	24; 76	18; 82
5	5	19; 36	17; 38	15; 40
	6	20; 40	18; 42	16; 44
	7	21; 44	20; 45	16; 49
	8	23; 47	21; 49	17; 53
	9	24; 51	22; 53	18; 57
	10	26; 54	23; 57	19; 61
	11	27; 58	24; 61	20; 65
	12	28; 62	26; 64	21; 69
	13	30; 65	27; 68	22; 73
	14	31; 69	28; 72	22; 78

n_1	n_2	0·10	0·05	0·01	n_1	n_2	0·10	0·05	0·01
		P (approx.)					*P* (approx.)		
	15	33; 72	29; 76	23; 82	8	11	59; 101	55; 105	49; 111
	16	34; 76	30; 80	24; 86		12	62; 106	58; 110	51; 117
	17	35; 80	32; 83	25; 90					
	18	37; 83	33; 87	26; 94		13	64; 112	60; 116	53; 123
	19	38; 87	34; 91	27; 98		14	67; 117	62; 122	54; 130
						15	69; 123	65; 127	56; 136
	20	40; 90	35; 95	28; 102		16	72; 128	67; 133	58; 142
6	6	28; 50	26; 52	23; 55		17	75; 133	70; 138	60; 148
	7	29; 55	27; 57	24; 60		18	77; 139	72; 144	62; 154
	8	31; 59	29; 61	25; 65		19	80; 144	74; 150	64; 160
	9	33; 63	31; 65	26; 70		20	83; 149	77; 155	66; 166
	10	35; 67	32; 70	27; 75					
	11	37; 71	34; 74	28; 80	9	9	66; 105	62; 109	56; 115
	12	38; 76	35; 79	30; 84		10	69; 111	65; 115	58; 122
	13	40; 80	37; 83	31; 89		11	72; 117	68; 121	61; 128
	14	42; 84	38; 88	32; 94		12	75; 123	71; 127	63; 135
	15	44; 88	40; 92	33; 99		13	78; 129	73; 134	65; 142
	16	46; 92	42; 96	34; 104		14	81; 135	76; 140	67; 149
	17	47; 97	43; 101	36; 108		15	84; 141	79; 146	69; 156
	18	49; 101	45; 105	37; 113		16	87; 147	82; 152	72; 162
	19	51; 105	46; 110	38; 118		17	90; 153	84; 159	74; 169
	20	53; 109	48; 114	39; 123		18	93; 159	87; 165	76; 176
7	7	39; 66	36; 69	32; 73		19	96; 165	90; 171	78; 183
	8	41; 71	38; 74	34; 78		20	99; 171	93; 177	81; 189
	9	43; 76	40; 79	35; 84	10	10	82; 128	78; 132	71; 139
	10	45; 81	42; 84	37; 89		11	86; 134	81; 139	73; 147
	11	47; 86	44; 89	38; 95		12	89; 141	84; 146	76; 154
	12	49; 91	46; 94	40; 100		13	92; 148	88; 152	79; 161
	13	52; 95	48; 99	41; 106		14	96; 154	91; 159	81; 169
	14	54; 100	50; 104	43; 111					
	15	56; 105	52; 109	44; 117		15	99; 161	94; 166	84; 176
	16	58; 110	54; 114	46; 122		16	103; 167	97; 173	86; 184
						17	106; 174	100; 180	89; 191
	17	61; 114	56; 119	47; 128		18	110; 180	103; 187	92; 198
	18	63; 119	58; 124	49; 133		19	113; 187	107; 193	94; 206
	19	65; 124	60; 129	50; 139					
	20	67; 129	62; 134	52; 114		20	117; 193	110; 200	97; 213
8	8	51; 85	49; 87	43; 93	11	11	100; 153	96; 157	87; 166
	9	54; 90	51; 93	45; 99		12	104; 160	99; 165	90; 174
	10	56; 96	53; 99	47; 105		13	108; 167	103; 172	93; 182

n_1	n_2	P (approx.)			n_1	n_2	P (approx.)		
		0·01	0·05	0·01			0·10	0·05	0·01
11	14	112; 174	106; 180	96; 190	14	19	192; 284	183; 293	168; 308
	15	116; 181	110; 187	99; 198		20	197; 293	188; 302	172; 318
	16	120; 188	113; 195	102; 206	15	15	192; 273	184; 281	171; 294
	17	123; 196	117; 202	105; 214		16	197; 283	190; 290	175; 305
	18	127; 203	121; 209	108; 222		17	203; 292	195; 300	180; 315
	19	131; 210	124; 217	111; 230		18	208; 302	200; 310	184; 326
	20	135; 217	128; 224	114; 238		19	214; 311	205; 320	189; 336
12	12	120; 180	115; 185	105; 195		20	220; 320	210; 330	193; 347
	13	125; 187	119; 193	109; 203					
	14	129; 195	123; 201	112; 212	16	16	219; 309	211; 317	196; 332
	15	133; 203	127; 209	115; 221		17	225; 319	217; 327	201; 343
	16	138; 210	131; 217	119; 229		18	231; 329	222; 338	206; 354
						19	237; 339	228; 348	210; 366
	17	142; 218	135; 225	122; 238		20	243; 349	234; 358	215; 377
	18	146; 226	139; 233	125; 247					
	19	150; 234	143; 241	129; 255	17	17	249; 346	240; 355	223; 372
	20	155; 241	147; 249	132; 264		18	255; 357	246; 366	228; 384
						19	262; 367	252; 377	234; 395
13	13	142; 209	136; 215	125; 226		20	268; 378	258; 388	239; 407
	14	147; 217	141; 223	129; 235					
	15	152; 225	145; 232	133; 244	18	18	280; 386	270; 396	252; 414
	16	156; 234	150; 240	136; 254		19	287; 397	277; 407	258; 426
	17	161; 242	154; 249	140; 263		20	294; 408	283; 419	263; 439
	18	166; 250	158; 258	144; 272	19	19	313; 428	303; 438	283; 458
	19	171; 258	163; 266	147; 282		20	320; 440	309; 451	289; 471
	20	175; 267	167; 275	151; 291					
					20	20	348; 472	337; 483	315; 505
14	14	166; 240	160; 246	147; 259					
	15	171; 249	164; 256	151; 269					
	16	176; 258	169; 265	155; 279					
	17	182; 266	172; 276	159; 289					
	18	187; 275	179; 283	163; 299					

TABLE A4

The Wilcoxon signed ranks test for two related samples

See § 10.4. The number of pairs of observations is n. The table gives the values of T (defined as the sum of positive ranks, or the sum of negative ranks, whichever is the smaller) for various values of P (the probability of a value of T equal to or less than the tabulated value if the null hypothesis is true). If there are more than 25 pairs of observations, use the method described at the end of § 10.4. Adapted from Wilcoxon and Wilcox (1964), with permission.

n	P value (two tail)		
	0·05	0·02	0·01
4	†0($P = 0\cdot125$) —		—
5	†0($P = 0\cdot0625$) —		—
6	0	—	—
7	2	0	—
8	4	2	0
9	6	3	2
10	8	5	3
11	11	7	5
12	14	10	7
13	17	13	10
14	21	16	13
15	25	20	16
16	30	24	20
17	35	28	23
18	40	33	28
19	46	38	32
20	52	43	38
21	59	49	43
22	66	56	49
23	73	62	55
24	81	69	61
25	89	77	68

† It is not possible to reach a value of P as small as 0·05 with such small samples (see §§ 6.2 and 10.4). The values of P for $T = 0$ are given.

TABLE A5

The Kruskal–Wallis one way analysis of variance on ranks (independent samples)

See § 11.5. For each value of H, the table gives the exact value of P (the probability of observing a value of H equal to or greater than the tabulated value if the null hypothesis is true, found from the randomization distribution of rank sums). This table deals only with $k = 3$ groups, the number of observations (n_1, n_2, and n_3) in each being up to 5. For larger samples or more groups use the method described at the end of § 11.5. From Kruskal and Wallis (1952, *J. Amer. Statist. Ass.* **47**, 614; **48**, 910) with permission of the author and publisher.

n_1	n_2	n_3	H	P	n_1	n_2	n_3	H	P
2	1	1	2·7000	0·500	4	2	1	4·8214	0·057
								4·5000	0·076
2	2	1	3·6000	0·200				4·0179	0·114
2	2	2	4·5714	0·067					
			3·7143	0·200	4	2	2	6·0000	0·014
								5·3333	0·033
								5·1250	0·052
3	1	1	3·2000	0·300				4·4583	0·100
								4·1667	0·105
3	2	1	4·2857	0·100					
			3·8571	0·133	4	3	1	5·8333	0·021
								5·2083	0·050
3	2	2	5·3572	0·029				5·0000	0·057
			4·7143	0·048				4·0556	0·093
			4·5000	0·067				3·8889	0·129
			4·4643	0·105					
					4	3	2	6·4444	0·008
3	3	1	5·1429	0·043				6·3000	0·011
			4·5714	0·100				5·4444	0·046
			4·0000	0·129				5·4000	0·051
								4·5111	0·098
3	3	2	6·2500	0·011				4·4444	0·102
			5·3611	0·032					
			5·1389	0·061					
			4·5556	0·100	4	3	3	6·7455	0·010
			4·2500	0·121				6·7091	0·013
								5·7909	0·046
3	3	3	7·2000	0·004				5·7273	0·050
			6·4889	0·011				4·7091	0·092
			5·6889	0·029				4·7000	0·101
			5·6000	0·050					
			5·0667	0·086	4	4	1	6·6667	0·010
			4·6222	0·100				6·1667	0·022
4	1	1	3·5714	0·200				4·9667	0·048
								4·8667	0·054

Sample sizes			H	P	Sample sizes			H	P
n_1	n_2	n_3			n_1	n_2	n_3		
			4·1667	0·082	5	3	3	7·0788	0·009
			4·0667	0·102				6·9818	0·011
4	4	2	7·0364	0·006				5·6485	0·049
			6·8727	0·011				5·5152	0·051
			5·4545	0·046				4·5333	0·097
			5.2364	0·052				4·4121	0·109
			4·5545	0·098					
			4·4455	0·103	5	4	1	6·9545	0·008
								6·8400	0·011
4	4	3	7·1439	0·010				4·9855	0·044
			7·1364	0·011				4·8600	0·056
			5·5985	0·049				3·9873	0·098
			5·5758	0·051				3·9600	0·102
			4·5455	0·099					
			4·4773	0.102	5	4	2	7·2045	0·009
								7·1182	0·010
4	4	4	7·6538	0·008				5·2727	0·049
			7·5385	0·011				5·2682	0·050
			5·6923	0·049				4·5409	0·098
			5·6538	0·054				4·5182	0·101
			4·6539	0·097					
			4·5001	0·104	5	4	3	7·4449	0·010
								7·3949	0·011
5	1	1	3·8571	0·143				5·6564	0·049
								5·6308	0·050
5	2	1	5·2500	0·036				4·5487	0·099
			5·0000	0·048				4·5231	0·103
			4·4500	0·071					
			4·2000	0·095	5	4	4	7·7604	0·009
			4·0500	0·119				7·7440	0·011
								5·6571	0·049
5	2	2	6·5333	0·008				5·6176	0·050
			6·1333	0·013				4·6187	0·100
			5·1600	0·034				4·5527	0·102
			5·0400	0·056					
			4·3733	0·090	5	5	1	7·3091	0·009
			4·2933	0·122				6·8364	0·011
								5·1273	0·046
5	3	1	6·4000	0·012				4·9091	0·053
			4·9600	0·048				4·1091	0·086
			4·8711	0·052				4·0364	0·105
			4·0178	0·095					
			3·8400	0·123	5	5	2	7·3385	0·010
								7·2692	0·010
5	3	2	6·9091	0·009				5·3385	0·047
			6·8218	0·010				5·2462	0·051
			5·2509	0·049				4·6231	0·097
			5·1055	0·052				4·5077	0·100
			4·6509	0·091					
			4·4945	0·101					

Sample sizes			H	P	Sample sizes			H	P
n_1	n_2	n_3			n_1	n_2	n_3		
5	5	3	7·5780	0·010				5·6429	0·050
			7·5429	0·010				4·5229	0·099
			5·7055	0·046				4·5200	0·101
			6·6264	0·051					
			4·5451	0·100	5	5	5	8·0000	0·009
			4·5363	0·102				7·9800	0·010
								5·7800	0·049
5	5	4	7·8229	0·010				5·6600	0·051
			7·7914	0·010				4·5600	0·100
			5·6657	0·049				4·5000	0·102

TABLE A6

The Friedman two way analysis of variance on ranks for randomized block experiments

See § 11.7. For each value of S the table gives the exact value of P (the probability of observing a value of S equal to or greater than the tabulated value if the null hypothesis is true, found from the randomization distribution of rank sums). Approximate P values are given at the head of the column. If the number of treatments, k, or the number of observations per treatment = number of blocks, n, is too large for this table, use the method described at the end of § 11.7. From Friedman, M. (1937, *J. Amer. Statist. Ass.* **32**, 688), by permission of the author and publisher.

No. of blocks n	k = 3 P≃0·05 S	P	P≃0·01 S	P	P≃0·001 S	P	k = 4 P≃0·05 S	P	P≃0·01 S	P	P≃0·001 S	P	k = 5 P≃0·05 S	P	P≃0·01 S	P	P≃0·001 S	P
2	—	—	—	—	—	—	20	0·042	—	—	—	—	64	0·045	76	0·0078	86	0·0009
3	18	0·028	—	—	—	—	37	0·033	—	—	—	—						
4	26	0·042	32	0·0046	—	—	52	0·036	64	0·0069	74	0·0009						
5	32	0·039	42	0·0085	50	0·0008	65	0·044	83	0·0087	105	0·0006						
6	42	0·029	54	0·0081	72	0·0001	76	0·043	100	0·0100	128	0·0009						
7	50	0·027	62	0·0084	86	0·0003												
8	50	0·047	72	0·0099	98	0·0009												
9	56	0·048	78	0·0100	114	0·0007												
10	62	0·046	96	0·0075	126	0·0008												

Table of the critical range (difference between rank sums for any two treatments) for comparing all pairs in the Kruskal–Wallis nonparametric one way analysis of variance (see §§ 11.5 and 11.9)

Values for which an exact P is given are abridged from the tables of McDonald and Thompson (1967), the remaining values are abridged from Wilcoxon and Wilcox (1964). Reproduction by permission of the authors and publishers. †Not attainable. Number of treatments (samples) = k. Number of observation (replicates) per treatment = n.

		P (approximate)						P (approximate)			
		0·01		0·05				0·01		0·05	
		crit.		crit.				crit.		crit.	
k	n	range	P	range	P	k	n	range	P	range	P
3	2	†		8	0·067	5	2	16	0·016	15	0·048
	3	17	0·011	15	0·064		3	32	0·007	28	0·060
	4	27	0·011	24	0·045		4	50	0·010	44	0·056
	5	39	0·009	33	0·048		5	75·8		63·5	
	6	51	0·011	43	0·049		6	99·3		83·2	
	7	67·6		54·4			7	124·8		104·6	
	8	82·4		66·3			8	152·2		127·6	
	9	98·1		78·9			9	181·4		152·0	
	10	114·7		92·3			10	212·2		177·8	
	11	132·1		106·3			11	244·6		205·0	
	12	150·4		120·9			12	278·5		233·4	
	13	169·4		136·2			13	313·8		263·0	
	14	189·1		152·1			14	350·5		293·8	
	15	209·6		168·6			15	388·5		325·7	
	16	230·7		185·6			16	427·9		358·6	
	17	252·5		203·1			17	468·4		392·6	
	18	275·0		221·2			18	510·2		427·6	
	19	298·1		239·8			19	553·1		463·6	
	20	321·8		258·8			20	597·2		500·5	
4	2	†		12	0·029	6	2	20	0·010	19	0·030
	3	24	0·012	22	0·043		3	39	0·009	35	0·055
	4	38	0·012	34	0·049		4	67·3		57·0	
	5	58·2		48·1			5	93·6		79·3	
	6	76·3		62·9			6	122·8		104·0	
	7	95·8		79·1			7	154·4		130·8	
	8	116·8		96·4			8	188·4		159·6	
	9	139·2		114·8			9	224·5		190·2	
	10	162·8		134·3			10	262·7		222·6	
	11	187·6		154·8			11	302·9		256·6	
	12	213·5		176·2			12	344·9		292·2	
	13	240·6		198·5			13	388·7		329·3	
	14	268·7		221·7			14	434·2		367·8	
	15	297·8		245·7			15	481·3		407·8	
	16	327·9		270·6			16	530·1		449·1	
	17	359·0		296·2			17	580·3		491·7	
	18	391·0		322·6			18	632·1		535·5	
	19	423·8		349·7			19	685·4		580·6	
	20	457·6		377·6			20	740·0		626·9	

TABLE A8

Table of the critical range (difference between rank sums for any two treatments) for comparing all pairs in the Friedman nonparametric two way analysis of variance (see §§ 11.7 and 11.9)

Values for which an exact P is given are abridged from McDonald and Thompson (1967), the remaining values are abridged from Wilcoxon and Wilcox (1964). Reproduction by permission of the authors and publishers. †Not attainable. Number of treatments $= k$. Number of replicates ($=$ number of blocks) $= n$.

| | | P (approximate) | | | | | | P (approximate) | | | |
| | | 0·01 | | 0·05 | | | | 0·01 | | 0·05 | |
k	n	crit. range	P	crit. range	P	k	n	crit. range	P	crit. range	P
3	3	†		6	0·028	5	2	†		8	0·050
	4	8	0·005	7	0·042		3	12	0·002	10	0·067
	5	9	0·008	8	0·039		4	14	0·006	12	0·054
	6	10	0·009	9	0·029		5	16	0·006	14	0·040
	7	11	0·008	9	0·051		6	17	0·013	15	0·049
	8	12	0·007	10	0·039		7	19	0·009	16	0·052
	9	12	0·013	10	0·048		8	20	0·012	18	0·036
							9	22	0·008	19	0·037
	10	13	0·010	11	0·037		10	23	0·009	20	0·038
	11	14	0·008	11	0·049		11	24	0·010	21	0·038
	12	14	0·012	12	0·038		12	25	0·011	22	0·038
	13	15	0·009	12	0·049		13	26	0·011	23	0·035
	14	16	0·007	13	0·038		14	27	0·011	24	0·034
	15	16	0·010	13	0·047		15	28	0·010	24	0·045
	16	16·5		13·3			16	29·1		24·4	
	17	17·0		13·7			17	30·0		25·2	
	18	17·5		14·1			18	30·9		25·9	
	19	18·0		14·4			19	31·7		26·6	
	20	18·4		14·8			20	32·5		27·3	
4	2	†		6	0·083	6	2	†		10	0·033
	3	9	0·007	8	0·049		3	14	0·008	13	0·030
	4	11	0·005	10	0·026		4	17	0·006	15	0·047
	5	12	0·013	11	0·037		5	19	0·010	17	0·047
	6	14	0·006	12	0·037		6	21	0·010	19	0·040
	7	15	0·008	13	0·037		7	23	0·010	20	0·049
	8	16	0·009	14	0·034		8	25	0·008	22	0·039
	9	17	0·010	15	0·032		9	26	0·012	23	0·043
	10	18	0·010	15	0·046		10	28	0·009	24	0·047
	11	19	0·009	16	0·041		11	29	0·012	26	0·036
	12	20	0·008	17	0·038		12	31	0·009	27	0·039
	13	21	0·008	18	0·032		13	32	0·010	28	0·039
	14	21	0·011	18	0·042		14	33	0·011	29	0·040
	15	22	0·010	19	0·037		15	34	0·012	30	0·040
	16	22·7		19			16	35·6		30·2	
	17	23·4		19·3			17	36·7		31·1	
	18	24·1		19·9			18	37·8		32·0	
	19	24·8		20·4			19	38·8		32·9	
	20	25·4		21·0			20	39·8		33·7	

Rankits (expected normal order statistics)

The use of Rankits to test for a normal (Gaussian) distribution is described in § 4.6. The observations are ranked, the rankit is found from the table, and plotted against the value of the observation (or any desired transformation of the observation). Negative values are omitted for samples larger than 10. By analogy with the smaller samples the rankit for the seventh observation in a sample of 11 is clearly −0·225 and that for the seventh in a sample of 12 is −0·103. The table is Bliss's (1967) adaptation of that of Harter (1961, *Biometrika* **48**, 151–65). Reproduced with permission.

Rank order	\multicolumn{9}{c}{Size of sample = N}
	2	3	4	5	6	7	8	9	10
1		0·564	0·864	1·029	1·163	1·267	1·352	1·424	1·485	1·539
2		−0·564	0·000	0·297	0·495	0·642	0·757	0·852	0·932	1·001
3			−0·864	−0·297	0·000	0·202	0·353	0·473	0·572	0·656
4				−1·029	−0·495	−0·202	0·000	0·153	0·275	0·376
5					−1·163	−0·642	−0·353	−0·153	0·000	0·123
6						−1·267	−0·757	−0·473	−0·275	−0·123
7							−1·352	−0·852	−0·572	−0·376
8								−1·424	−0·932	−0·656
9									−1·485	−1·001
10										−1·539

Rank order	11	12	13	14	15	16	17	18	19	20
1	1·586	1·629	1·668	1·703	1·736	1·766	1·794	1·820	1·844	1·867
2	1·062	1·116	1·164	1·208	1·248	1·285	1·319	1·350	1·380	1·408
3	0·729	0·793	0·850	0·901	0·948	0·990	1·029	1·066	1·099	1·131
4	0·462	0·537	0·603	0·662	0·715	0·763	0·807	0·848	0·886	0·921
5	0·225	0·312	0·388	0·456	0·516	0·570	0·619	0·665	0·707	0·745
6	0·000	0·103	0·191	0·267	0·335	0·396	0·451	0·502	0·548	0·590
7			0·000	0·088	0·165	0·234	0·295	0·351	0·402	0·448
8					0·000	0·077	0·146	0·208	0·264	0·315
9							0·000	0·069	0·131	0·187
10									0·000	0·062

Rank order	21	22	23	24	25	26	27	28	29	30
1	1·889	1·910	1·929	1·948	1·965	1·982	1·998	2·014	2·029	2·043
2	1·434	1·458	1·481	1·503	1·524	1·544	1·563	1·581	1·599	1·616
3	1·160	1·188	1·214	1·239	1·263	1·285	1·306	1·327	1·346	1·365
4	0·954	0·985	1·014	1·041	1·067	1·091	1·115	1·137	1·158	1·179
5	0·782	0·815	0·847	0·877	0·905	0·932	0·957	0·981	1·004	1·026
6	0·630	0·667	0·701	0·734	0·764	0·793	0·820	0·846	0·871	0·894
7	0·491	0·532	0·569	0·604	0·637	0·668	0·697	0·725	0·752	0·777
8	0·362	0·406	0·446	0·484	0·519	0·553	0·584	0·614	0·642	0·669
9	0·238	0·286	0·330	0·370	0·409	0·444	0·478	0·510	0·540	0·568
10	0·118	0·170	0·218	0·262	0·303	0·341	0·377	0·411	0·443	0·473
11	0·000	0·056	0·108	0·156	0·200	0·241	0·280	0·316	0·350	0·382
12		0·000	0·000	0·052	0·100	0·144	0·185	0·224	0·260	0·294
13					0·000	0·048	0·092	0·134	0·172	0·209
14							0·000	0·044	0·086	0·125
15									0·000	0·041

TABLE A9 (Continued)

Rank order	Size of sample = N									
	31	32	33	34	35	36	37	38	39	40
1	2·056	2·070	2·082	2·095	2·107	2·118	2·129	2·140	2·151	2·161
2	1·632	1·647	1·662	1·676	1·690	1·704	1·717	1·729	1·741	1·753
3	1·383	1·400	1·416	1·432	1·448	1·462	1·477	1·491	1·504	1·517
4	1·198	1·217	1·235	1·252	1·269	1·285	1·300	1·315	1·330	1·344
5	1·047	1·067	1·087	1·105	1·123	1·140	1·157	1·173	1·188	1·203
6	0·917	0·938	0·959	0·979	0·998	1·016	1·034	1·051	1·067	1·083
7	0·801	0·824	0·846	0·867	0·887	0·906	0·925	0·943	0·960	0·977
8	0·694	0·719	0·742	0·764	0·786	0·806	0·826	0·845	0·863	0·881
9	0·595	0·621	0·646	0·670	0·692	0·714	0·735	0·755	0·774	0·793
10	0·502	0·529	0·556	0·580	0·604	0·627	0·649	0·670	0·690	0·710
11	0·413	0·442	0·469	0·496	0·521	0·545	0·568	0·590	0·611	0·632
12	0·327	0·358	0·387	0·414	0·441	0·466	0·490	0·514	0·536	0·557
13	0·243	0·276	0·307	0·336	0·364	0·390	0·416	0·440	0·463	0·486
14	0·161	0·196	0·228	0·259	0·289	0·317	0·343	0·369	0·393	0·417
15	0·080	0·117	0·151	0·184	0·215	0·245	0·273	0·300	0·325	0·350
16	0·000	0·039	0·076	0·110	0·143	0·174	0·203	0·232	0·258	0·284
17			0·000	0·037	0·071	0·104	0·135	0·165	0·193	0·220
18					0·000	0·035	0·067	0·099	0.128	0·156
19							0·000	0·033	0·064	0·094
20									0·000	0·031

Rank order	41	42	43	44	45	46	47	48	49	50
1	2·171	2·180	2·190	2·199	2·208	2·216	2·225	2·233	2·241	2·249
2	1·765	1·776	1·787	1·797	1·807	1·817	1·827	1·837	1·846	1·855
3	1·530	1·542	1·554	1·565	1·577	1·588	1·598	1·609	1·619	1·629
4	1·357	1·370	1·383	1·396	1·408	1·420	1·431	1·442	1·453	1·464
5	1·218	1·232	1·246	1·259	1·272	1·284	1·296	1·308	1·320	1·331
6	1·099	1·114	1·128	1·142	1·156	1·169	1·182	1·194	1·207	1·218
7	0·993	1·009	1·024	1·039	1·054	1·068	1·081	1·094	1·107	1·119
8	0·898	0·915	0·931	0·946	0·961	0·976	0·990	1·004	1·017	1·030
9	0·811	0·828	0·845	0·861	0·877	0·892	0·907	0·921	0·935	0·949
10	0·729	0·747	0·764	0·781	0·798	0·814	0·829	0·844	0·859	0·873
11	0·651	0·671	0·689	0·707	0·724	0·740	0·757	0·772	0·787	0·802
12	0·578	0·598	0·617	0·636	0·654	0·671	0·688	0·704	0·720	0·735
13	0·507	0·528	0·548	0·568	0·586	0·604	0·622	0·639	0·655	0·671
14	0·439	0·461	0·482	0·502	0·522	0·540	0·559	0·576	0·593	0·610
15	0·373	0·396	0·418	0·439	0·459	0·479	0·498	0·516	0·534	0·551
16	0·309	0·333	0·355	0·377	0·398	0·419	0·438	0·457	0·476	0·494
17	0·246	0·270	0·294	0·317	0·339	0·360	0·381	0·400	0·419	0·438
18	0·183	0·209	0·234	0·258	0·281	0·303	0·324	0·345	0·364	0·384
19	0·122	0·149	0·175	0·200	0·224	0·247	0·269	0·290	0·310	0·330
20	0·061	0·089	0·116	0·142	0·167	0·191	0·214	0·236	0·257	0·278
21	0·000	0·030	0·058	0·085	0·111	0·136	0·160	0·183	0·205	0·227
22			0·000	0·028	0·055	0·081	0·106	0·130	0·153	0·176
23					0·000	0·027	0·053	0·078	0·102	0·125
24							0·000	0·026	0·051	0·075
25									0·000	0·025

References

ANON. (1932). *Illustrated London News* **180**, 1057.

BAILEY, N. T. J. (1964). *The elements of stochastic processes.* Wiley, New York.

—— (1967). *The mathematical approach to biology and medicine.* Wiley, London.

BAIN, W. A. and BATTY, J. E. (1956). Inactivation of adrenaline and noradrenaline by human and other mammalian liver *in vitro. Br. J. Pharmacol.* **11**, 52–7.

BARTLETT, M. S. (1947). The use of transformations. *Biometrics* **3**, 39–52.

BAYES, T. (1763). An essay towards solving a problem in the doctrine of chances. *Phil. Trans. Soc.* **53**, 370.

BERNARD, C. (1965). *An introduction to the study of experimental medicine.* Collier Books edition (1961). Collier Books, New York.

BLISS, C. I. (1947). 2 × 2 Factorial experiments in incomplete groups for use in biological assays. *Biometrics* **3**, 69–88.

BLISS, C. I. (1967). *Statistics in biology,* Vol. I. McGraw-Hill.

BOYD, I. A. and MARTIN, A. R. (1956). The end-plate potential in mammalian muscle. *J. Physiol., Lond.* **132**, 74–91.

BOX, G. E. P. and COX, D. R. (1964). An analysis of transformations. *Jl R. statist. Soc.* **B26**, 211–43.

BROWNLEE, K. A. (1965). Statistical theory and methodology in science and engineering, 2nd edn. Wiley, New York.

BURN, J. H., FINNEY, D. J., and GOODWIN, L. G. (1950). *Biological standardization,* 2nd edn. Oxford University Press.

BURNSTOCK, G. and HOLMAN, M. E. (1962). Spontaneous potentials at sympathetic nerve endings in smooth muscle. *J. Physiol., Lond.* **160**, 446–60.

COCHRAN, W. G. (1952). The χ^2 test of goodness of fit. *Ann. math. Statist.* **23**, 315–45.

—— and COX, G. M. (1957). *Experimental designs,* 2nd end. Wiley, New York; Chapman and Hall, London.

COLQUHOUN, D. (1963). Balanced incomplete block designs in biological assay illustrated by the assay of gastrin using a Youden square. *Br. J. Pharmac. Chemother.* **21**, 67–77.

—— (1968). The rate of equilibration in a competitive *n* drug system and the auto-inhibitory equations of enzyme kinetics: some properties of simple models for passive sensitization. *Proc. R. Soc.* **B170**, 135–54.

—— (1969). A comparison of estimators for a two-parameter hyperbola. *Jl R. statist. Soc.* Ser. C (Applied Statistics) **18**, 130–40.

—— and TATTERSALL, M. (1970). Rapid histamine assays: a method and some theoretical considerations. *Br. J. Pharmac. Chemother.* **38**,   .

COX, D. R. (1962). *Renewal theory.* Methuens, London, Science Paperback (1967).

—— and LEWIS, P. A. W. (1966). *The statistical analysis of a series of events.* Methuen, London.

CUSHNY, A. R. and PEEBLES, A. R. (1905). The action of optical isomers. II. Hyoscines. *J. Physiol., Lond.* **32**, 501–10.

416   *References*

DEWS, P. B. and BERKSON, J. (1954). *Statistics and mathematics in Biology*, (Eds. O. Kempthorne, Th.A. Bancroft, J. W. Gowen, and J. L. Lush), pp. 361–70. Iowa State College Press.

Documenta Geigy scientific tables, 6th edn (1962). J. R. Geigy, S. A. Basle, Switzerland.

DOWD, J. E. and RIGGS, D. S. (1965). A comparison of estimates of Michaelis–Menten kinetic constants from various linear transformations. *J. biol. Chem.* **240**, 863–9.

DRAPER, N. R. and SMITH, H. (1966). *Applied regression analysis*. Wiley, New York.

DUNNETT, C. W. (1964). New tables for multiple comparisons with a control. *Biometrics* **20**, 482–91.

DURBIN, J. (1951). Incomplete blocks in ranking experiments. *Br. J. statist. Psychol.* **4**, 85–90.

FELLER, W. (1957). *An introduction to probability theory and its applications*, Vol. 1, 2nd edn. Wiley, New York.

—— (1966). *An introduction to probability theory and its applications*, Vol. 2, 2nd edn. Wiley, New York.

FINNEY, D. J. (1964). *Statistical method in biological assay*, 2nd edn. Griffin, London.

—— LATSCHA, R., BENNETT, B. M., and HSU, P. (1963). *Tables for testing significance in a* 2×2 *table*. Cambridge University Press.

FISHER, R. A. (1951). *The design of experiments*, 6th edn. Oliver and Boyd, Edinburgh.

—— and YATES, F. (1963). *Statistical tables for biological, agricultural and medical research*, 6th edn. Oliver and Boyd, Edinburgh.

GOULDEN, C. H. (1952). *Methods of statistical analysis*, 2nd edn. Wiley, New York.

GUILFORD, J. P. (1954). *Psychometric methods*, 2nd edn. McGraw-Hill, New York.

HEMELRIJK, J. (1961). Experimental comparison of Student's and Wilcoxon's two sample tests. In *Quantitive methods in pharmacology* (Ed. H. de Jonge). North Holland, Amsterdam.

HOOKE, R. and JEEVES, T. A. (1961). 'Direct search' solution of numerical and statistical problems. *J. Ass. comput. Mach.* **8**, 212–29.

KATZ, B. (1966). *Nerve, muscle and synapse*. McGraw-Hill, New York.

KEMPTHORNE, O. (1952). *The design and analysis of experiments*. Wiley, New York.

KENDALL, M. G. and STUART, A. (1961). *The advanced theory of statistics*, Vol. 2. Griffin, London.

—— —— (1963). *The advanced theory of statistics*, Vol. 1, 2nd ed. Griffin, London.

—— —— (1966). *The advanced theory of statistics*, Vol. 3. Griffin, London.

LINDLEY, D. V. (1965). *Introduction to probability and statistics from a Bayesian viewpoint*, Part 1. Cambridge University Press.

—— (1969). In his review of "*The structure of inference*" by D. A. S. Fraser. *Biometrika* **56**, 453–6.

MAINLAND, D. (1963). *Elementary medical statistics*, 2nd edn. Sauders, Philadelphia.

—— (1967a). Statistical ward rounds—1. *Clin. Pharmac. Ther.* **8**, 139–46.

—— (1967b). Statistical ward rounds—2. *Clin. Pharmac. Ther.* **8**, 346–55.

MARLOWE, C. (1604). *The tragicall history of Doctor Faustus*. London: Printed by V. S. for Thomas Bushell.

MARTIN, A. R. (1966). Quantal nature of synaptic transmission. *Physiol. Rev.* **46**, 51–66.

MASSEY, H. S. W. and KESTELMAN, H. (1964). *Ancillary mathematics*, 2nd edn. Pitman, London.

MATHER, K. (1951). *Statistical analysis in biology*, 4th edn. Methuen, London.

MCDONALD, B. J. and THOMPSON, W. A. JR. (1967). Rank sum multiple comparisons in one- and two-way classifications. *Biometrika*, **54**, 487–97.

MOOD, A. M. and GRAYBILL, F. A. (1963). *Introduction to the theory of statistics*, 2nd edn. McGraw-Hill Kogakusha, New York.

NAIR, K. R. (1940). Table of confidence intervals for the median in samples from any continuous population. *Sankhya* **4**, 551–8.

OAKLEY, C. L. (1943). He-goats into young men: first steps in statistics. *Univ. Coll. Hosp. Mag.* **28**, 16–21.

OLIVER, F. R. (1970). Some asymptotic properties of Colquhoun's estimators of a rectangular hyperbola. *J. R. statist. Soc.* (Series C, Applied statistics) **19**, 269–73.

PEARSON, E. S. and HARTLEY, H. O. (1966). *Biometrika tables for statisticians*, *Vol.* 1, 3rd edn. Cambridge University Press.

POINCARÉ, H. (1892). *Thermodynamique*. Gauthier-Villars, Paris.

RANG, H. P. and COLQUHOUN, D. (1973). *Drug Receptors: Theory and Experiment*. In preparation.

SCHOR, S. and KARTEN, I. (1966). Statistical evaluation of medical journal manuscripts. *J. Am. med. Ass.* **195**, 1123–8.

SEARLE, S. R. (1966). *Matrix algebra for the biological sciences*. Wiley, New York.

SIEGEL, S. (1956*a*). *Nonparametric statistics for the behavioural sciences*. McGraw-Hill, New York.

—— (1956*b*). A method for obtaining an ordered metric scale. *Psychometrika* **21**, 207–16.

SNEDECOR, G. W. and COCHRAN, W. G. (1967). *Statistical methods*, 6th edn. Iowa State University Press, Iowa.

STONE, M. (1969). The role of significance testing: some data with a message. *Biometrika* **56**, 485–93.

STUDENT (1908). The probable error of a mean. *Biometrika* **6**, 1–25.

TAYLOR, D. (1957). *The measurement of radioisotopes*, 2nd edn. Methuen, London.

THOMPSON, SILVANUS, P. (1965). *Calculus made easy*. Macmillan, London.

TIPPETT, L. H. C. (1944). *The methods of statistics*, 4th edn. Williams and Norgate, London; Wiley, New York.

TREVAN, J. W. (1927). The error of determination of toxicity. *Proc. R. Soc.* **B101**, 483–514.

TUKEY, J. W. (1954). Causation, regression and path analysis. In *Statistics and mathematics in biology* (Eds. O. Kempthorne, Th. A. Bancroft, J. W. Gowen, and J. L. Lush), p. 35. Iowa State College Press, Iowa.

WILCOXON, F. and WILCOX, ROBERTA, A. (1964). *Some rapid approximate statistical procedures*. Published and distributed by Lederle Laboratories, Pearl River, New York.

WILDE, D. J. (1964). *Optimum seeking methods*. Prentice-Hall, Englewood Cliffs, N. J.

WILLIAMS, E. J. (1959). *Regression analysis*. Wiley, New York, Chapman and Hall, London.

Index